Mixed-Signal CMOS for Wireline Communication

Get up to speed with the fundamentals of complementary metal oxide semiconductor (CMOS) integrated circuit (IC) design for wireline communication with this practical introduction, from short-reach optical links to various electrical links. It presents practical coverage of the state of the art, equipping readers with all the tools needed to understand these circuits and then design their own. A comprehensive treatment of components, including details for front-end circuits, equalizers, oscillators, phase-locked loops and clock and data recovery systems, accompanies significant coverage of inverter-based circuits, preparing the reader for modern designs in nano-scale CMOS. Numerous inline examples demonstrate concepts and solutions, allowing readers to absorb the theory and confidently apply concepts to new scenarios. Suitable for graduate students and professional engineers working in mixed-signal integrated circuit design for high-speed interconnects, and including over 100 end-of-chapter problems to extend learning (with online solutions for instructors), this versatile book will equip readers with an unrivalled understanding of exactly what goes into a modern wireline link – and why.

Glenn Cowan is Professor of Electrical and Computer Engineering at Concordia University and has been teaching courses on mixed-signal integrated circuits for over ten years. His research interests include low-power mixed-signal circuits for wireline communication, biomedical signal acquisition and computation.

Mixed-Signal CMOS for Wireline Communication

Transistor-Level and System-Level Design Considerations

GLENN COWAN

Concordia University

Shaftesbury Road, Cambridge CB2 8EA, United Kingdom

One Liberty Plaza, 20th Floor, New York, NY 10006, USA

477 Williamstown Road, Port Melbourne, VIC 3207, Australia

314–321, 3rd Floor, Plot 3, Splendor Forum, Jasola District Centre,
New Delhi – 110025, India

103 Penang Road, #05–06/07, Visioncrest Commercial, Singapore 238467

Cambridge University Press is part of Cambridge University Press & Assessment,
a department of the University of Cambridge.

We share the University's mission to contribute to society through the pursuit of
education, learning and research at the highest international levels of excellence.

www.cambridge.org
Information on this title: www.cambridge.org/9781108490009

DOI: 10.1017/9781108779791

First published 2024

A catalogue record for this publication is available from the British Library.

A Cataloging-in-Publication data record for this book is available from the Library of Congress

ISBN 978-1-108-49000-9 Hardback

Contents

Preface

Topic Area

This book introduces readers to high-speed wireline communication, both electrical links and short-reach optical links, from the perspective of complementary metal oxide semiconductor (CMOS) integrated circuit (IC) design. Although a broad topic, the fundamentals of electrical and optical channels as well as the electronics for transmitting and receiving data using these channels are presented. The material in this book grew out of ELEC413/6071: Mixed-Signal VLSI for Communication Systems (Concordia University) and a few tutorials I have given over the years. ELEC413/6071 focuses exclusively on wireline communication (as opposed to wireless communication), teaching final-year undergraduate and graduate students about the IC design concepts needed to build transmitters (Tx) and receivers (Rx), so-called transceivers.

Starting from fundamentals of communication theory and the basics of transmission lines, the book explains electrical-link transceiver circuitry, optical-link transceiver circuitry, oscillators, phase-locked loops (PLLs) and clock and data recovery (CDR) systems. Over the past decade or two, CMOS inverter-based circuits have become much more common in high-speed design. These are presented along with an introduction to how PLLs are made digital and how conventional mixed-signal links are moved into digital signal processor-based links.

Some features of the book include:

- Coverage of both electrical and optical links starting from the basics.
- A detailed introduction to thermal noise analysis of transistor circuits as an appendix.
- Discussion of modern CMOS implementation of transceiver circuits using inverter-based approaches.
- Analysis of optical receivers that use equalization to enable higher gain and lower noise operation, known as "low-bandwidth" receivers.
- Introduction of the fundamentals of PAM4 signal transmission and digital techniques in PLLs and equalizers.
- Baud-rate CDR systems.

- Examples in each chapter that start with introductory-level concepts illustrating the material.
- Back-of-the-chapter problems throughout.

Who Should Read This Book?

The target audience for this book is a graduate student, though it is fairly accessible to a well-qualified final-year undergraduate student. The ideal student has learned about metal–oxide–semiconductor field-effect transistors (MOSFETs), their high-frequency models and analyzing circuits for dc and ac signals. The fundamentals of differential pairs and current mirrors are required; the material in a typical first graduate course in analog IC design dealing with operational amplifier design would be helpful. Undergraduate courses in digital communications and electromagnetics are also helpful. Some familiarity with computer-aided design tools such as Cadence is an asset.

Another target audience is the IC designer working in another application area. Anyone already thinking about dc operating points, gain/bandwidth, noise, mismatch, choosing the bias currents, widths and lengths of transistors for circuits in wireless communication, low-speed analog, ADC/DAC design or biomedical IC design can read this book and consider a career shift into IC design for wireline communication.

One Book?

This book aims to address IC design for both electrical and optical links. This is a broad scope; however, whether a link's channel is a printed circuit board (PCB) trace or an optical fibre, high-speed, low-power CMOS ICs are required. Both types of links require phase-locked loops and other synchronization circuitry as well as high-speed latches, multiplexers and demultiplexers. In the past, channel equalization using transmitter-side feed-forward equalizers and/or receiver-side continuous-time linear equalizers or decision feedback equalizers was limited to electrical links. Nowadays, equalization is widely deployed in laser-diode drivers in optical links and increasingly in optical receivers that use emerging low-bandwidth techniques to improve receiver sensitivity. Finally, with optical transceivers residing on a separate die from the host CMOS chip, electrical links between processors and optical transceivers must be implemented. In light of the shared building blocks and similarity in circuit approaches, this book presents design considerations for both electrical and optical links.

Book Organization and Use

In a 13-week course such as ELEC413/6071, I use material from the entire book. However, to allow a more in-depth treatment, the material in this book can be subdivided in different ways. For example, Chapters 1 to 10 address channels and mainly "front-end" circuitry. Chapter 11 introduces advanced topics related to PAM4 and DSP-based links. Clock generation and distribution along with Tx/Rx synchronization and CDR are presented in Chapters 12 to 15.

The best way for instructors to use this book depends on how many wireline courses their institution offers, and whether topics such as voltage-controlled oscillator (VCO) design and PLL design are treated in separate courses that perhaps address VCOs and PLLs for both wireline and wireless communication. Some possible subdivisions include:

- **Option 1: Front-end IC design for electrical and optical links:** Chapters 1 to 9, extending to Chapter 10 (low-bandwidth optical receivers) or 11 (advanced topics: PAM4 and DAC/ADC-based links) time permitting.
- **Option 2: IC design for electrical links:** Chapters 1 to 6 which cover electrical link front-end circuit design, Chapter 11 (advanced topics: PAM4 and DAC/ADC-based links) and Chapters 12 to 15, which cover synchronization aspects.
- **Option 3: IC design for optical links:** Chapters 1 (introduction), 3 (decision circuits), 5 (electrical link Tx, needed as an introduction to optical Tx circuitry), 7 (optical channels and devices), 8 (optical Tx circuits), 9 (conventional optical Rx circuits), and 12 to 15, which cover synchronization. Depending on time, also Chapter 10 and/or 11. If low-bandwidth optical Rxs are presented (Chapter 10), equalization is needed as prerequisite material (Chapter 4).

Please bring any typos or other errors that are inevitably discovered to my attention. Suggestions for additional examples or conceptual clarifications are also welcome.

The back-of-the-chapter problems warrant some remarks. Some topics are amenable to paper-and-pencil problems. Apply this equation to compute that quantity. Some of these simpler concepts are illustrated in examples. Familiarity with this book's material benefits greatly from instructor-led project work, involving system-level simulation and transistor-level schematic and layout design. Design targets need to be technology and application dependent. The reader will see some problems as being a little vague. These are meant to be suggestions to instructors of possible areas to go into. There will not be a solution manual to these. It would be great if we can share our course assignments and design projects.

Acknowledgements

Many thanks are due to many people. The team at Cambridge and its partners supported this project from day one and has been helpful along the way. Thanks are due to Laura Blake, Naomi Chopra, Ali Clark, Collette Forder, Elizabeth Horne, Suresh Kumar, Holly Paveling, Bharathan Sankar, Dinesh Singh Negi, and Sarah Strange. My expertise in this area has grown out of my work with numerous MASc and PhD students working in wireline IC design over the years. Four recent PhD graduates and one current PhD student should be singled out: Abdullah Ibn Abbas, Diaaeldin

Abdelrahman, Monireh Moayedi Pour Fard, Christopher Williams and Di Zhang. Several of their papers are referenced. Diaa's and Monir's work provided several figures. Di helped with material related to his research area. Sara Granata provided invaluable feedback on the text from the perspective of the fourth-year undergrad. The journey into wireline was launched by the IBM Thomas J. Watson Research Center taking a chance on a new PhD grad who had worked on kHz-range analog computers.

The approach to presenting transmission lines closely follows material by Professor Chris Trueman (Concordia), who kindly reviewed the chapter. Noise and mismatch is presented in the way I learned from Professor Peter Kinget (Columbia) when I replaced him in a course while he was on a parental leave. Everything I know about digital VLSI I learned from Professor Kenneth L. Shepard (Columbia). Professor Yannis Tsividis (Columbia), my PhD supervisor, got me thinking about impedance scaling, noise/power trade-offs, and the merits of getting explanations right.

I am grateful for the encouragement throughout this project from my wife and frequent co-author/collaborator, Odile Liboiron-Ladouceur (McGill).

Notation and Acronyms

ADC	Analog-to-digital converter
BER	Bit error rate
CDR	Clock and data recovery
CID	Consecutive identical digit
CMOS	Complementary metal oxide semiconductor
CP	Charge pump
CPU	Central processing unit
CTLE	Continuous-time linear equalizer
DAC	Digital-to-analog converter
DFE	Decision feedback equalizer
DSP	Digital signal processor or processing
f_b, f_{bit}	Data rate
FFE	Feed-forward equalizer
FIR	Finite impulse response
GPU	Graphics processing unit
IC	Integrated circuit
IIR	Infinite impulse response
ILO	Injection-locked oscillator
ISI	Intersymbol interference
k	Boltzmann constant: $1.38064852 \times 10^{-23}$ JK^{-1}
LPF	Low-pass filter
LSB	Least-significant bit
LTI	Linear time invariant
MOSFET	Metal–oxide–semiconductor field-effect transistor
MSB	Most-significant bit
NRZ	Non-return to zero
PAM	Pulse-amplitude modulation
PCB	Printed circuit board
PD	Phase detector/photodiode/photodetector
PFD	Phase-frequency detector

PLL	Phase-locked loop
PSD	Power spectral density
PVT	Process, voltage and temperature
RLM	Ratio level mismatch
T	Absolute temperature in Kelvin
TDECQ	Transmitter and dispersion eye closure quaternary
q	Electron charge: $1.6021765 \times 10^{-19}$C
Q	Quality factor
UI	Unit interval
V-to-I	Voltage-to-current
VCO	Voltage-controlled oscillator
VEO	Vertical eye opening
ζ	Damping factor

1 Introduction to Wireline Communication

Digital systems, be they smartphones, laptops, desktops or data centres, have components that process vast amounts of data, but also transfer data both within the system or over longer distances. For example, a desktop computer will have a combination of central processing units (CPUs) and graphics processing units (GPUs) with communication busses connecting them to memory and to each other. A data centre has thousands of links within its constituent racks and from rack to rack. The shortest of these are implemented electrically whereas beyond a few metres, links are usually fibre-based optical links. Optical links also connect data centres to one another and most of the connections in the internet, although the final connection to an end-user's residence or workplace (the so-called "last mile") can be optical or electrical over a twisted pair of wires.

Near the end of the writing of this text, Gordon Moore's long and impactful life came to a peaceful end. His observation that the number of transistors on an integrated circuit (IC) doubled every two years or so, carved into the lexicon as "Moore's Law," is universally known within the IC industry and well-known among those less technical. The original observation spoke to the number of transistors on a chip, although the ensuing decades have seen comparable trends in the shrinking of transistors' dimensions and, until the early 2000s, geometric increases in microprocessor clock speeds. A reader might assume that decades of complementary metal oxide semiconductor (CMOS) technology scaling, which have brought us multi-core processors, issuing multiple instructions per sub-nanosecond clock period, would naturally give rise to similar year-over-year improvements in wired links, perhaps even making wireline link design relatively straightforward and an unworthy subject for a textbook. Although CMOS scaling allows for terabytes per second (TB/s) processing ability from a modern microprocessor or GPU, getting those data from one chip to another is still a pinch point in system design. In fact, as discussed in Section 1.1.2, the doubling of data rates in wired links is considerably slower than every two years. Electrical links are limited by the frequency response of the electrical channel connecting transmitter and receiver, presented in detail in Chapter 2, which no matter how fast the transmitter operates, distorts the data. Fast transistors help, but clever techniques to mitigate the channel's low-pass response (known as equalization, see Chapter 4) are still needed and have their own limitations. The bandwidth of optical devices such as photodetectors, modulators and lasers can limit the speed of optical links. Thus, wired links are indeed challenging to design and their performance is critical to ensuring high-performance digital systems. Also, their power dissipation can be a non-trivial percentage of an overall system's power dissipation.

Over the past decades, data rates have increased from 10s to 100s of Mb/s in the 1980s to 112 Gb/s (and higher) line rates for electrical and >100 Gb/s per wavelength in optical links. Looking forward, we expect higher data rates achieved by higher symbol rates and higher-order modulation sending more bits per symbol.

To reduce cost it can be favourable to integrate more transceiver blocks on the same chip, usually a CMOS chip. Circuitry that was implemented using bipolar technology is increasingly implemented using CMOS. Not only has this changed the underlying transistor technology, but also the topology of circuits. We see increased use of inverter-based designs, where the well-known two-transistor logic circuit is used as a transmitter, or an amplifier. Another important trend is the increased use of digital techniques in aspects of the system that were once analog or mixed-signal. Channel equalization is increasingly performed in the digital domain. On the clocking side, phase-locked loops now use digitally controlled oscillators and digital loop filters rather than analog tuning and analog filters.

Wireline communication continues to play an important commercial role. The optical transceiver market is worth approximately 10 billion USD as of 2023 (American billion, i.e., 10^9). Electrical links usually sit on larger chips so calculating their market share is difficult. However, large companies such as Xilinx and other makers of field-programmable gate arrays (FPGAs) can differentiate their products by the performance and versatility of the input/output (I/O) offerings. Successful demonstration of high-end I/O is also a necessary step for any foundry supporting the fabless market. Hence, the mixed-signal systems covered in this book play an important role in the current semiconductor market and will continue to drive sales in the future.

This book provides an introduction to integrated circuit design for wired communication. This includes electrical links for various serial link standards and short-reach optical links. Data are modulated using pulsed amplitude modulation (PAM), that is, discrete amplitude of voltage or optical power. Emphasis in this text is on the two-valued signals (PAM2) but PAM4 concepts are also introduced. The discrete multitone schemes used in digital subscriber line communication are not presented. Optical links that use intensity modulation and direct detection (IMDD) of PAM2 and PAM4 signals are presented, including links for multi-mode and single-mode fibre. Longer reach links that use coherent modulation are not presented in detail, but what distinguishes them from IMDD links is mainly in the optical components. Thus, the fundamentals of their circuit building blocks are presented.

1.1   Overview of the State of the Art

This section gives an overview of the state of the art in wireline communication. Before doing so, some important metrics for wireline communication are defined.

1.1.1 Wireline Communication Metrics

Before taking a look at technology trends in more detail, and discussing where they have arrived in the 2020s, the primary metrics of a wireline link must be defined. These include:

- **Data rate:** This is the number of bits of information transmitted per second over a single lane of a wireline standard. Nowadays this is in the range of Gb/s. In some cases, a standard can be described with an aggregate data rate transmitted over multiple lanes. For example 100 Gb Ethernet can be implemented as 10 lanes of 10 Gb/s or 4 lanes of 25 Gb/s. In the context of this book, these scenarios are described as having data rates of 10 and 25 Gb/s, respectively. Finally, since more than one bit can be encoded at a time (per symbol), the data rate can be higher than the symbol rate or baud rate. We will discuss this in more detail when modulation formats are presented in Section 1.4.
- **Energy efficiency:** The performance of many large ICs is constrained by a maximum power density, above which air cooling is no longer viable. Hence, power dissipation is a critical specification in all I/O standards. To make fair comparisons across various data rates, power dissipation is normalized to data rate. Therefore, energy efficiency is computed as $\frac{\text{Power (mW)}}{\text{Data rate (Gb/s)}}$ giving energy per transmitted bit, usually reported in units of pJ/bit.
- **Channel loss:** Although copper is a great conductor for dc current, as presented in Chapter 2 copper traces on printed circuit boards (PCBs) show significant high-frequency loss, making high-speed links challenging, particularly over longer reach. If two electrical links operate at the same data rate, we would expect the one that can support more channel loss at the Nyquist frequency ($f_{baud}/2$) to dissipate more energy per bit and require more circuitry.

1.1.2 The Current State of the Art in Wired Links

Each year, during the lead-up to the International Solid-State Circuits Conference (ISSCC), an overview of the state of the art in integrated circuits is published in an article titled, "Through the Looking Glass" [1]. Figure 1.1 (a) shows the data rate of various standards from 2000 through to 2021. In this context, the vertical axis is the "per-lane" data rate as opposed to an aggregate data rate achieved through a group of lanes. All standards show a consistent upward trend, doubling in approximately four years, a trend much slower than the 18 month to two-year doubling of transistor density predicted by Moore's Law. The underlying CMOS nodes are captured in Figure 1.1 (b) where different markers show the year of publication of the ICs. As expected, the fastest line rates are implemented with smaller feature-size technologies that offer both high-speed operation and reasonable power dissipation when equalization is implemented digitally.

Figure 1.2 (a) shows the energy efficiency of links as a function of channel loss at the Nyquist frequency. Over time, the dots have moved down vertically. The trend

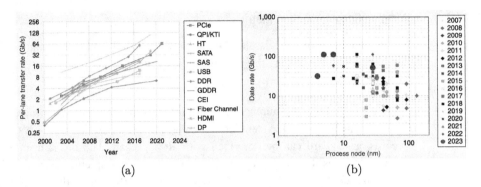

Figure 1.1 Trends in wireline: (a) data rate increase for various wireline standards, 2000 to present; (b) data rate for various CMOS process technology nodes published at ISSCC from 2007 to present. © 2023 IEEE. Reprinted, with permission, from [1].

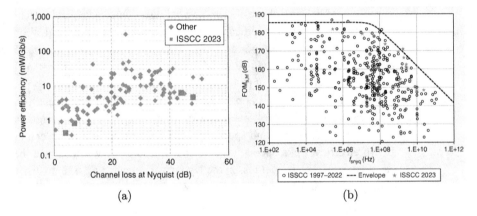

Figure 1.2 Trends in wireline and analog-to-digital converters: (a) energy cost per transmitted bit vs channel loss; (b) analog-to-digital converter performances vs Nyquist frequency. © 2023 IEEE. Reprinted, with permission, from [1].

toward better efficiency with lower loss is also clearly visible. Long-reach optical links have employed high-speed analog-to-digital converters (ADC), driving high-speed ADC development for the last two decades. Recently, we see ADCs deployed in electrical links and short-reach optical links. This is viable with reduced ADC and digital signal processor (DSP) power dissipation. The vertical axis of Figure 1.2 (b) is a figure of merit (FOM) computed by normalizing the signal to noise and distortion to the power dissipation of the converter. The horizontal axis is the Nyquist frequency of the converter. For wireline applications, we are interested in converters with Nyquist frequencies above 10 GHz. Over the past decade, the computed FOM has improved by about 10 dB [2], making ADC-based links more viable.

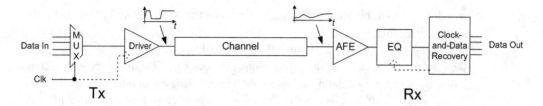

Figure 1.3 Generic block diagram of a wireline link.

1.2 Exemplary Links

In this section, block diagrams of wireline links are presented, starting with a very generic link, capturing key features common to both electrical and optical links. Examples of electrical and optical links are described and contrasted.

1.2.1 Generic Wireline Link

Figure 1.3 shows a generic block diagram of a wireline link. At the centre is a channel, which can represent either a PCB trace or coaxial cable of an electrical link or a fibre optic cable. A scenario often considered as the worst case electrical channel over which a high-speed electrical link is to operate is the backplane connection. The lower portion of Figure 2.3 shows two horizontal boards connected to a vertical board (known as a backplane). The overall channel consists of PCB traces and two connectors, not to mention the chip packages. Returning to Figure 1.3, "Data In" are generated by a CPU, GPU or some other "data-producing" chip. Regardless of the context, the clock speed of large digital chips is usually in the range of 1–3 GHz. To maximize the I/O capability of such a chip, each I/O connection is used to its fullest by launching data out of the chip at as high a data rate as possible. Regardless of the type of link, the transmitter, denoted as Tx, aggregates several lower-speed data streams (Data In) into a high-speed data stream, using a multiplexer (MUX). The driver launches binary data down the channel using a signal that swings between two levels (or more), encoding binary 1s and 0s. The driver may or may not retime data using a clock signal, as implied by the dashed "Clk" line. At the output of the channel, the data signal is degraded, having a smaller amplitude and possibly intersymbol interference (defined in Section 1.5). The receiver (Rx) of the link includes an analog front-end (AFE) and possibly equalization (EQ). The equalizer's job is to undo some of the effects of the channel so that the data can be properly detected. Ultimately, the data are sampled using high-speed latches in the clock and data recovery block. This block aligns an Rx clock to appropriately sample the data and produce a large amplitude data signal. Data are demultiplexed into Data Out. These lower data rate signals are available to the "data-using" chip on the other end of the link. In an electrical link, the MUX and driver are on the "data-producing" chip while the AFE, EQ and CDR are on the "data-using" chip.

This block diagram also captures the functionality of an optical link, although, in the case of an optical link, the driver block includes electrical to optical conversion, either by directly modulating a laser diode or by using an optical modulator. This functionality is often outside of the "data-producing" chip. Likewise the AFE will include a photodiode to convert the optical signal back to an electrical signal as well as a transimpedance amplifier. Similar to the Tx side, this AFE functionality is usually outside of the "data-receiving chip."

1.2.2   Electrical Link Example

To give the circuits presented in subsequent chapters some context, this section intro-duces a detailed example of an electrical link transceiver. The block diagram in Figure 1.4 [3] shows a link targeting the CEI-56G-VSR standard, which, as presented later in Figure 1.7, operates over a short channel that introduces a modest amount of loss. The transceiver supports PAM4 signalling, meaning two bits of information are sent at a time, by creating a four-valued signal. The bits are referred to as the most-significant bit (MSB) and the least-significant bit (LSB). Although the stand-ard operates at 56 Gb/s, the design was evaluated up to 64 Gb/s. An overview of the transceiver is as follows:

- The transmitter, shown in Figure 1.4 (c), is supplied with 40 parallel data streams for each of the MSB and LSB at 800 Mb/s, which get multiplexed together to give MSB and LSB data at 32 Gb/s. The 40:8 block produces 8 data streams at 4 Gb/s. The rest of the circuitry up to the Output Network is used to generate a signal that is the convolution of the 32 Gb/s data stream and the impulse response of a finite-impulse response filter, known as a feed-forward equalizer (FFE).
- The summation of the convolution is done in the Output Network, producing a differential output OUT_P/OUT_N.
- The transmitter, shown in Figure 1.4 (c) receives a clock from the clock generator portion, presented in Figure 1.4 (b). The block with the circular arrow indicates a phase rotator, which can modify the phase of the clock. DCD refers to duty cycle distortion. This block restores the clock signal to have close to 50% duty cycle. This is critical to do when data processing occurs on both edges of the clock.
- The operations in the various stages of multiplexing need different clock frequen-cies derived from the quarter-rate clock generated in the clock generator. When operating at 64 Gb/s PAM4, each data stream is at 32 Gb/s and hence the quarter-rate clock is at 8 GHz.
- Electrostatic discharge protection (ESD) is added near the outputs. The coupled coils, denoted by the curved double-ended arrow, extend the bandwidth of the circuit that would otherwise be limited by the parallel combination of the ESD's capacitance and the channel impedance. This technique, known as T-coil bandwidth extension, is presented in detail in Section 5.2.2.

- The receiver, shown in Figure 1.4 (a), also uses a T-coil to extend the bandwidth of the signal path and improve input impedance matching.
- Although the transmitted data are binary, the received signal is first processed by an AFE.
- Blocks VGA1 and VGA2 are variable gain amplifiers. These are used to increase the signal amplitude to a desired level. Variable gain is needed because input amplitude will vary based on the actual channel used.
- Also inside the analog front-end is a continuous-time linear equalizer (CTLE), in this case composed of three sub-circuits, which each modify the frequency content of the incoming signal in a particular range (low, mid and high).
- The AFE enables a direct loop back (LPBK) from the transmitter (Tx). This is useful for testing the transceiver or to calibrate various circuits.
- The receiver then samples the equalized input signal using so-called Data and Edge samplers in an Alexander phase detector (see Section 15.2.4). The former sample the data and make a decision on whether the transmitted bit was a 1 or a 0. The latter sample the signal where it changes from one symbol to another. Sampling during a transition can be used to determine if the clock signals are properly synchronized with the data.
- The samplers are driven by eight phases of the 8 GHz clock. These are all operating at 8 GHz, but are spaced by $45°$.
- The output of the Edge samplers, processed by the CLK Recovery block, control more phase rotators and DCD blocks to align the clock signals to the incoming data.
- Incoming data are also sampled by the Eye Monitor block. By sampling the signal with varying time offsets and thresholds, a detailed picture of the incoming signal can be generated. The Adaptation Controller block uses this information to adjust settings in the VGAs and the CTLE.
- The clock generator, shown in Figure 1.4 (b), has an Integer-N phase-locked loop (PLL) and an injection locked ring oscillator (ILO). The PLL locks an internal oscillator to a low-frequency reference clock signal.
- The PLL uses two oscillators to cover the entire 4–8 GHz range, producing a differential output. This output drives the ILO which produces the required eight output phases, spaced by $45°$.
- The outputs of the ILO are used in the transmitter and receiver.
- The power dissipation required to generate clock signals in the clock generator block can be shared by several transceivers. It is common for the output of the clock generator block to be used by four or eight transceivers.

This book aims to explain the operation of all of the blocks in this link.

1.2.3   Exemplary Optical Link

A block diagram of an optical link is shown in Figure 1.5. This is a generic representation of the links used in short-reach applications such as those in data centres. The link includes several blocks, each discussed below:

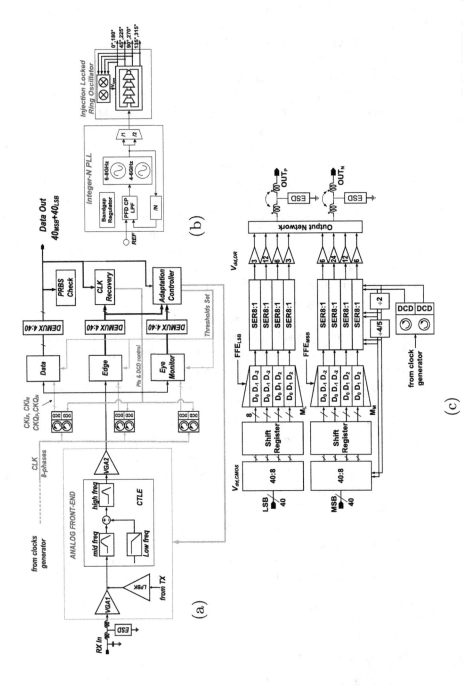

Figure 1.4 Example of a transceiver for a 64 Gb/s PAM4 electrical link: (a) receiver; (b) clock generation circuitry; (c) transmitter. © 2019 IEEE. Reprinted, with permission, from [3].

- The optical transmitter module (OTx) and optical receiver module (ORx) are frequently outside of the "data-producing"/ "data-using" chip packages because conventional packaging techniques do not allow bringing in an optical fibre. On the other hand, the techniques used to package the OTx and ORx do allow for the fibre.
- Short electrical channels separate the OTx/ORx from the "data-producing/using" chips.
- Since the I/O count of the data-producing chip is limited, data are serialized before being transmitted by the electrical link transmitter (ETx) across the short electrical channel with no additional multiplexing in the OTx.
- The OTx includes a driver and one of the following:

  - a laser diode whose current is directly modulated by the driver or
  - a laser diode biased with a constant current (referred to as a CW laser) whose optical output is modulated using an external modulator such as a Mach–Zehnder interferometer.

- As data rates increase, the OTx may include an electrical link Rx to properly recover the data transmitted down the short electrical channel.
- Optical fibre serves as the channel. In data centres most links use multi-mode fibre, frequently bundled such that many fibres are combined together. When links are implemented as a number of parallel links, the laser diodes are in an array, matched in pitch to the cores of the fibre.
- The ORx has a photodiode (PD) that converts the optical signal to an electrical current, known as a photocurrent. The PDs are also in an array matched to the fibre array.
- A transimpedance amplifier (TIA) amplifies the photocurrent. Its output voltage signal is still too small to drive the clock and data recovery (CDR) circuit and is hence amplified further.
- A main amplifier further amplifies the signal for the CDR.
- Separating the main amplifier from the CDR is frequently a short electrical channel.
- The electrical link receiver (ERx) has a CDR that will align a sampling clock to the middle of the data bits and capture the high-speed stream. In this diagram the CDR is also performing serial-to-parallel conversion, outputting four lower rate streams. As data rates increase, the ERx may have equalization.

In Figure 1.5, many details are omitted, such as the biasing circuitry for the lasers and PDs. Also, the clock generation circuitry is highly simplified. In some cases, the ORx block includes a CDR which retimes the data before retransmitting to the data-using chip. Nevertheless, circuit building blocks used in optical links bear strong resemblance to those in electrical links. The next example highlights these similarities.

In this example a transceiver (Figure 1.6 [4]) supporting IEEE802.3b 100GBASE-LR4 is described. This is a standard for an aggregate data rate of 100 Gb/s across four single-mode fibres with one signal per fibre. It has a single PLL that supplies a clock to four Txs and Rxs. What is not shown in this block diagram on the Tx side are the

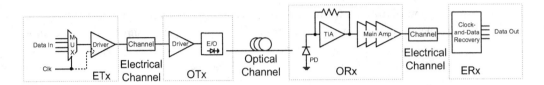

Figure 1.5 Example of a short-reach optical fibre link.

lasers and drivers. On the Rx side, the photodiodes and the transimpedance amplifiers are also not shown.

What we notice from Figure 1.6 is that, on the Tx side, data are multiplexed using two stages of multiplexers. An FFE is used before the driver. The FFE in this case is used to equalize the short electrical channel between the driver and the modulator, rather than a long electrical channel as in Figure 1.4. On the Rx side, once the optical signal has been converted to an electrical current by way of a photodiode and then amplified by a TIA, the rest of the receiver is very similar to the electrical links in Figure 1.4. The decision feedback equalizer (DFE) in the optical link is there not to compensate for the optical channel, but to compensate for the short electrical link between the TIA and the rest of the receiver.

When we look at the left side of Figure 1.6 we see the interface between this chip and a larger digital system. This interface is an electrical link. Notice that the top left is the receiver from the digital system and this chip. It is identical to the circuitry in the bottom right which is the receiver for signals coming in from the optical link. That the same circuit can be used for these two parts illustrates the similarity between electrical and optical links.

1.2.4   Commonalities and Differences Between Electrical and Optical Links

These examples of electrical and optical links have many common building blocks. For example, both include gain blocks, clock and data recovery and latches.

There are several important differences between these links that will influence IC design. Short-reach optical links tend to be receiver noise limited, with transmitter power set such that a noise floor at the receiver is overcome. Receiver design focuses heavily on minimizing input-referred noise and achieving high gain. On the other hand, electrical links are often channel limited such that receiver design focuses on equalization of the received signal.

Short-reach optical and electrical links can be synchronized using the same approach. However, because optical links can reach farther for a given data rate, optical links are more likely to operate transmitters and receivers from separate reference frequencies, giving rise to small differences in the reference clock's frequency, a scenario referred to as "plesiochronous." Links of a few centimetres are usually implemented electrically, and may use a common reference clock for Tx and Rx circuits or a source-synchronous clocking style in which a clock is forwarded from transmitter to receiver along with the data stream(s).

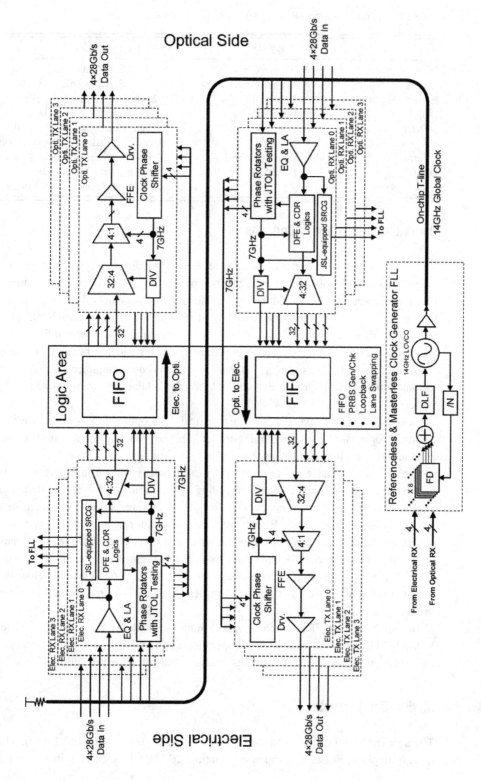

Figure 1.6 Example of a 25 Gb/s optical link supporting 100G. © 2015 IEEE. Reprinted, with permission, from [4].

Table 1.1 Summary of various electrical- and optical-link standards.

Name	Type	Notes
PCIe	Mostly electrical	Serial interface
USB	Mostly electrical	Serial interface
Hypertransport	Electrical	No new data rates since 2008
DDR/GDDR	Electrical	Parallel memory interface
SONET	Optical	Synchronous standard where all Tx/Rx are synched using atomic clocks
Ethernet	Optical and electrical	Many flavours of standards
OIF-CEI	Electrical	Multiple standards for various electrical channels up to 1 m

1.3   Overview of Wireline Standards

Wireline standards enable transmitters and receivers from different vendors to function together. Consumers take for granted that two products with a USB port can communicate with each other, provided the right cable is used. The standards also dictate the testing required to determine that a portion of the link is "standard compliant." Data rate, Tx signal swing, channel loss and impedance, modulation format and jitter tolerance are some of the characteristics specified in a typical standard; these are discussed in this book. Standards also specify higher-layer aspects of communication (i.e., from the data link layer or higher) which will not directly dictate how the front-end ICs are designed.

Table 1.1 presents several common electrical- and optical-link standards whose datarate progressions are shown in Figure 1.1 (a). The circuits presented in this book are relevant to each of these standards.

A nice summary table of various OIF-CEI standards is shown in Figure 1.7. The reason for having several standards operating at the same symbol rate is that if an electrical channel is shorter, inducing less loss, the transceiver needs less equalization and can dissipate less power. By subdividing the design space, transceivers can be optimized for different applications. Of course, if a transceiver that meets the most lossy channel can be reconfigured to dissipate less power when servicing a lower loss channel, then it is possible for one design to meet the needs of all links. However, reconfiguration will also have a power overhead and unused circuitry still occupies valuable chip area.

High-performance FPGAs often support multiple standards. For example a Virtex VU27P FPGA supports up to 48 links operating up to 58 Gb/s. The same transceiver can operate with NRZ or PAM4 signals across a range of data rates and standards.

1.4   Modulation Formats

The simplest way to transmit data (i.e., 1s and 0s) is shown in Figure 1.8 (a). This is an example of pulsed amplitude modulation (PAM), where an amplitude level of

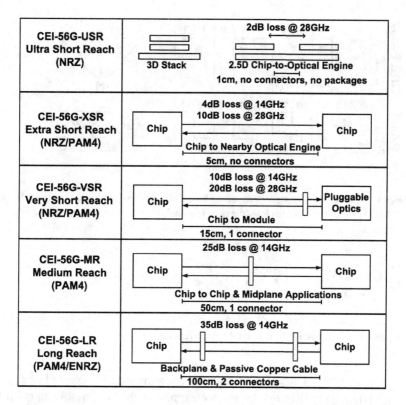

Figure 1.7 Examples of 56 Gb/s electrical link standards for various applications. © 2019 IEEE. Reprinted, with permission, from [3].

L_1 represents a 1 and L_0 represents a 0. Each symbol in this case encodes one bit. Since there are only two signal levels, this is PAM2. In the case of an optical link the quantity that is modulated is the optical power, whereas in an electrical link it is usually the voltage that is modulated, though current-mode signalling is also possible. The signal level is held for the bit period (T_b) or unit interval (UI). Notice that where consecutive 1s are present, the signal remains at the L_1 level, giving rise to the term "non-return to zero" (NRZ).

While imprecise, PAM2 is often referred to as simply NRZ.

An NRZ signal is decoded at the receiver by comparing the received signal to a decision threshold, shown as L_d. This comparison can be interpreted as sampling the signal once per UI, where the dots represent the samples to be compared against L_d.

Rather than continuing to reduce the UI in order to build links with higher data rate, the data rate can be increased by transmitting more than one bit per symbol. An example of this is PAM4, shown in Figure 1.8 (b), in which two bits are transmitted per symbol. The full-scale voltage or optical power range is therefore subdivided into four transmission levels (L_{00}, L_{01}, L_{10} and L_{11}). With more signal levels, the spacing between levels decreases. For a given data rate encoding more bits per symbol relaxes the bandwidth of the system but requires lower noise and better linearity.

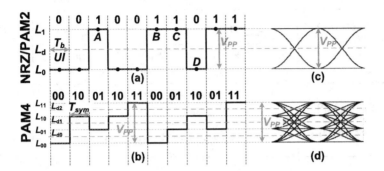

Figure 1.8 Time-domain signals encoding binary data: (a) pulsed amplitude modulation two (PAM2) example signal; (b) PAM4 example signal; (c) PAM2 eye diagram; and (d) PAM4 eye diagram.

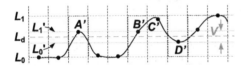

Figure 1.9 NRZ signal with intersymbol interference.

Decoding a PAM4 signal requires three comparison levels denoted as L_{d0}, L_{d1} and L_{d2}. When more than one bit is encoded per symbol we must distinguish between bit rate (Gb/s) and symbol rate. Symbol rate is equivalent to baud rate, measured in GBd/s. The unit interval is a measure of symbol duration and is the reciprocal of baud rate. For example, a 28 GBd/s PAM4 link operates at 56 Gb/s. The UI denoted by T_{sym} is 35.7 ps.

1.5   Intersymbol Interference

Under ideal channel characteristics, the received signal at one decision point is independent of the signal arriving at the previous or following decision points. However, channel and circuit limitations, such as insufficient bandwidth, impedance mismatch or dispersion, can give rise to intersymbol interference (ISI), where the symbol at one decision point is influenced by the symbol transmitted in a previous symbol or in a future symbol. In Figure 1.8 (a) the signal has negligible ISI. For example, points B and C have the same level. On the other hand, in Figure 1.9 the signal level at points B' and C' differs despite both encoding transmitted 1s. The signal in Figure 1.9 encodes the same bit stream as in Figure 1.8 (a), but due to insufficient bandwidth in the system, the signal after the first 1 is lower than the value to which the signal settles after two 1s. As a side note, we refer to repeated binary symbols as consecutive identical digits (CIDs). Put another way, the zero transmitted before point B' interferes with the first 1 that gives rise to B'.

ISI is important to consider because, as we will quantify in Section 1.10, it can lead to bit errors. In Figure 1.8 (a) all transmitted 1s and 0s are equally far ($\frac{V_{pp}}{2}$) from the decision threshold L_d. In Figure 1.9, on the other hand, points A', B' and D' are much closer to the decision threshold. The voltage V shows the peak-to-peak minimum distance of data samples around L_d: ISI has reduced the distance from L_d of some samples from $\frac{V_{pp}}{2}$ to $\frac{V}{2}$. Even though some samples remain $\frac{V_{pp}}{2}$ away from L_d, that some samples are much closer will greatly increase the probability of bit errors.

In this example the effect of ISI persists for only one or two UIs. However, in electrical links with high channel loss or reflections, ISI can persist for many UIs. The next section presents a commonly used tool that gives a synoptic view of the impact of ISI on a data stream. Rather than view long data streams as a function of time, an eye diagram can be constructed, allowing the straightforward viewing of ISI.

ISI also introduces variation in the zero crossing of data, a phenomenon known as jitter, discussed in more detail in Section 1.11.

1.6 Eye Diagrams

There are several perspectives from which signals in wired communication can be viewed, the eye diagram being one frequently used. Figure 1.10 (a) shows an NRZ signal, where each two-UI segment is drawn with a different line style. These segments are then overlaid in Figure 1.10 (b) giving rise to an eye diagram. Notice how the signal at the decision time $t_d = 1.0$ UI associated with transmitted 1s is clustered around a single level. Since the level is independent of the preceding or following bits, it is concluded that the system introduces no ISI. This is contrary to the case in Figure 1.11 (b) where a 1 following several 0s reaches a reduced amplitude compared to a 1 following a 1.

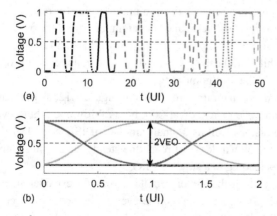

(a)

(b)

Figure 1.10 Construction of an eye diagram by subdividing a signal in sections of two UIs in length: (a) NRZ signal; (b) eye diagram. This signal has negligible ISI.

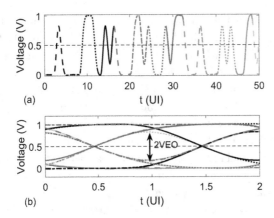

Figure 1.11 Construction of an eye diagram by subdividing a signal in sections of two UIs in length: (a) NRZ signal; (b) eye diagram. This signal has non-negligible ISI.

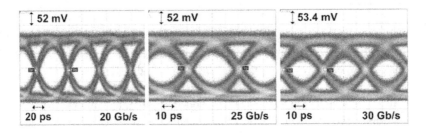

Figure 1.12 Eye diagrams measured at the output of an optical receiver. © 2018 IEEE. Reprinted, with permission, from [5].

In Figure 1.11 (b) the distance between the minimum 1 value (or maximum 0) value and the decision threshold L_d is the vertical eye opening or VEO. The overall distance from maximum 0 value to minimum 1 value is 2 VEO. This can be interpreted as a peak-to-peak voltage, while VEO can be viewed as a peak voltage. In this discussion, the decision threshold is in the mid-point of the eye.

Examples of measured eye diagrams are shown in Figure 1.12, as measured on an oscilloscope. The lefthand eye measured at 20 Gb/s is fairly open, but nevertheless shows some jitter and noise as well as a small amount of ISI. It is clear that when there is a bit transition, the signal does not reach the steady-state value of the signal. At 30 Gb/s, ISI increases leading to a smaller VEO.

When the ISI introduced by a channel or circuit is larger, the eye opening decreases, as shown in Figure 1.13 (a). In some cases, the eye can be completely closed: i.e., VEO = 0. Figure 1.13 (b) shows the eye when equalization techniques, introduced in Chapter 4 and adapted for laser drivers in Chapter 8, are used.

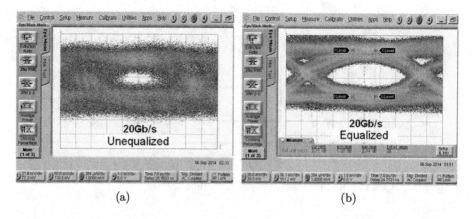

(a)   (b)

Figure 1.13 Measured eye diagrams: (a) with significant ISI and (b) after equalization. © 2016 IEEE. Reprinted, with permission, from [5].

1.7   Definitions Specific to Optical Signals

Figure 1.14 shows an example of optical power in the time domain for a given sequence of transmitted bits. If we consider the optical power associated with 1 and 0 as P_1 and P_0, respectively, we can define the following quantities:

$$P_{avg} = \frac{P_1 + P_0}{2}, \tag{1.1}$$

$$ER = \frac{P_1}{P_0}, \tag{1.2}$$

$$OMA = P_1 - P_0 = 2P_{avg}\frac{ER - 1}{ER + 1}, \tag{1.3}$$

$$ER_{dB} = 10\log\frac{P_1}{P_0}, \tag{1.4}$$

where P_{avg} is the average optical power, assuming a balanced data stream (i.e., an equal number of 1s and 0s), ER is the extinction ratio of the signal and OMA is the optical modulation amplitude. Finally, ER_{dB} is the extinction ratio expressed in dB. Unlike voltage signals which can have negative values, optical signals are always positive. That is, $P_0 \geq 0$.

Since the data is carried in the OMA, any increase in both P_1 and P_0 wastes optical power. However, Chapter 7 explains that lasers operate more slowly as $P_0 \to 0$. In terms of signal definitions, as $P_0 \to 0$, $OMA \to 2P_{avg}$ and $ER \to \infty$.

1.8   Unit Pulse Input and Unit Pulse Response

If it is assumed that the channel and the circuits in the signal path are linear and time invariant, the effect of bandwidth limitations on a data stream can be easily quantified

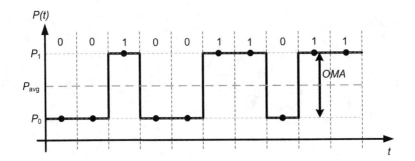

Figure 1.14 Idealized optical signal in the time domain.

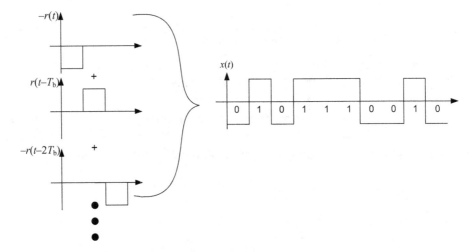

Figure 1.15 Example of a data stream represented by scaled and shifted rectangular pulses.

if we decompose the input signal into a sum of pulses. This is depicted in Figure 1.15. For example, the input signal $x(t)$ can be written as:

$$x(t) = \sum_{k=-\infty}^{\infty} d_k r(t - kT_b), \tag{1.5}$$

where $r(t)$ is a rectangular pulse lasting one UI (T_b) and d_k represents the kth transmitted bit, with the proviso that when the transmitted bit is a zero, $d_k = -1$. The input signal may also include a dc offset term, not shown in (1.5). However, this can be ignored for the time being. The input pulse $r(t)$ is given by

$$r(t) = \begin{cases} 1, & \text{for } 0 \le t < T_b \\ 0, & \text{otherwise.} \end{cases} \tag{1.6}$$

Using the unit-step function, $u(t)$, as a building block, the input pulse becomes

$$r(t) = u(t) - u(t - T_b). \tag{1.7}$$

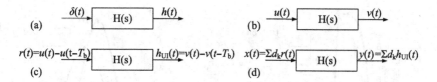

Figure 1.16 Relationship between unit impulse, $\delta(t)$, impulse response, $h(t)$, unit-step input, $u(t)$, step response, $v(t)$, unit input pulse, $r(t)$, pulse response, $h_{UI}(t)$, overall input, $x(t)$ and overall output $y(t)$. Linear time-invariant system assumed.

Figure 1.15 shows the first three bits of $x(t)$ decomposed into $-r(t)$, $r(t - T_b)$ and $-r(t - 2T_b)$.

By applying the principle of superposition, we can find the response of a system to a data sequence $x(t)$ if we know the system's response to the unit pulse $r(t)$. The response of a system to the unit pulse is referred to as the unit-pulse response, denoted as $h_{UI}(t)$. This notation is used to conjure up the notion of impulse response from linear time-invariant (LTI) system theory; however, unlike the traditional impulse response that is due to an input of infinite amplitude, but vanishing duration (i.e., $\delta(t)$), the pulse response is due to a unit-amplitude signal lasting one UI.

Figure 1.16 shows an LTI system, $H(s)$, with various input signals applied. In (a) the unit impulse $\delta(t)$ at the input gives rise to the impulse response $h(t)$. When the input is the unit step $u(t)$, the output is the step response, denoted here as $v(t)$. Since $u(t) = \int_{-\infty}^{t} \delta(\tau)d\tau$, $v(t) = \int_{-\infty}^{t} h(\tau)d\tau$. Figure 1.16 (c) shows that $h_{UI}(t)$, which is the system's response to the unit pulse $r(t)$, can be computed as:

$$h_{UI}(t) = v(t) - v(t - T_b),\tag{1.8}$$

where $v(t)$ is the unit-step response. Finally, as shown in Figure 1.16 (d), if we know $h_{UI}(t)$, we can find the output of the system $y(t)$ as a summation of shifted and scaled instances of $h_{UI}(t)$:

$$y(t) = \sum_{k=-\infty}^{\infty} d_k h_{UI}(t - kT_b).\tag{1.9}$$

The pulse response shows the gain of the system as well as the severity of ISI introduced by the system. Figure 1.17 shows examples of pulse responses. Each continuous-time pulse response is overlaid with a discrete-time sequence of points, constructed by sampling the continuous-time signal at intervals of T_b relative to the peak of each $h_{UI}(t)$. We refer to the peak as the main cursor (denoted as h_0) and other samples as ISI terms ($h_k, k \neq 0$). Samples to the left of h_0 ($k < 0$) are referred to as pre-cursor ISI whereas samples to the right ($k > 0$) are post-cursor ISI.

In Figure 1.17 (a), pulse responses of first-order systems are presented. For signal $h_{UI1}(t)$ the data rate (f_b) and the bandwidth (f_{bw}) are both equal to 25 GHz, giving a pulse response that settles to its final value within one UI ($h_0 = 1$). This pulse response also decays to zero within the second UI ($h_1 = 0$). Intuitively, this system at this data rate will exhibit negligible ISI. If the bandwidth were decreased to half the data rate the pulse response, as depicted by h_{UI2}, reaches to a slightly lower level

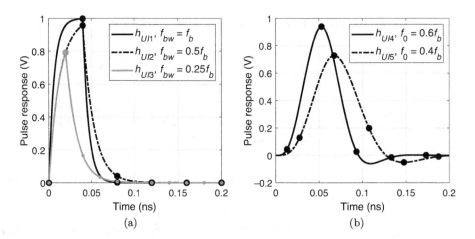

Figure 1.17 Various pulse responses: (a) pulse response of first-order systems. Data rate, $f_b = 25$ Gb/s for h_{UI1} and h_{UI2}, whereas for $h_{UI3}, f_b = 50$ Gb/s; (b) pulse response of fourth-order systems consisting of two second-order systems with $Q = 0.7$ for $f_b = 25$ Gb/s.

in the first UI. This gives rise to a lower effective gain. Also, the pulse response will not decay to zero in the second UI. These two effects indicate the presence of ISI, which will degrade the link's performance. However, the lower system bandwidth will be easier to implement and will also add less noise to the signal. Hence, we usually select bandwidths in the range of 50 to 70% of the data rate.

Since the pulse response is constructed based on the step response of a system, it is characteristic of the system, but it also depends on the targeted data rate. Whereas h_{UI1} and h_{UI2} had a UI of 40 ps corresponding to $f_{bit} = 25$ Gb/s, to construct h_{UI3} the unit interval was decreased to 20 ps, leaving the system unchanged, relative to that used for h_{UI2}. Compared to h_{UI2}, h_{UI3} has a smaller main cursor and more ISI. In all cases, the samples of the pulse response before the main cursor (i.e., $h_k, k < 0$) are zero. This is a property of a first-order system: The main cursor occurs at $t = T_b$. Since the pulse response is zero for $t \le 0$, pre-cursor ISI is zero.

Figure 1.17 (b) shows the pulse response of fourth-order systems, given by:

$$H(s) = \left(\frac{\omega_0^2}{s^2 + \frac{\omega_0}{Q}s + \omega_0^2} \right)^2, \tag{1.10}$$

where $Q = 0.7$ and $\omega_0 = 2\pi f_0$. The system giving rise to h_{UI4} has $f_0 = 0.6f_b$, resulting in modest ISI. In this case, the main cursor occurs for $t > T_b$, leading to $h_{-1} \ne 0$. Also, the pulse response dips below zero, indicating that ISI can be both positive and negative. When the magnitude of poles (f_0) is reduced, h_{UI5} has larger ISI terms and a reduced main cursor.

Figure 1.18 shows the pulse response of an electrical channel where the load resistance is not matched to the transmission line's characteristic impedance. The modelling of transmission lines is discussed in detail in Chapter 2. This pulse response, evaluated for 10 Gb/s, has significant ISI around the main cursor. Both h_{-1} and h_1 are greater

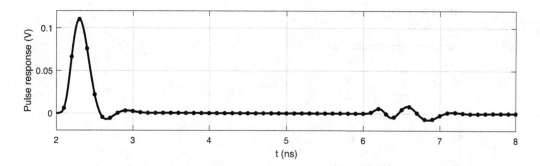

Figure 1.18 Pulse response of an electrical channel showing ISI due to channel loss and reflections.

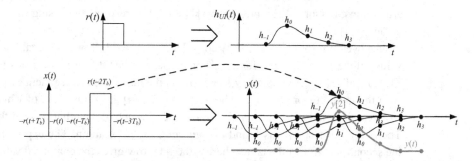

Figure 1.19 Output decomposed into summation of unit pulse responses. Top: unit pulse $r(t) \rightarrow h_{UI}(t)$. Bottom: $x(t) \rightarrow y(t)$. Output is a summation of scaled and shifted unit pulse responses $h_{UI}(t)$. Lone 1 in $x(t)$ assumed to be preceded and followed by many 0s.

than 50% of h_0. This is due to frequency-dependent loss in the transmission line. The main cursor, occurring at $t \approx 2.3$ ns, is delayed with respect to the input pulse due to the finite propagation delay of the signal along the line. Due to impedance discontinuities, signal reflections occur leading to an echo of the pulse between $t = 6$ ns and $t = 7$ ns.

Figure 1.19 illustrates how a system's response to a data stream can be constructed by summing shifted and scaled unit pulse responses. The values of the sampled pulse response are summarized in Table 1.2. The lower portion of the figure shows an input signal, $x(t)$, applied to the system. We can view this as a discrete-time sequence, $x[n]$, where $x[n]$ is the value of the signal at time t given by $nT_b < t < (n + 1)T_b$. It is assumed that many zeros have been applied before the isolated 1 starting at $t = 2T_b$, referred to as $x[2] = 1$. Each unit input pulse gives rise to an output pulse response. The output at each time point can be found by summing all pulse responses that overlap. For example, where the main cursor of the isolated one appears, the overall output is given by:

$$y[2] = h_0 - h_{-1} - h_1 - h_2 \qquad (1.11)$$

Table 1.2 Pulse response amplitude.

Sample	Value
$h_{k<-1}$	0
h_{-1}	0.02
h_0	1.0
h_1	0.5
h_2	0.2
$h_{k>2}$	0

where h_{-1} is the pre-cursor ISI from the zero ($x[3]$) following the one and h_1 and h_2 are the post-cursor ISI samples from $x[1]$ and $x[0]$, respectively. Using the values in Table 1.2, $y[2] = 0.28$.

The overall response is found by considering Table 1.3. Rows in the table show the values of the pulse response for the kth bit aligned such that h_0 is in the $n = k$ position. The row of Table 1.3 denoted by $k = 2$ shows the pulse response sequence h_k centred with the main cursor at $n = 2$. The other bits in $x[n]$ are zeros, each of which gives rise to an inverted pulse response $(-h_k)$ as seen in each of the other rows of the table. The output at each $y[n]$ is found by summing the nth column. The expression is in the second-last row of the table whereas the final row gives the computed value of the output given this particular pulse response and input sequence. The output sequence $y[n]$ and a continuous-time output $y(t)$ are shown in Figure 1.19 in grey. During the long run of zeros, ISI pushes the steady-state value of $y[n]$ down from $-h_0 = -1$ to $-\Sigma_k h_k$. Due to ISI, the output due to the isolated 1 does not reach $h_0 = 1$, but reaches only $h_0 - ISI = 0.28$. This is much closer to $L_d = 0$ than h_0.

In the absence of ISI, a transmitted 1 leads to a value of $y(t) = h_0$ and a transmitted 0 leads to a value of $y(t) = -h_0$. In this scenario a decision threshold of 0 is assumed since it is the midpoint between the zero and one signal values. Since we assume all combinations of transmitted bits are possible, the ISI from preceding and following bits can degrade a given main cursor. Therefore, the worst-case ISI is computed as:

$$ISI_{MAX} = \sum_{k \neq 0} |h_k|. \tag{1.12}$$

Given an eye diagram, the vertical eye opening (VEO) is the most important amplitude metric if we assume that the signal will be processed without further equalization to remove ISI. Thus, VEO is computed as

$$VEO = (h_0 - ISI_{MAX}). \tag{1.13}$$

The peak to peak VEO would be $2(h_0 - ISI_{MAX})$.

The polarity of ISI has no bearing on the VEO calculation since we assume all combinations of bits are possible, giving rise to a combination for which all ISI combines destructively with the main cursor. Interestingly, the values associated with long runs

Table 1.3 Summation of pulse responses to find system output $y[n]$ by summing shifted, scaled pulse response sequences h_k.

k	d_k	$y[0]$	$y[1]$	Contribution to $y[2]$	$y[3]$	$y[4]$	$y[5]$
-3	-1	0	0	0	0	0	0
-2	-1	$-h_2$	0	0	0	0	0
-1	-1	$-h_1$	$-h_2$	0	0	0	0
0	-1	$-h_0$	$-h_1$	$-h_2$	0	0	0
1	-1	$-h_{-1}$	$-h_0$	$-h_1$	$-h_2$	0	0
2	1	0	h_{-1}	h_0	h_1	h_2	0
3	-1	0	0	$-h_{-1}$	$-h_0$	$-h_1$	$-h_2$
4	-1	0	0	0	$-h_{-1}$	$-h_0$	$-h_1$
5	-1	0	0	0	0	$-h_{-1}$	$-h_0$
6	-1	0	0	0	0	0	$-h_{-1}$
7	-1	0	0	0	0	0	0
Total		$-(h_{-1}+$ h_0+h_1+ $h_2)$	$-(-h_{-1}+$ h_0+h_1+ $h_2)$	$-(h_{-1}-$ h_0+h_1+ $h_2)$	$-(h_{-1}+$ h_0-h_1+ $h_2)$	$-(h_{-1}+$ h_0+h_1- $h_2)$	$-(h_{-1}+$ h_0+h_1+ $h_2)$
Value		-1.72	-1.68	0.28	-0.72	-1.32	-1.72

of 1s or 0s do not correspond simply to h_0 and $-h_0$. When several 1s are transmitted, ISI from previous and following bits adds up, possibly boosting the signal value above h_0. Then, L_1 and L_0 are given by

$$L_1 = h_0 + \sum_{k \neq 0} h_k,$$

$$L_0 = -L_1. \tag{1.14}$$

Notice that L_1 is not equal to the main cursor plus the worst-case ISI. In the case of a long run of 1s, ISI is combined such that positive values of ISI increase L_1 and negative values decrease L_1. In systems with negative ISI, due to reflections or insufficient damping, the maximum value for a one, L_{1max} is larger than L_1. Thus, L_{1max} and L_{0min} are given as

$$L_{1max} = h_0 + \sum_{k \neq 0} |h_k| = h_0 + ISI_{MAX},$$

$$L_{0min} = -L_{1max}. \tag{1.15}$$

If all ISI is positive, then $L_1 = L_{1max}$ and $L_0 = L_{0min}$.

The eye diagram can also indicate the presence of jitter in a data signal. Jitter is discussed in the following section alongside noise and bit error rate.

Example 1.1   Find the unit pulse response of a unity dc gain first-order system with a -3 dB bandwidth of 4 GHz for a unit interval corresponding to 10 Gb/s. Determine the worst-case ISI relative to the main cursor.

Solution

The transfer function of this system is:

$$H(s) = \frac{2\pi f_{-3dB}}{s + 2\pi f_{-3dB}}. \tag{1.16}$$

Using (1.9) the pulse response can be found from the step response, where the step response is:

$$v(t) = 1 - \exp\left(-2\pi f_{-3dB} t\right), \text{ for } t \geq 0. \tag{1.17}$$

This gives rise to:

$$h_{UI}(t) = v(t) - v(t - T_b) \tag{1.18}$$

$$= \begin{cases} 1 - \exp\left(-2\pi f_{-3dB} t\right), & \text{for } 0 \leq t < T_b \\ (1 - \exp\left(-2\pi f_{-3dB} T_b\right))\exp\left(-2\pi f_{-3dB}(t - T_b)\right), & \text{for } t \geq T_b. \end{cases} \tag{1.19}$$

With $T_b = 100$ ps and $f_{-3dB} = 4$ GHz, the pulse response becomes:

$$h_{UI}(t) = \begin{cases} 1 - \exp\left(-2\pi 4\text{GHz}t\right), & \text{for } 0 \leq t < T_b \\ (1 - \exp\left(-0.8\pi\right))\exp\left(-2\pi 4\text{GHz}(t - T_b)\right), & \text{for } t \geq T_b. \end{cases} \tag{1.20}$$

Table 1.4 First-order pulse response amplitude.

Sample	Value
$h_{k<0}$	0
h_0	0.919
h_1	0.0744
h_2	0.0060

The peak of (1.20) occurs at $t = T_b$. Thus, the sampled pulse response will be found by evaluating (1.20) at multiples of T_b as summarized in Table 1.4. Beyond h_2 the sampled pulse response decays by $\exp(0.8\pi) = 0.081$ relative to the previous sample. The worst-case ISI is given by the sum of $h_{k>0} = 0.081$. Relative to the main cursor, worst-case ISI is 8.8% of h_0. The vertical eye opening is:

$$VEO = (h_0 - ISI_{MAX}) = (0.919 - 0.081) = 0.838 \tag{1.21}$$

In the absence of ISI, the vertical eye opening would be 1 (for a system with unity dc gain). Due to the reduced bandwidth, the VEO has been reduced to 0.838. Further reduction in bandwidth reduces the main cursor and increases the worst-case ISI, further reducing the VEO. For a first-order system, a bandwidth of 50 to 70% of the data rate is usually targeted to ensure low ISI but also low noise operation.

1.9   Frequency Content of NRZ Signals

The preceding sections have introduced time-domain techniques for considering the response of a system to random data. In this section, the frequency content of NRZ signals is presented. If we assume the bits d_k in (1.5) are random, we cannot discuss the Fourier transform of $x(t)$ per se, but rather we can describe its power spectral density (PSD). Given the formulation in (1.5) it can be shown that the PSD of $x(t)$ denoted as $S_x(f)$ is given by:

$$S_x(f) = \frac{1}{T_b}|R(f)|^2, \tag{1.22}$$

where $R(f)$ is the Fourier transform of the rectangular pulse, $r(t)$. The Fourier transform of the rectangular pulse is the well-known *sinc* function, given by:

$$R(f) = T_b \frac{\sin(\pi f T_b)}{\pi f T_b} = T_b \text{sinc}(\pi f T_b). \tag{1.23}$$

Therefore, the PSD of NRZ data ($S_x(f)$) simplifies to:

$$S_x(f) = T_b \text{sinc}^2(\pi f T_b). \tag{1.24}$$

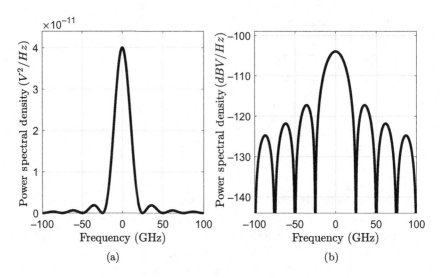

Figure 1.20 Power spectral density of 25 Gb/s random NRZ data: (a) linear vertical axis; (b) vertical axis in dBV/Hz.

The PSD of a data stream consisting of random NRZ data at 25 Gb/s is shown in Figure 1.20 (a) with a linear vertical scale and in part (b) with a logarithmic vertical scale. The PSD is 0 at non-zero multiples of $f_{bit} = \frac{1}{T_b}$. Most of the power is within the central lobe extending from $-f_{bit}$ to f_{bit}. This motivates bandwidths on the order of 50–70 % of the data rate to process these signals. Nevertheless, the time-domain response remains the preferred method to judge signal integrity.

1.10   Noise and Bit Error Rate

A critical metric of a communication link is its bit error rate (BER). Once a target BER is achieved, link designs operating at a given data rate are compared based on other metrics such as power dissipation or chip area. The BER is defined as the number of bits that are erroneously received divided by the total number of bits transmitted. Of course, to make a useful statement about a link's BER, enough bits must be transmitted. While this sounds obvious, since 10^{-12} is a commonly targeted BER, one must receive at least 10^{12} bits before a remotely accurate measure of BER can be made. One approach that is often taken is to test for "error-free" operation among a number of bits slightly larger than $1/BER$. For example, a BER of better than 10^{-12} can be claimed with 95% confidence if error-free operation for 3×10^{12} bits is observed. Note that at 10 Gb/s, 10^{12} bits requires 100 s to transmit.

Bit errors are due to several factors including noise, jitter, power-supply fluctuations and signal crosstalk. In order to quantify how noise effects BER, consider the eye diagram in Figure 1.21 where the thick black line shows an eye diagram without noise. The system processing the data adds negligible ISI since the signal reaches $\pm V_P$ at the decision time, t_d. The thin grey line shows the signal when noise is added.

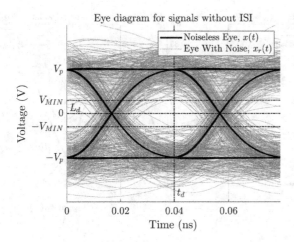

Figure 1.21 Eye diagram for signals without ISI: $x(t)$ is noiseless whereas $x_r(t)$ is the received signal with noise added.

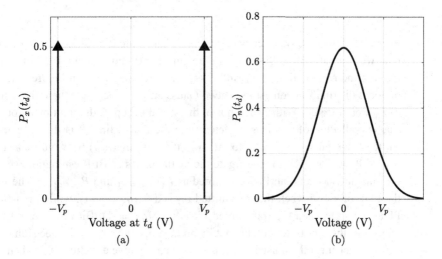

Figure 1.22 Probability density function of: (a) the signal $x(t_d)$ and (b) additive noise $n(t_d)$.

We will start with an idealized scenario in which the signal is applied to an ideal decision circuit. The decision circuit is said to be ideal because it always outputs a logic high when a signal larger than a fixed decision threshold L_d is applied at the decision time t_d and outputs a logic low when the input signal is lower than the threshold. The probability density function (PDF) of $x(t)$ at the decision time t_d is shown in Figure 1.22 (a), assuming equal probability of 1s and 0s. Since the signal is either $-V_P$ or V_P, each with 50% probability, the PDF consists of δ-functions located at $\pm V_P$.

The noise $n(t)$ that was added to the noiseless signal in Figure 1.21 to give $x_r(t)$ has a Gaussian PDF with zero mean and a variance of σ_n^2. The PDF of $n(t)$ is shown in

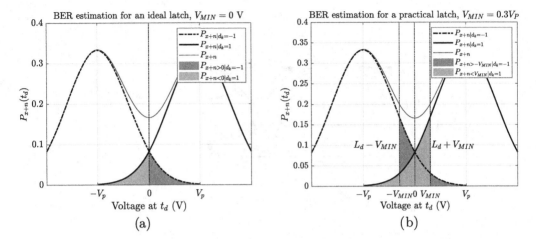

Figure 1.23 Probability density function of the signal plus noise $x(t_d) + n(t_d)$ without ISI: (a) area under the curve for calculating BER with an ideal latch and (b) area under the curve for calculating BER with a practical latch having $V_{MIN} = 0.3V_P$.

Figure 1.22 (b). The variance is larger than what would typically occur to demonstrate how we estimate BER. Since x_r is the sum of $n(t)$ and $x(t)$, the PDF of x_r is found through the convolution of P_n and $P_x(t_d)$. The result, shown in Figure 1.23 (a), is a bimodal distribution consisting of two Gaussian PDFs centred where the δ-functions of P_x were located. With added noise, the signal occupies distributions around $\pm V_P$. The overall PDF of $x + n$ is depicted with the dotted line. It is the sum of the two shifted Gaussians. If we were to measure the PDF of $x_r(t_d)$ we would construct this overall PDF. However, to investigate the mathematics of BER estimation we consider the constituent Gaussian PDFs, denoted as $P_{x+n|d_k=-1}$ and $P_{x+n|d_k=1}$. The former is the PDF of $x + n$ when the transmitted bit is a 0, ($d_k = -1$), while the latter is the PDF of $x + n$ when the transmitted bit is a 1, ($d_k = 1$). Bit errors result from the regions of the constituent PDFs that lie on the opposite side of the decision threshold. Bit errors occur if the noise is large enough in a negative direction that, when added to a 1, the overall signal is less than the decision threshold. Likewise, a bit error occurs if the noise increases a zero-level signal above the decision threshold. The regions of the PDF that give rise to errors for an ideal decision circuit are shaded in grey in Figure 1.23 (a). The probability of an error is the area within these shaded regions. Mathematically, the probability of a bit error is given by:

$$P_{ERR} = P_{d_k=1}P_{x+n<0|d_k=1} + P_{d_k=-1}P_{x+n>0|d_k=-1}, \tag{1.25}$$

where $P_{d_k=1}$ and $P_{d_k=-1}$ are the probability that a given transmitted bit is a 1 or 0, respectively, usually both equal to 0.5. Then, $P_{x+n<0|d_k=1}$ is the probability that the noise, $n(t)$, is less than $-V_P$ and $P_{x+n>0|d_k=-1}$ is the probability that the noise is larger than V_P. Although the shaded regions lie on opposite sides of each PDF's mean, their areas are equal and can be found by finding the probability that $n(t) > V_P$.

Table 1.5 Required $\frac{V_P}{\sigma_N}$ for various BER.

Target BER	Required $\frac{V_P}{\sigma_N}$ using (1.27)	Estimated $\frac{V_P}{\sigma_N}$ using (1.28)
10^{-6}	4.7534	4.7615
10^{-9}	5.9978	6.0020
10^{-12}	7.0344	7.0371
10^{-15}	7.9413	7.9432

The probability of a bit error becomes:

$$P_{ERR} = P_{n>V_P} = Q\left(\frac{V_P}{\sigma_N}\right). \tag{1.26}$$

Here, $Q(x)$ gives the area under the tail of a standard Gaussian distribution beyond $\frac{V_P}{\sigma_N}$. $Q(x)$ is given by:

$$Q(x) = \int_x^\infty \frac{1}{\sqrt{2\pi}} \exp\left(\frac{-u^2}{2}\right) du. \tag{1.27}$$

The result of integration can be looked up in tables or, for $x > 3$, can be approximated as:

$$Q(x) \approx \frac{1}{x\sqrt{2\pi}} \exp\left(\frac{-x^2}{2}\right). \tag{1.28}$$

Example 1.2   Find the BER if $\sigma_n = 4$ mV and $V_P = 28$ mV.

Solution
Substituting $x = \frac{V_P}{\sigma_n}$ into (1.28) evaluates to:

$$Q(7) \approx \frac{1}{7\sqrt{2\pi}} \exp\left(\frac{-7^2}{2}\right) = 1.305 \times 10^{-12}. \tag{1.29}$$

The required $\frac{V_P}{\sigma_N}$ for various BER is summarized in Table 1.5. Due to the shape of a Gaussian distribution, the BER decreases significantly for small increases in SNR. The righthand column in Table 1.5 shows the estimated SNR required for each BER using (1.28). These values are close to but larger than those computed using (1.27), meaning that the approximation is slightly pessimistic.

A real latch requires a minimum voltage swing in order to reliably resolve its input value in the required amount of time. For example, if operated with a unit interval of 100 ps, a given latch might require an input differential voltage of 40 mV. Input voltages within the range of -40 mV to $+40$ mV may lead to metastability and ultimately errors. The horizontal lines in Figure 1.21 at $\pm V_{MIN}$ define a region within which

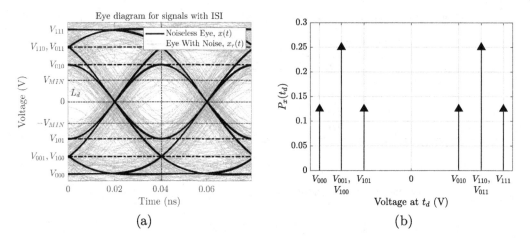

(a)   (b)

Figure 1.24 Eye diagram and PDF of a signal with ISI: (a) eye diagram showing levels at $t = t_d$ associated with different bit patterns centred around t_d; (b) PDF of the signal at $t = t_d$.

bit errors will occur. The impact on BER is visible in Figure 1.23 (b) where an error occurs if noise leads to an x_r falling between $-V_{MIN}$ and $+V_{MIN}$. The latch may also have some equivalent input noise, but this can be combined with the noise in the signal to give an overall noise. If we are pessimistic and assume all inputs between $-V_{MIN}$ and $+V_{MIN}$ lead to bit errors, the BER can be calculated as:

$$P_{ERR} = Q\left(\frac{V_P - V_{MIN}}{\sigma_N}\right). \tag{1.30}$$

A BER of 10^{-12} requires an argument of $Q() \approx 7$. For the case of $V_{MIN} = 0$, this corresponds to a ratio of V_P to σ_N of 7, or a peak-to-peak signal to noise ratio of 14. With $V_{MIN} \neq 0$, for a BER of 10^{-12} we require:

$$V_P = 7\sigma_N + V_{MIN}. \tag{1.31}$$

Another way of interpreting (1.30) is as follows: suppose for an ideal latch a given BER requires an SNR of x. This means that V_p must be a distance of $x\sigma_N$ from the decision threshold. With a real latch, V_p must increase so that it is a distance of $x\sigma_N$ from V_{MIN}.

The ISI degrades the BER because for some bit combinations the signal will be closer to the decision threshold than V_P. Consider the eye diagram in Figure 1.24 (a). Due to ISI the noiseless signal does not reach to $\pm V_P$ when there is a bit transition. The levels reached by the signal are shown on the y-axis by a three-bit sequence where the middle number is the transmitted bit at the decision time. For example, V_{110} indicates that the current and previous bits were 1 while the next bit is a 0. The PDF of the signal is shown in Figure 1.24 (b). At the decision time, the signal is $\pm V_P$ only one-quarter of the time unlike in the ISI-free case where the signal is always $\pm V_P$. The closest the signal is to L_d at t_d is V_{010} or V_{101}. For the noiseless signal, these voltages determine the VEO.

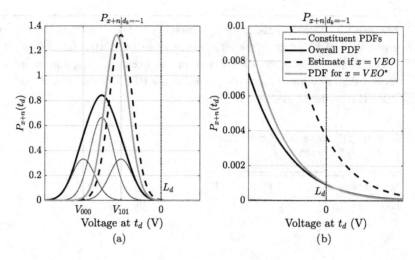

Figure 1.25 Probability density function of the signal plus noise $x(t_d) + n(t_d)$ with ISI: (a) PDFs for calculating BER with ISI and (b) expanded view of PDFs for calculating BER with ISI.

The overall PDF once noise is introduced is the convolution of the Gaussian noise PDF with the collection of δ-functions in Figure 1.24 (b). The result for $d_k = -1$ is shown in Figure 1.25. Each Gaussian noise PDF must be scaled by the amplitude of the δ-functions. The thin dotted lines show the PDFs centred at each possible noiseless voltage. The thick solid line shows the overall PDF. Figure 1.25 (b) shows an expanded view of the PDFs around the decision threshold. We make the following observations for this particular case:

1. The overall PDF (solid line) coincides with the constituent PDF centred at V_{101} – i.e., the PDF closest to L_d. Here this constituent PDF took into account that $P_{x=V_{101}} = 0.125$.
2. A PDF centred at V_{101} but with a weight assuming $P_{x=V_{101}} = 0.5$ would pessimistically estimate BER. This is shown in the thick dashed line ($x = $ VEO). This is what the PDF would be for an ISI-free signal with the same VEO as the signal with ISI.
3. A pessimistic estimate of BER might lead a designer to increase the amplitude of the signal to achieve a desired BER, giving rise to a VEO of VEO*. The slight leftward shift from VEO to VEO* shows that the pessimistic BER estimation from assuming $P_{x=V_{101}} = 0.5$ leads to a small over-design, which is generally acceptable.

As alluded to above, rather than construct a PDF with six shifted Gaussians, a pessimistic estimate of BER can be found by assuming $P_{x=V_{101}} = 0.5$; or, put another way, that the signal is $\pm V_{VEO}$ at t_d. Equation (1.30) can be modified to include ISI as:

$$P_{ERR} = Q\left(\frac{V_{VEO} - V_{MIN}}{\sigma_N}\right). \tag{1.32}$$

Example 1.3   Use the pessimistic approach in (1.32) to determine the required V_P to give a BER of 10^{-12} given the estimates in Table 1.6 of $x(t_d)$ for various bit combinations. Assume that $\sigma_n = 14$ mV and for this exercise, V_{MIN} can be ignored. As a second step, find the value of V_P needed if the BER associated with the constituent PDFs is considered, as shown in Figure 1.25.

Table 1.6 $x(t_d)$ for various bit combinations; middle bit is the symbol at t_d.

Bit combination	Fraction of V_P
000	−1
001	−0.76
100	−0.76
101	−0.515
010	0.515
011	0.76
110	0.76
111	1

Solution
Solving (1.32) for x, such that $Q(x) = 10^{-12}$ gives $x = 7.037$. This means that $VEO = x\sigma_n = 98.52$ mV. This in turn means that $V_P = \frac{VEO}{0.515} = 191.3$ mV.

Table 1.7 summarizes the estimation of the BER taking into account the probability of each signal level in the eye diagram. The PDF centred at $-0.515\,V_P$ contributes the most. However, since it only occurs with 0.25 probability, rather than giving a BER of 10^{-12} as expected, we get a BER of 0.25×10^{-12}. Therefore the required value of V_P could be reduced very slightly to about 186 mV. Here, $Q(x = 6.842) = 4 \times 10^{-12}$, which when scaled by $P_{x=V_{101}} = 0.25$ gives a BER of 10^{-12}. The top two entries in the righthand column of Table 1.7 are ignored because they are several orders of magnitude below the target BER.

Table 1.7 Bit error rate estimation from constituent PDFs. $V_P = 191.3$ mV.

Bits	$P_{x\vert d_k=-1}$	Fraction of V_P	V (mV)	$x = \frac{V}{\sigma_n}$	$Q(x)$	P_{ERR}
000	0.25	−1	−191.3	13.664	8.34×10^{-43}	−
001, 100	0.5	−0.76	−145.4	10.385	1.47×10^{-25}	−
101	0.25	−0.515	−98.52	7.037	1.00×10^{-12}	2.5×10^{-13}

The pessimistic approach taken by applying (1.32) estimated a required V_P to be 191.3 mV, only 5 mV larger than the more detailed analysis. Thus, the simplicity of (1.32) outweighs the slight over-design in required V_P.

The aforementioned analysis assumes that the latch captures the input signal's value instantly and is always activated at the same location in the eye. In reality the latch responds to an input signal over an interval of time, a phenomenon discussed in Chapter 3. Secondly, due to clock jitter, actual decision times vary relative to uniformly spaced decision times.

In the above analysis, the noise associated with 1s and 0s was assumed to be equal. This is usually denoted as N_1 and N_0, respectively. Generally, N_1 and N_0 have identical variance; however, in certain optical links where amplification is done in the optical domain, noise is signal-dependent, leading to more noise for the L_1 level than the L_0 level.

1.11   Jitter

To define and discuss jitter, consider an ideal jitter-free periodic signal CLK_I having a uniform period of T_b, shown in Figure 1.26. Zero crossings are uniformly spaced. On the other hand CLK_J has jitter, manifested as deviations in its zero crossings near T_b, $2T_b$, $3T_b$ ... at zero-crossings 1, 2, 3... Assuming CLK_J differs from CLK_I only in how quickly it evolves as a function of time, we can write:

$$CLK_J(t) = CLK_I(t + J(t)), \tag{1.33}$$

where $J(t)$ is the difference in the arrival time of the rising edges of the two clock signals. In this case, $J(t)$ is the absolute jitter of $CLK_J(t)$ relative to $CLK_I(t)$. Further discussion of the frequency content of jitter and its relationship to phase noise is presented in Chapter 13. For the time being, we will consider random and deterministic jitter. Random jitter results from thermal and flicker noise sources in resistors and transistors. These give rise to a Gaussian distribution of $J(t)$. Deterministic jitter (DJ) results from power-supply noise and other signal coupling. The distribution of deterministic jitter may have only discrete values, but given that it may originate from several sources, it may have an intractably large number of values. Random jitter is unbounded, due to its Gaussian distribution while deterministic jitter is bounded.

Consider a deterministic jitter bounded by $+/-$DJ and random jitter with σ_{RJ}. Conceptually, a bit error arises if, at the actual decision time given by $J(t)$, the signal noise moves a transmitted 1 to a received 0, or moves a transmitted 0 to a received 1. A consequence of moving a decision time left or right relative to the ideal decision time is that invariably the eye closes, reducing the distance between the threshold and signal, and thereby increasing the likelihood that noise moves the signal to the wrong side. Mathematically the joint probability distribution between noise and jitter must be considered. A simpler, but more pessimistic approach is to require that, for the target BER, the eye is open enough to yield the target BER even when the jitter is within its worst-case range as set by the target BER. In reality, the zero crossings of both a clock signal and a data signal will deviate from their ideal locations due to jitter. The preceding discussion can be adapted to account for this by assuming the data signal has no jitter and combining the data jitter with the clock jitter.

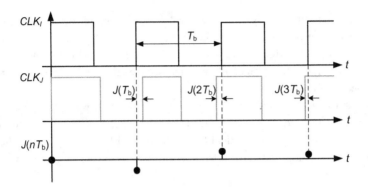

Figure 1.26 Clock signals without (CLK_I) and with (CLK_J) jitter. Discrete-time jitter signal ($J(nT_b)$).

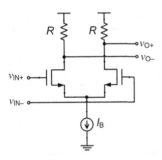

Figure 1.27 Example of a current-mode logic buffer with resistive loads.

1.12 Full-Rail CMOS vs CML Philosophy

The traditional approach to design mixed-signal wireline circuits has been to use resistively or actively loaded differential pairs, possibly with inductors, giving wireline circuits a look of analog circuits. Circuits of this class are often referred to as current-mode logic circuits, or CML. An example of a CML circuit is shown in Figure 1.27. It can operate as a small-signal amplifier or as a logic buffer, depending on the input signal swing. With degeneration resistors added to help with linearity, we see this circuit form the core of equalizers and amplifiers. Without any extra elements to linearize it, this circuit can be used as a clock buffer, or as part of a ring oscillator. Two CML buffers can be combined to form a high-speed latch, which in turn can be used to build a flip-flop.

As emphasized in undergraduate electronics courses, the output signal is taken differentially: i.e., as the difference between the two outputs: $v_O = v_{O+} - v_{O-}$. This, along with the use of a current source, improves power-supply and common-mode rejection. By drawing constant current from the supply, switching activity does not modulate the supply voltage. This reduction in induced power-supply noise further reduces coupling between circuits. However, CML circuits have a few disadvantages, some of which become more significant as the supply voltage is reduced in smaller feature size technologies.

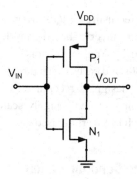

Figure 1.28 CMOS inverter.

The first challenge with CML circuits is the current source. As supply voltage is reduced, the dc voltage across each transistor must reduce. With lower V_{DS} a larger width transistor must be used, which leads to larger capacitance at the output of the current source. This capacitance reduces the output impedance of the current source at frequencies above dc, reducing common-mode rejection of the circuit, and thereby reducing the potential benefit of CML. By allocating more voltage across the current source, the overdrive voltage on the switching pair is reduced. This reduces the f_T of the switching pair, limiting the speed of the circuit.

An approach that is becoming more widespread since the early 2000s is the use of CMOS inverter-based circuits, an example of which is shown in Figure 1.28. The transistors in a CMOS inverter can have a V_{GS} of V_{DD} when the circuit switches. However, they can also be used as small-signal amplifiers when biased around the inverter's switching threshold. CML circuits tend to be larger than CMOS circuits due to the current source transistor, often designed with larger-than-minimum length and due to the loads, often implemented as polysilicon resistors. Although much smaller, without a current source, CMOS circuits are much more prone to power-supply noise, since variations in V_{DD} lead to delay variations when used as logic circuits. When used as amplifiers, power-supply noise couples into the output signal and modulates the circuit's bias point. Also, CMOS logic circuits induce more power-supply noise since current is drawn intermittently, or only when the circuit switches. Nevertheless, with careful regulator design, decoupling capacitance and the use of pseudo-differential configurations, CMOS circuits are being used more widely in wired communication systems.

IC designers might assume that an inverter was always used as a logic circuit with full-rail inputs before this relatively recent trend. This has not always been the case; in fact Frank Wanless's 1967 patent (filed in 1963, assigned to Fairchild) describing the use of CMOS transistors in logic circuits [7] only slightly predates Joseph Burns's inverter-based amplifier patent [8] that was filed in 1967 and issued in 1968. The 1970s brought a flurry of activity aimed at improving the common-mode and power-supply rejection of inverter-based CMOS amplifiers with patents from Sheng Teng Hsu [9], Andrew Dingwall [10] and Richard Pryor [11]. Interesting from a historical perspective, the four inverter-based amplifier patents were all assigned to Radio Corporation of America (RCA). No sooner had engineers started using CMOS inverters for logic than they used them for amplifiers. This interest in inverter-based amplifiers waned as CMOS operational amplifiers were pioneered by Yannis Tsividis in 1976 [12], though they enjoyed a second wave in the late 1980s and early 1990s through the work of Bram Nauta [13] [14]. Since the early 2000s, designers recognize that for high-speed operation their simplicity makes them compatible with highly scaled CMOS technology. This book presents both CML and CMOS inverter-based circuits.

1.13   Summary and Future Directions in Wireline Communication

This chapter introduced much of the terminology needed for discussing wireline links, namely: eye diagram, pulse response, intersymbol interference, bit error rate, extinction ratio, optical modulation amplitude and jitter. Block diagrams of some wireline link examples were presented showing similarities and differences between electrical and optical transceivers, and CMOS and CML circuit building blocks were introduced.

Per-lane data rates have increased steadily over the past decades and will continue to rise. It is unclear what will happen if CMOS scaling stops at a future technology node. However, further improvements in data rate can come from improved packaging giving rise to better channels. To increase data rate, without increasing the symbol rate, PAM4 will be deployed more widely. In fact as early as ISSCC 2019, most transceivers for 56 Gb/s and higher used PAM4 signalling. This is significant because PAM4 receivers usually require additional forward-error control codes (FEC) to achieve low BER. While most NRZ links can operate at 10^{-12} BER, with the smaller VEO, PAM4 receivers rely on FEC to take the BER from a pre-FEC level of around 10^{-3} to a post-FEC 10^{-12}. At ISSCC 2020, all transceivers operating at 112 Gb/s used PAM4 signalling. Naturally, one might wonder if even higher-order modulation will make inroads into short-reach links.

Long-reach telecom links that use coherent receivers have leveraged complex modulation schemes for decades. However, these require high-speed ADCs in the receivers. CMOS scaling has improved the energy efficiency of ADCs considerably to the point where today, ADCs are used in backplane electrical links. Once a link has an ADC, the system designer can consider transmitting more bits per symbol since detection is done in the digital domain. Although high-speed digital signal processing may still use considerable power, its speed and power dissipation improves with CMOS scaling. Thus, wider use of DACs and ADCs is expected.

In optical links a few trends are expected. In data centres where low-cost vertical cavity surface emitting lasers (VCSELs) and multi-mode fibre (MMF) have been the preferred technology, increasing distances and data rates will favour the deployment of single-mode fibre. Within the photonics community, advances in photonic integration, notably silicon photonics, have been shown to drastically reduce the size of transceivers for telecom links. However, we do not yet see widespread use of silicon photonic links between nearby chips. Looking ahead, we expect to see electrical to optical and optical to electrical conversions being moved closer to the large digital chips so as to reduce the impact of the electrical interfaces. We also expect to see optical links being deployed over physically shorter links.

Problems

1.1 Estimate the BER if a signal with a noiseless peak-to-peak VEO of 40 mV and a root-mean-squared noise of 4 mV is applied to an ideal latch. To what level must the noise be reduced to achieve a BER of 10^{-12}?

1.2 Estimate the BER if a signal with a noiseless peak-to-peak VEO of 40 mV and a root-mean-squared noise of 4 mV is applied to a decision circuit requiring a minimum peak-to-peak voltage of 15 mV. To what level must the signal's amplitude be increased to achieve a BER of 10^{-12}?

1.3 Consider the system with a transfer function of $H(s) = \frac{2\pi f_p}{s+2\pi f_p}$:

(a) Find the pulse response of the system if $f_p = 5$ GHz and the data rate is 10 Gb/s.
(b) How much ISI relative to the main cursor is present?
(c) What is the vertical eye opening assuming an input signal of 1 V_{PP}?
(d) Repeat (a) to (c) if f_p is reduced to 3 GHz?
(e) At what value of f_p does the eye become closed?

1.4 From the step response of a channel shown in Figure 1.29:

(a) Sketch the pulse response for 5 Gb/s data.
(b) Determine the amount of ISI relative to the main cursor.
(c) Indicate whether random data sent through this channel would have a closed or open eye.

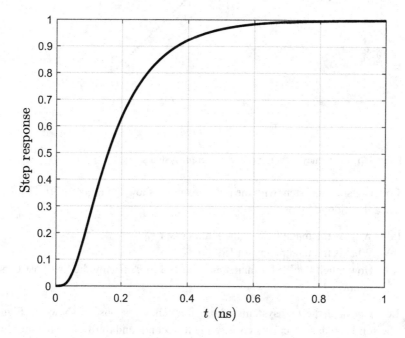

Figure 1.29 Channel step response for problem 1.4.

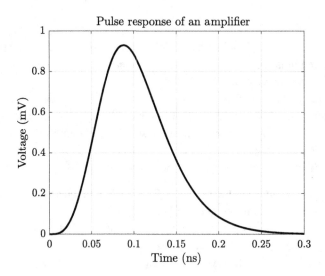

Figure 1.30 Amplifier pulse response for 20 Gb/s data.

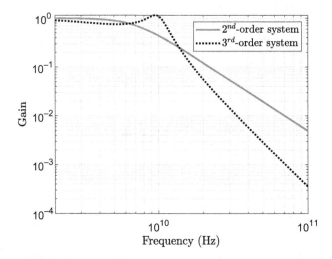

Figure 1.31 Frequency response of two different systems.

1.5 Consider a system with the pulse response shown in Figure 1.30. The input pulse had a duration of 50 ps.

(a) What is the magnitude of the main cursor h_0?
(b) What is the total precursor ISI?
(c) How much total ISI remains, normalized to h_0 if only the first post-cursor were removed?

1.6 Consider the two systems whose frequency response is shown in Figure 1.31. The solid and dashed curves correspond to second- and third-order transfer functions, respectively, given by:

$$H_{2nd} = \frac{(2\pi 7 \text{ GHz})^2}{s^2 + \sqrt{2}(2\pi 7 \text{ GHz})s + (2\pi 7 \text{ GHz})^2}, \qquad (1.34)$$

$$H_{3rd} = \frac{2\pi 3.5 \text{ GHz}(2\pi 10 \text{ GHz})^2}{(s + 2\pi 3.5 \text{ GHz})(s^2 + 0.30(2\pi 10 \text{ GHz})s + (2\pi 10 \text{ GHz})^2)}. \qquad (1.35)$$

The second-order transfer function has a quality factor of $\frac{1}{\sqrt{2}}$, leading to the maximally flat (Butterworth) response. It has complex conjugate poles, but no peaking. Its bandwidth is equal to the pole magnitude (7 GHz in this case). The third-order transfer function has a bandwidth of 11.2 GHz and 1.2 dB of amplitude peaking. This question investigates whether the larger bandwidth of the third-order transfer function translates to a better pulse response and reduced ISI.

(a) Using Matlab, plot the pulse response of the two systems for 10 Gb/s data.
(b) Determine the amount of ISI and resulting VEO.
(c) Based on your findings in (b) which system is better?

1.7 Sketch the power spectral density of an NRZ data stream of random bits at a bit rate of 56 Gb/s. Compute the power spectral density at dc, 28 GHz, 56 GHz and 84 GHz. Be sure to give appropriate units.

2 Electrical Channels

Electrical-link design is challenging due to frequency-dependent loss and reflections associated with electrical channels as well as crosstalk between nearby channels. The equations describing lossless and lossy transmission lines are introduced in this chapter, followed by a brief discussion of loss mechanisms. The characteristics of various channels are presented, along with the effect of wirebonds and packages.

The goal of this chapter is to provide link designers with a methodology to estimate the overall pulse response of a channel consisting of a lossy transmission line and the relevant package and chip parasitic elements. Knowing the pulse response, a link designer can then contemplate and model the equalization (discussed in Chapter 4) needed to properly detect transmitted bits. This chapter is organized to discuss transmission line fundamentals and then overall channels. Readers coming from electronics may have forgotten their undergraduate electromagnetic wave theory; thus, material in Section 2.1 serves as a review.

Figure 2.1 shows the frequency response of two representative channels as quantified by their s_{21} response. Part (a) shows a lossy channel where the attenuation increases slowly at lower frequency before increasing more steeply at higher frequency. Such a channel gives rise to a reduced pulse height as well as a wider pulse with some pre-cursor and possibly significant post-cursor ISI. Unlike Figure 2.1 (b), part (a) has a monotonically decreasing s_{21}, leading to a relatively "well-behaved" pulse response. Channels such as this can be equalized, with increasing power dissipation up to about 45 dB of loss at the Nyquist frequency (i.e., $f_{bit}/2$ for NRZ signalling).

The response in Figure 2.1 (b) also has increased attenuation at higher frequency. However, there is a notch, typically caused by an impedance discontinuity, such as a via stub in a printed circuit board (PCB) or a connector. Impedance discontinuities lead to reflections. Rather than producing a pulse response with a tail of gradually decreasing ISI, a notch can predict "echos" in the pulse response, meaning that large post-cursor ISI can appear 10s of UIs after the main cursor, whose exact location varies based on the channel length. This complicates equalization.

In addition to the input-output transmission of a given electrical channel, link designers must also contend with crosstalk between nearby channels. This is shown in Figure 2.2. Here, it is assumed that Tx 1 sends a one that appears at Rx 1, with smaller amplitude and spread out in time. However, a copy of this pulse appears, with additional attenuation at Rx 2. We can view it as an additional ISI term, further reducing

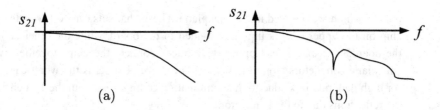

Figure 2.1 Frequency response of electrical channels: (a) well-behaved channel and (b) channel with impedance discontinuities.

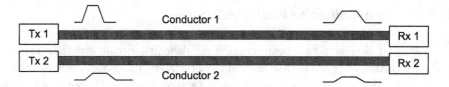

Figure 2.2 Crosstalk in parallel conductors.

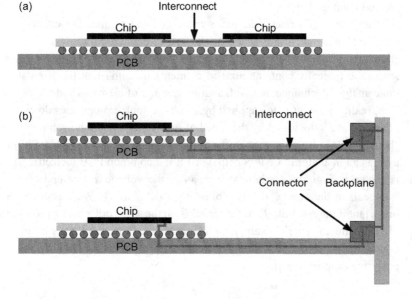

Figure 2.3 Electrical channel examples: (a) chip-to-chip interconnect through a package; (b) interconnect through a backplane.

Rx 2's vertical eyeopening (VEO). However, unlink ISI due to a given channel, equalizing crosstalk onto channel 2, requires knowledge of link 1's operation. Of course, crosstalk is reciprocal; pulses from channel 2 are added to the signal on link 1.

A cross-sectional view of example channels is shown in Figure 2.3, taken from the OIF-CEI working group. The upper link connects two chips within the same package. Such a short-reach link will present 5–10 dB of loss allowing high-speed transmission with modest equalization. The lower link connects two chips that sit on different PCBs, both of which are plugged into a backplane. These channels can be 1 m in length. The chip on the upper card is flip-chip bonded to the board, while the other connects to the board via a package. Copper PCB traces connect the chips together, with two backplane connectors along the way. This type of channel is the worst-case channel for high data rate links due to the combination of high loss from the 1 m channel and the reflections due to the connectors.

2.1   Transmission Line Fundamentals

Chip designers have been trained from their first circuit analysis course to analyze circuits using traditional, lumped circuit analysis techniques, such as nodal analysis. In using such a technique, elements are assumed to be connected to one another using ideal conductors that present no resistance, inductance or capacitance. The voltages at opposite ends of such a conductor are always equal, regardless of how quickly the applied voltage changes.

In an "on-chip" context the series resistance and parallel capacitance of intercon-nects are often taken into account when an integrated-circuit layout is "extracted," allowing a simulation to include said parasitic components. However, the simulated schematic typically contains lumped elements meaning that the simulator uses the same analysis technique, but with a larger number of elements and nodes.

In reality, a voltage step applied by a voltage source propagates down a conductor as a wave at a velocity set by the speed of light in the materials around it. In air this is about 30 cm/ns dropping to $\approx$15 cm/ns for a conductor surrounded by SiO_2 (the oxide used as an insulator in CMOS chips) or FR4, a common PCB dielectric. An example is shown in Figure 2.4 where voltage v_1 is the voltage at one end of a conductor. It appears at the other end as v_2 following a delay of T_d. Wave phenomena become important if the rise time T_r of the signal is on the same order as the propagation delay of the conductor T_d. For example, if a signal has a 10 ps rise time, we can use lumped circuit analysis only if T_d is much less than 10 ps. According to [15], lumped circuit analysis is appropriate if:

$$\frac{T_r}{T_d} > 6. \tag{2.1}$$

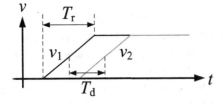

Figure 2.4 Delay and rise time example.

In Figure 2.4, $T_d < T_r$ but the criterion in (2.1) is not met. For a 10 ps rise time, lumped analysis is appropriate if $T_d < 1.6$ ps. This corresponds to a length (L) of:

$$L = vT_d = 15\ \frac{\text{cm}}{\text{ns}} 1.6\ \text{ps} = 0.024\ \text{cm} = 240\ \mu\text{m}, \tag{2.2}$$

where v is the speed of light in the material surrounding the conductor; 240 μm is a considerable distance on a chip. Most wires will be shorter than this length allowing lumped analysis, although their series resistance and parallel capacitance may nevertheless be important. In the case of a longer line, a distributed model that accounts for wave propagation and reflections must be used. It is important to remember that (2.1) relates to the signal's rise/fall times (T_r) and not the bit period (T_b).

The rest of this section reviews the following topics fundamental to understanding transmission line operation:

1. Wave propagation along lossless lines.
2. Characteristic impedance.
3. Reflections and reflection coefficients.
4. Lossy transmission lines.

The presentation of this material was helped greatly by [16].

2.1.1   Distributed Model of a Lossless Transmission Line

A model of a lossless transmission line is shown in Figure 2.5. The transmission line is assumed to be uniform having a per-unit length inductance of l and a per-unit length capacitance of c. Therefore, a section of line of length Δz has inductance and capacitance of $\Delta z l$ and $\Delta z c$, respectively. One could imagine dividing the transmission line into several sections and use lumped-element circuit analysis (or circuit simulation tools), although this usually requires many sections, particularly for a lossless line, and never accurately predicts the step response of the line or its response to frequencies above $\frac{1}{\Delta z \sqrt{lc}}$. Instead, in this section, a distributed model is derived, by writing circuit-analysis equations for the lumped section. This gives equations parameterized by Δz which can be rewritten by letting $\Delta z \to 0$.

The single section in Figure 2.5 (b) can be analyzed with a Kirchhoff's voltage law (KVL) and Kirchhoff's current law (KCL) equation giving:

$$v(z) = v(z + \Delta z) + l\Delta z \frac{\partial i(z)}{\partial t}, \tag{2.3}$$

$$i(z) = i(z + \Delta z) + c\Delta z \frac{\partial v(z + \Delta z)}{\partial t}. \tag{2.4}$$

Each can be rearranged to isolate the spatial dependence, leading to:

$$\frac{v(z + \Delta z) - v(z)}{\Delta z} = -l\frac{\partial i(z)}{\partial t}, \tag{2.5}$$

$$\frac{i(z + \Delta z) - i(z)}{\Delta z} = -c\frac{\partial v(z + \Delta z)}{\partial t}. \tag{2.6}$$

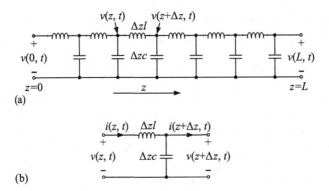

Figure 2.5 Lossless transmission line with per-unit length inductance l and capacitance c: (a) overall line; (b) expanded view of a section of length Δz. The ellipse draws attention to a small, but long-lasting ISI tail.

Taking the limits of each left-hand side as $\Delta z \to 0$ yields partial differential equations in space and time:

$$\frac{\partial v(z)}{\partial z} = -l\frac{\partial i(z)}{\partial t}, \tag{2.7}$$

$$\frac{\partial i(z)}{\partial z} = -c\frac{\partial v(z)}{\partial t}. \tag{2.8}$$

Equations (2.7) and (2.8) can each be rearranged to give an equation in either current or voltage forming the well-known wave equations:

$$\frac{\partial^2 v}{\partial z^2} = lc\frac{\partial^2 v}{\partial t^2} = \frac{1}{v_0^2}\frac{\partial^2 v}{\partial t^2}, \tag{2.9}$$

$$\frac{\partial^2 i}{\partial z^2} = lc\frac{\partial^2 i}{\partial t^2} = \frac{1}{v_0^2}\frac{\partial^2 i}{\partial t^2}, \tag{2.10}$$

where $v_0 = \frac{1}{\sqrt{lc}}$ is the phase velocity in units m/s. The solution to the wave equations is very general. In fact, any function (f) in which the argument appropriately combines space (z) and time (t) is a solution, giving rise to this general solution:

$$v(z,t) = f^+\left(t - \frac{z}{v_0}\right) + f^-\left(t + \frac{z}{v_0}\right). \tag{2.11}$$

Therefore we expect the lossless line to support a wave travelling toward increasing z given by f^+ and a wave travelling in the negative direction (f^-). The actual functional form of f is determined by the voltage generator connected to the line. In the case of a generator at $z = 0$ the amplitude and shape of f^- depends on the termination of the transmission line at $z = L$.

For a given f the current can be found. It also consists of two waves, scaled by the characteristic impedance of the transmission line:

$$i(z,t) = \sqrt{\frac{c}{l}} f^+ \left(t - \frac{z}{v_0} \right) - \sqrt{\frac{c}{l}} f^- \left(t + \frac{z}{v_0} \right) = \frac{1}{Z_0} f^+ \left(t - \frac{z}{v_0} \right) - \frac{1}{Z_0} f^- \left(t + \frac{z}{v_0} \right),$$

(2.12)

where $Z_0 = \sqrt{\frac{l}{c}}$ is the characteristic impedance of the transmission line. Notice that the f^- term in the current is subtracted from the f^+ term. Due to the linearity of the transmission line, the shape of the current wave is the same as the shape of the voltage wave, in spite of the line being composed of reactive elements.

Although (2.11) shows $v(z,t)$ as a superposition of only two terms, due to reflections at the ends of the transmission line, each of f^+ and f^- can be the sum of several waves. To illustrate this, we first define the reflection coefficient at the load end as:

$$\Gamma_L = \frac{f^-}{f^+} = \frac{R_L - Z_0}{R_L + Z_0},$$

(2.13)

where R_L is the resistance of a load connected to the transmission line at $z = L$, and Γ_L represents the ratio between the reflected wave f^- and incident wave f^+. Equation (2.13) is derived by setting the ratio of (2.11) and (2.12) (at $z = L$) to R_L and solving for $\frac{f^-}{f^+}$. To consider how a wave travelling in the negative direction is reflected at the generator end (2.13) applies but the source resistance (R_S) replaces the load resistance to give Γ_S.

Example 2.1   Consider the transmission line shown in Figure 2.6. It has per-unit capacitance and inductance of $c = 100$ pF/m and $l = 0.25$ μH/m, respectively. Compute the characteristic impedance and phase velocity. Find the reflection coefficient at the source and load. Determine the voltage at the source (v_I) and load (v_L) ends of the line over a few ns when the input V_S is a 4 V step.

Solution

The characteristic impedance (Z_0) is:

$$Z_0 = \sqrt{\frac{l}{c}} = \sqrt{\frac{0.25 \ \mu H/m}{100 \ pF/m}} = 50 \ \Omega.$$

(2.14)

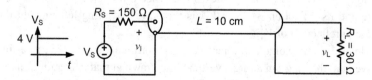

Figure 2.6 Lossless transmission line example.

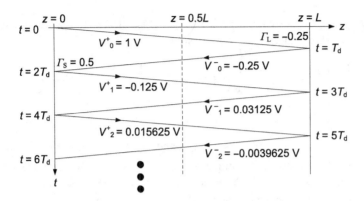

Figure 2.7 Lossless transmission line bounce diagram example.

The phase velocity is:

$$v_0 = \frac{1}{\sqrt{lc}} = \frac{1}{\sqrt{(0.25\ \mu H/m)(100\ pF/m)}} = 20\ cm/ns. \tag{2.15}$$

Therefore, the 10 cm line has a time delay (T_d) of 0.5 ns. The source and load reflection coefficients are:

$$\Gamma_S = \frac{R_S - Z_0}{R_S + Z_0} = \frac{150 - 50}{150 + 50} = 0.50, \tag{2.16}$$

$$\Gamma_L = \frac{R_L - Z_0}{R_L + Z_0} = \frac{30 - 50}{30 + 50} = -0.25. \tag{2.17}$$

The step input launches a travelling wave down the transmission line, the amplitude of which (V_0^+) is calculated using voltage division between Z_0 and R_S:

$$V_0^+ = \frac{Z_0}{Z_0 + R_S}V_S = \frac{50}{50 + 150}4\ V = 1\ V. \tag{2.18}$$

The voltage along the line at any position and time can be found using the bounce diagram shown in Figure 2.7. The horizontal axis is the distance, z, along the line and the vertical axis is time (t). The diagonal lines are the step input wave fronts. Thus, V_0^+, as calculated in (2.18), arrives at $z = L$ at $t = T_d$, seeing a reflection coefficient of $\Gamma_L = -0.25$. Therefore, the reflected wave at the load $V_0^- = -0.25$ V. Since the line is lossless, the step input travels back and forth an infinite number of times, being reflected at load and source by the corresponding reflection coefficient. If either R_L or R_S were matched to Z_0, one reflection coefficient would be 0, preventing additional reflections. The subscript of V indexes the reflection cycles; $V_0^{+/-}$ are the incident voltage and its reflection whereas V_1^+ is the wave travelling left to right due to the source-end reflection of V_0^-.

To determine the voltage at a particular z-coordinate, say $z = \frac{L}{2}$, as a function of time, we read the diagram along a vertical line drawn at that z location. The voltage is zero at $t = 0$ and stays zero until the first wave front crosses the dashed vertical line. Each time a diagonal line (indicating a wave) crosses the vertical line, the voltage

Table 2.1 Interpretation of the bounce diagram in Figure 2.7 at $z = \frac{L}{2}$.

t	Additional wave name	Wave amplitude (V)	$v(z = \frac{L}{2}, t)$ (V)
0	N/A	N/A	0
$\frac{T_d}{2}$	v_0^+	1	1
$\frac{3T_d}{2}$	v_0^-	-0.25	0.75
$\frac{5T_d}{2}$	v_1^+	-0.125	0.625
$\frac{7T_d}{2}$	v_1^-	0.03125	0.65625
$\frac{9T_d}{2}$	v_2^+	0.015625	0.671875
$\frac{11T_d}{2}$	v_2^-	-0.0039625	0.667969

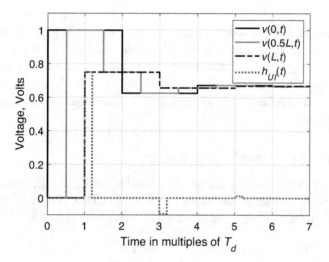

Figure 2.8 Time-domain plots of $v(z = 0, t)$, $v(z = \frac{L}{2}, t)$, $v(z = L, t)$ and $h_{UI}(t)$.

of the wave adds to the cumulative voltage. Table 2.1 summarizes the calculation for $z = \frac{L}{2}$ assuming for $t < 0$, the voltage along the line is 0.

With each wavefront, the voltage at $z = \frac{L}{2}$ approaches its steady-state value. Intuitively, as $t \to \infty$ we approach *dc* operating conditions, under which the lossless line acts like a perfect conductor. This results in a voltage along the line constant with respect to z:

$$v(z, t \to \infty) = \frac{R_L}{R_L + R_S} V_S = \frac{30}{30 + 150} 4 = \frac{2}{3} \text{ V.} \tag{2.19}$$

Time-domain plots of $v(z = 0, t)$, $v(z = \frac{L}{2}, t)$ and $v(z = L, t)$ are shown in Figure 2.8. The voltage step at the source occurs at $t = 0$. The line has an overall propagation delay of T_d. Therefore, halfway down the line ($z = \frac{L}{2}$), the 1 V step appears at $\frac{T_d}{2}$. At the load end of the line ($z = L$) the wave arrives and is instantly reflected with $\Gamma_L = -0.25$, appearing as a voltage of 0.75 V. As seen in Figure 2.8 and Table 2.1, all three voltages settle to the calculated equilibrium voltage of $\frac{2}{3}$ V.

The pulse response at the load is calculated as:

$$h_{UI}(t) = v(L, t) - v(L, t - T_b). \tag{2.20}$$

The dotted line in Figure 2.8 shows the pulse response for 10 Gb/s data. It is composed of a 1-UI-wide main pulse arriving at $t = T_d$, followed by 1-UI-wide reflections occurring at $2T_d$ intervals. Each reflection is the result of the load-reflected pulse travelling to the source end and then back to load, a total time of $2T_d = 1$ ns, which for this example is $10T_b$. This long time delay between reflections can complicate ISI mitigation.

The example presented above is highly idealized since it is based on a lossless line. Additionally, the terminations of the line are real resistors. In practice, transmission lines have frequency-dependent loss mechanisms that spread out input pulses. Also, reactive parasitic components of the source and load impedances add additional frequency-dependent behaviour. Models of lossy transmission lines are discussed in the next subsection.

2.1.2   Distributed Model of a Lossy Transmission Line

Figure 2.9 shows a per-unit length model of a lossy transmission line. Per-unit series resistance r and per-unit shunt conductance g have been added. These can be added to the KVL (2.3) and KCL (2.4) equations of the lossless line to give:

$$v(z) = v(z + \Delta z) + l\Delta z \frac{\partial i(z)}{\partial t} + r\Delta z i(z), \tag{2.21}$$

$$i(z) = i(z + \Delta z) + c\Delta z \frac{\partial v(z + \Delta z)}{\partial t} + g\Delta z v(z + \Delta z). \tag{2.22}$$

As in the lossless case, each can be rearranged to isolate the spatial dependence, leading to:

$$\frac{v(z + \Delta z) - v(z)}{\Delta z} = -l\frac{\partial i(z)}{\partial t} - ri(z), \tag{2.23}$$

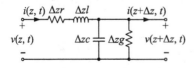

Figure 2.9 Lossy transmission line with per-unit length inductance l, capacitance c, resistance r and conductance g.

$$\frac{i(z+\Delta z) - i(z)}{\Delta z} = -c\frac{\partial v(z+\Delta z)}{\partial t} - gv(z+\Delta z). \tag{2.24}$$

Taking the limits of each left-hand side as $\Delta z \to 0$ yields partial differential equations in space and time:

$$\frac{\partial v(z)}{\partial z} = -l\frac{\partial i(z)}{\partial t} - ri(z), \tag{2.25}$$

$$\frac{\partial i(z)}{\partial z} = -c\frac{\partial v(z)}{\partial t} - gv(z). \tag{2.26}$$

Equations (2.25) and (2.26) are the lossy versions of the transmission line equations. Their solution for sinusoidal steady-state operation is found by first transforming them to the phasor domain, where the uppercase V and I denote phasor domain quantities:

$$\frac{\partial V(z)}{\partial z} = -(r+j\omega l)I(z), \tag{2.27}$$

$$\frac{\partial i(z)}{\partial z} = -(g+j\omega c)V(z). \tag{2.28}$$

Each can be rearranged to give an equation in either current or voltage forming the well-known wave equations:

$$\frac{\partial^2 V}{\partial z^2} = (r+j\omega l)(g+j\omega c)V(z) = \gamma^2 V(z), \tag{2.29}$$

$$\frac{\partial^2 I}{\partial z^2} = (r+j\omega l)(g+j\omega c)I(z) = \gamma^2 I(z), \tag{2.30}$$

where $\gamma = \sqrt{(r+j\omega l)(g+j\omega c)}$ is the propagation constant capturing both spatial dependence and frequency dependence. The general solution to (2.29) is:

$$V(z) = V^+e^{-\gamma z} + V^-e^{\gamma z} \tag{2.31}$$

where V^+ and V^- are phasors representing waves travelling in the +ve and -ve direction, respectively. To better understand (2.31), the propagation constant can be expanded as:

$$\gamma = \sqrt{(r+j\omega l)(g+j\omega c)} = \alpha + j\beta, \tag{2.32}$$

where α models the loss and β models the phase velocity of the line at a particular angular frequency, ω. Rewriting (2.31) using (2.32) for the particular case of $V^+ = A^+e^{j\theta^+}$ and $V^- = A^-e^{j\theta^-}$ gives:

$$V(z) = A^+e^{j\theta^+}e^{-\gamma z} + A^-e^{j\theta^-}e^{\gamma z}, \tag{2.33}$$

$$= A^+e^{j\theta^+}e^{-(\alpha+j\beta)z} + A^-e^{j\theta^-}e^{(\alpha+j\beta)z}, \tag{2.34}$$

$$= A^+e^{-\alpha z}e^{j(\theta^+ - \beta z)} + A^-e^{\alpha z}e^{j(\theta^- + \beta z)}. \tag{2.35}$$

When converted back to the time-domain the voltage along the transmission line becomes:

$$v(z,t) = A^+e^{-\alpha z}\cos(\omega t + \theta^+ - \beta z) + A^-e^{\alpha z}\cos(\omega t + \theta^- + \beta z). \tag{2.36}$$

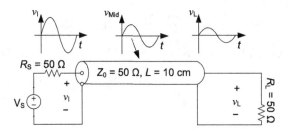

Figure 2.10 Decay in signal amplitude along a matched, lossy transmission line.

The source and load impedances determine the amplitudes $A^{+/-}$ and phase angles $\theta^{+/-}$. From (2.36) we see that the A^+ wave decays exponentially as it travels toward $+ve$ z. The A^- wave, which travels in the $-ve$ z direction gets smaller in amplitude as z decreases. For the matched case, Figure 2.10 shows the decaying amplitude of A^+ along the line. The voltage v_{Mid} is the voltage halfway down the line. The plotted voltages appear to have same phase, occurring for the particular case of $\beta = \frac{4\pi k}{L}$, where k is an integer greater than 0. The line, although it is distributed, is still linear. Sinusoidal excitation gives voltages at a given z location that are also sinusoidal.

The current in the phasor domain is:

$$I(z) = \frac{V^+}{Z_0}e^{-\gamma z} - \frac{V^-}{Z_0}e^{\gamma z}, \tag{2.37}$$

where:

$$Z_0 = \sqrt{\frac{r + j\omega l}{g + j\omega c}}. \tag{2.38}$$

The characteristic impedance of the lossy transmission has two important differences compared to the lossless case. Firstly, it is frequency-dependent even if r, g, l and c are assumed to be frequency-independent. As presented below, these parameters are themselves frequency-dependent, giving a more complicated overall dependency on frequency. Secondly, Z_0 is a complex number. Even for a resistive termination, a lossy transmission line presents a complex and frequency-dependent impedance.

With this preliminary discussion of lossy transmission lines completed, a framework for determining the input impedance and transfer functions of overall channels is needed. This is presented in the next section.

2.2   *ABCD* Parameter Descriptions of Two-Port Networks

ABCD parameters are one type of two-port parameter. Figure 2.11 shows a generic two-port network with port voltages and currents labelled. For most two-port parameter definitions the reference direction for I_2 is directed into port 2, whereas for *ABCD* parameters I_2 is flowing out of port 2. A feature of *ABCD* parameters is that cascades

Table 2.2 *ABCD* parameters of common elements.

Element	*ABCD* matrix
Shunt admittance (Y)	$\begin{bmatrix} 1 & 0 \\ Y & 1 \end{bmatrix}$
Series impedance (Z)	$\begin{bmatrix} 1 & Z \\ 0 & 1 \end{bmatrix}$
Transmission line $(Z_0, \gamma L)$	$\begin{bmatrix} \cosh(\gamma L) & Z_0\sinh(\gamma L) \\ \frac{1}{Z_0}\sinh(\gamma L) & \cosh(\gamma L) \end{bmatrix}$

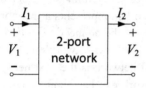

Figure 2.11 *ABCD*-parameter definition.

of networks see their *ABCD* parameters combined through matrix multiplication, making them useful for deriving overall transfer functions of complicated channels. The terminal characteristics are related as:

$$\begin{bmatrix} V_1 \\ I_1 \end{bmatrix} = \begin{bmatrix} A & B \\ C & D \end{bmatrix}\begin{bmatrix} V_2 \\ I_2 \end{bmatrix}, \tag{2.39}$$

where each term is defined as:

$$A = \frac{V_1}{V_2}\bigg|_{I_2=0} \qquad B = \frac{V_1}{I_2}\bigg|_{V_2=0}, \tag{2.40}$$

$$C = \frac{I_1}{V_2}\bigg|_{I_2=0} \qquad D = \frac{I_1}{I_2}\bigg|_{V_2=0}. \tag{2.41}$$

Some useful *ABCD* parameter descriptions of common elements are summarized in Table 2.2.

When two-port networks are cascaded, the overall network's *ABCD* parameters are found by multiplying the constituent *ABCD* matrices. Thus we can use the ABCD parameters of a transmitter, channel, receiver and parasitic elements from packages to build an overall model. Once an *ABCD* parameter description of an overall link is developed, the transfer function from input to output is found as follows:

1. Assume all load impedances are included in the overall *ABCD* description, giving an open-circuit connection to the right-hand side of the network.
2. Set $I_2 = 0$. Here, I_2 is the output current on the right-hand side of the overall channel.
3. With $I_2 = 0$, $V_1 = AV_2$, and V_2 is the output voltage of the overall channel.
4. Rearrange to give: $\frac{V_2}{V_1} = \frac{1}{A}$.

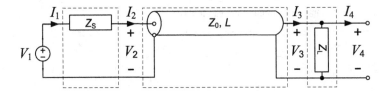

Figure 2.12 Example circuit showing how a transfer function is computed based on *ABCD* parameters of individual elements.

By a similar argument, the overall input impedance, $\frac{V_1}{I_1}$ is equal to $\frac{A}{C}$.

An example of a network consisting of a series source impedance and shunt load admittance is shown in Figure 2.12. Each dashed box represents a two-port network that will be described by its *ABCD* parameters. At the connection of two networks, the output port's voltage V_2 of the left network is the next network's input voltage V_1. Likewise, the output port's current I_2 of the left network is the next network's input current I_1. The *ABCD* parameters of each network can be found in Table 2.2. The overall *ABCD* parameters are:

$$\begin{bmatrix} A & B \\ C & D \end{bmatrix} = \begin{bmatrix} 1 & Z_S \\ 0 & 1 \end{bmatrix} \begin{bmatrix} \cosh(\gamma L) & Z_0\sinh(\gamma L) \\ \frac{1}{Z_0}\sinh(\gamma L) & \cosh(\gamma L) \end{bmatrix} \begin{bmatrix} 1 & 0 \\ Y_L & 1 \end{bmatrix}. \tag{2.42}$$

Then, (2.42) is simplified to give:

$$\begin{bmatrix} A & B \\ C & D \end{bmatrix} = \begin{bmatrix} 1 & Z_S \\ 0 & 1 \end{bmatrix} \begin{bmatrix} \cosh(\gamma L) + \frac{Z_0}{Z_L}\sinh(\gamma L) & Z_0\sinh(\gamma L) \\ \frac{1}{Z_0}\sinh(\gamma L) + \frac{1}{Z_L}\cosh(\gamma L) & \cosh(\gamma L) \end{bmatrix}. \tag{2.43}$$

Finally, *A* is:

$$A = \cosh(\gamma L) + \frac{Z_0}{Z_L}\sinh(\gamma L) + Z_S\left(\frac{1}{Z_0}\sinh(\gamma L) + \frac{1}{Z_L}\cosh(\gamma L)\right) \tag{2.44}$$

$$= \left(1 + \frac{Z_S}{Z_L}\right)\cosh(\gamma L) + \left(\frac{Z_0}{Z_L} + \frac{Z_S}{Z_0}\right)\sinh(\gamma L). \tag{2.45}$$

The transfer function from V_1 to V_4, given by $\frac{1}{A}$ is:

$$\frac{V_4}{V_1} = \frac{1}{A} = \frac{1}{\left(1 + \frac{Z_S}{Z_L}\right)\cosh(\gamma L) + \left(\frac{Z_0}{Z_L} + \frac{Z_S}{Z_0}\right)\sinh(\gamma L)}. \tag{2.46}$$

Z_0 and γ are frequency-dependent, meaning that (2.46) must be evaluated across the frequency range of interest. This is easily plotted using software such as MAT-LAB. However, to assess the channel characteristics for binary data, a time-domain representation is also needed, such as the step response or pulse response.

The relatively simple example in the preceding paragraphs included a uniform transmission line, a load impedance and a source impedance. However, the approach can be extended to model more elaborate package parasitics, transmission lines interrupted by connectors, or other impedance discontinuities.

2.2.1   Computing Time-Domain Behaviour from an *ABCD* Parameter Description

The approach described above computes the frequency response of an overall channel. The impulse response of a channel is computed as:

$$h(t) = \mathcal{F}^{-1}\{H(f)\} = \mathcal{F}^{-1}\left\{\frac{1}{A(f)}\right\}, \tag{2.47}$$

where $\mathcal{F}^{-1}$ denotes an inverse Fourier transform. Unlike lumped elements which give rise to rational polynomial transfer functions in "s," transmission lines give frequency responses involving hyperbolic trigonometric functions. Rational polynomials have known inverse transform pairs allowing the relatively straightforward (albeit possibly cumbersome) calculation of closed-form time-domain responses. This process is not possible for lossy transmission lines. Instead, we evaluate (2.47) numerically at discrete values of f thereby giving $h(t)$ at discrete values of t. MATLAB code to facilitate *ABCD* parameter analysis of channels, used to generate the plots in Examples 2.2, 2.3 and 2.4, can be found here [17].

2.2.2   Channel Response for a Lossy Line with Constant *r*, *l*, *g*, *c*

As a simple example to illustrate the effect of loss on the frequency-domain and time-domain response of a channel, consider (2.46) for the case where source and load are matched to the line:

$$Z_L = Z_S = Z_0. \tag{2.48}$$

Therefore (2.46) simplifies to:

$$\frac{V_4}{V_1} = \frac{1}{2}\frac{1}{\cosh(\gamma L) + \sinh(\gamma L)} = \frac{1}{2}e^{-\gamma L}. \tag{2.49}$$

Before investigating the frequency dependence of the r, l, g, c parameters, we consider a few particular cases with constant values of transmission line parameters by revisiting (2.32) above. In the lossless case (i.e., $r = g = 0$), γ reduces to:

$$\gamma_{\alpha=0} = \sqrt{(j\omega l)(j\omega c)} = j\omega\sqrt{lc}. \tag{2.50}$$

Purely imaginary, $\gamma_{\alpha=0}$ increases linearly with frequency, meaning that the line is an ideal delay. A voltage division of $\frac{1}{2}$ exists due to the matched source and load terminations. For the case of non-zero r and g, consider the special case of $\frac{r}{l} = \frac{g}{c}$:

$$\gamma = \sqrt{(r + j\omega l)(g + j\omega c)} = \sqrt{lc}\sqrt{\left(\frac{r}{l} + j\omega\right)\left(\frac{g}{c} + j\omega\right)} = v_0\left(\frac{r}{l} + j\omega\right). \tag{2.51}$$

With the propagation constant simplified to a constant real part and an imaginary part increasing linearly with ω, the line, with matched source and load, acts as an ideal delay with a *frequency-independent attenuation*. Equation (2.49) can be rewritten as:

$$\frac{V_4}{V_1} = \frac{1}{2}e^{-(v_0(\frac{r}{l} + j\omega)L)} = \frac{1}{2}\exp\left(-\frac{r}{Z_0}L\right)\exp\left(-jv_0\omega L\right). \tag{2.52}$$

The line acts as an ideal delay, with an additional attenuation determined by the length of the line. Although (2.38) gives a frequency-dependent expression for Z_0, the setting $\frac{r}{l} = \frac{g}{c}$ simplifies the characteristic impedance to:

$$Z_0 = \sqrt{\frac{r + j\omega l}{g + j\omega c}} = \sqrt{\frac{l(r/l + j\omega)}{c(g/c + j\omega)}} = \sqrt{\frac{l}{c}}, \tag{2.53}$$

becoming frequency-independent. When $\frac{r}{l} \neq \frac{g}{c}$, we must consider the real and imaginary parts of (2.32). Also, we must consider the frequency dependence of Z_0. For $\omega = 0$,

$$\gamma(\omega = 0) = \sqrt{(r + j(0)l)(g + j(0)c)} = \sqrt{rg}, \tag{2.54}$$

$$Z_0(\omega = 0) = \sqrt{\frac{r}{g}}. \tag{2.55}$$

Unlike rational polynominal transfer functions where we consider the overall magnitude of terms of the form $(\alpha + j\beta)$, to determine the magnitude of the response of $e^{-\gamma L}$ we need the real part of γ. It can be written as:

$$\Re(\gamma) = \frac{1}{\sqrt{2}}\sqrt{\sqrt{r^2 + \omega^2 l^2}\sqrt{g^2 + \omega^2 c^2} - \omega^2 lc + rg}, \tag{2.56}$$

which depends on ω. As $\omega \to \infty$ (2.56) can be simplified to:

$$\Re(\gamma)_{\omega \to \infty} = \frac{1}{\sqrt{2}}\sqrt{rg + \frac{r^2}{2Z_0(\infty)^2} + \frac{g^2 Z_0(\infty)^2}{2}}, \tag{2.57}$$

where $Z_0(\infty)$ is the characteristic impedance for large ω. Also, $Z_0(\infty) = \sqrt{\frac{l}{c}}$. Equation 2.57 allows us to estimate the high-frequency loss of the line.

The pulse response of a line with r, l, g, c parameters that are constant with respect to frequency is investigated in the example below.

Example 2.2 Find the pulse response of a transmission line with the following parameters:

- $L = 1$ m
- $v_0 = 15$ cm/ns
- $\sqrt{\frac{l}{c}} = 50\ \Omega$
- $r = 200\ \Omega/\text{m}$
- $g = 10^{-4}\ \Omega^{-1}/\text{m}$
- $Z_S = Z_L = 50\ \Omega$

Solution
The per-unit inductance and capacitance are calculated as:

$$l = \frac{50\ \Omega}{v_0}, \tag{2.58}$$

$$c = \frac{1}{(50\ \Omega)v_0}. \tag{2.59}$$

Note that the values of r and g have been chosen to provide significant loss over the line's 1 m length. Also, g has been chosen to be far from the criterion that leads to constant loss with respect to frequency outlined in (2.51). This gives a variation in the real part of γ.

In this case, for $\omega = 0$, $\gamma = \sqrt{rg} = 0.1414$. Rather than evaluate (2.49), we must revert to (2.46) since at low frequency, $Z_0 \rightarrow \sqrt{\frac{r}{g}} = 1414\ \Omega$. Evaluating (2.46) for $\omega = 0$ gives:

$$\frac{V_4}{V_1} = \frac{1}{\left(1 + \frac{50\ \Omega}{50\ \Omega}\right)\cosh(0.1414) + \left(\frac{1414\ \Omega}{50\ \Omega} + \frac{50\ \Omega}{1414\ \Omega}\right)\sinh(0.1414)} \tag{2.60}$$

$$. = \frac{1}{(2)(1.010) + (28.32)(0.1419)} = 0.1656. \tag{2.61}$$

As ω increases the real part of $\gamma \rightarrow 2.0025$ and $Z_0 \rightarrow \sqrt{\frac{l}{c}} = 50\ \Omega$. We can evaluate (2.49) since $Z_L = Z_S = \sqrt{\frac{l}{c}}$ giving:

$$\left|\frac{V_4}{V_1}\right| = \frac{1}{2}e^{\Re(-\gamma L)} = \frac{1}{2}e^{-2.0025} = 0.0675. \tag{2.62}$$

The overall magnitude response therefore varies from -15.6 to -23.4 dB.

Figure 2.13, generated using [17] shows the magnitude response of the terminated line along with its step and pulse responses. Although the transfer function varies with frequency, by 1 GHz it reaches a nearly constant attenuation. Therefore, the pulse does not spread out. The step response has an instantaneous rise time followed by a slower rise time to reach the steady-state gain. This is similar to what is seen when a 10x scope probe has mismatched resistive and capacitive divide ratios. The pulse response shown in Figure 2.13 and expanded in Figure 2.14 shows a sharp, square pulse followed by a long tail. Although none of the ISI terms is large, since the tail persists for so long, the worst-case ISI here leads to complete eye closure.

The preceding example shows how constant r, l, g, c parameters lead to some frequency-dependent attenuation, but without significant pulse spreading. Also, it highlighted the need to consider the frequency-dependent characteristic impedance and propagation constant when computing the transfer function in a simple transmission line problem. Note that several phenomena discussed in the next section lead to transmission line (r, l, g, c) parameters that vary significantly with frequency, rendering the preceding example unrealistic.

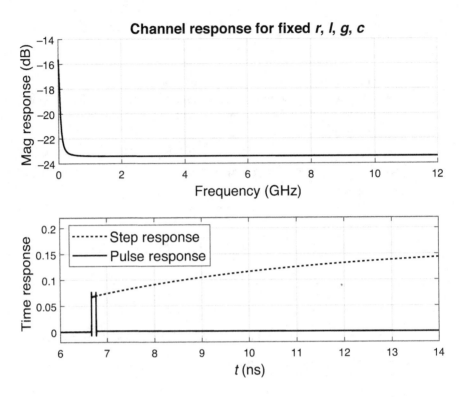

Figure 2.13 Behaviour of matched transmission line with frequency independent r, l, g, c parameters: (top) frequency response; (bottom) step and pulse responses for 100 ps UI.

2.3　Transmission Line Parameters

As seen in Example 2.2, a lossy line with constant r, l, g, c parameters shows relatively constant loss with respect to frequency and hence has a pulse response that shrinks as the line length is increased but does not spread out considerably. The pulse widening seen in real transmission lines is due to the parameters themselves being dependent on frequency.

There are several transmission line structures. Three configurations commonly used in PCBs are shown in Figure 2.15. In all cases, the signal-carrying conductor has a width w and thickness t. The dark grey layer along the bottom is a ground plane. Dielectric material is shown in light grey. Figure 2.15 (a) shows a microstrip line, where it is assumed that there is no conductor close to the signal trace on the top layer. Figure 2.15 (b) shows a coplanar waveguide with ground plane below. As $s \rightarrow \infty$ this type of transmission line behaves like a microstrip line. Figure 2.15 (c) shows a stripline where the conductor is sandwiched inside dielectric layers. For configurations (a) and (b) an electromagnetic wave propagates in the dielectric and air, seeing different phase velocity. This can lead to dispersion. On the other hand, striplines provide a uniform transmission medium eliminating this phenomena as well as providing better

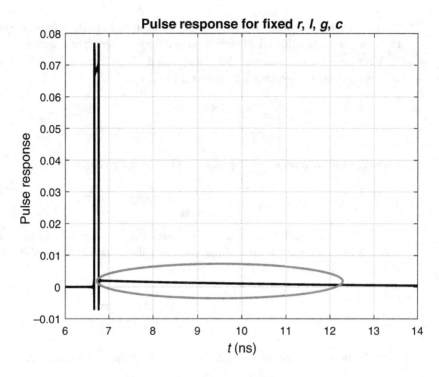

Figure 2.14 Expanded view of pulse response for 100 ps UI of a matched transmission line with frequency independent r, l, g, c parameters.

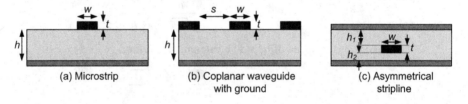

Figure 2.15 Cross sections of common transmission lines: (a) microstrip; (b) coplanar waveguide with ground plane; (c) asymmetrical stripline.

shielding against outside interference. Unfortunately, with more of the wave propagating in the PCB dielectric, this configuration suffers from more dielectric loss. This, combined with the added manufacturing complexity of accessing signals buried inside the board, has made this type of transmission line less common today.

This section focuses on microstrip lines, although online calculators provide a means to estimate Z_0 for all three configurations as well as coaxial cables.

2.3.1 Z_0, Low-Frequency Per-Unit Inductance and Capacitance

The equations that relate l and c to geometry are relatively complicated. The height dimension is typically dictated by the stack-up of the PCB, leaving the width of the line as the main design parameter. In this context, w is chosen to achieve the target Z_0. However, to model a line, the individual l and c parameters are needed.

A commonly used expression for calculating the high-frequency Z_0 ($Z_0(\infty)$) for the microstrip cross section shown in Figure 2.15 (a) is:

$$Z_0 = \begin{cases} \dfrac{Z_f}{2\pi\sqrt{\epsilon_{eff}}}\ln\left(8\dfrac{h}{w}+\dfrac{w}{4h}\right), & \text{for } \frac{w}{h} \leq 1 \\[3mm] \dfrac{Z_f}{\sqrt{\epsilon_{eff}}\left(\frac{w}{h}+1.393+0.667\ln\left(\frac{w}{h}+1.444\right)\right)}, & \text{for} \frac{w}{h} \geq 1, \end{cases}$$ (2.63)

where Z_f is the characteristic impedance of free space given by:

$$Z_f = \sqrt{\frac{\mu_0}{\epsilon_0}} \approx 120\pi.$$ (2.64)

The effective dielectric constant, ϵ_{eff}, captures the fact that the electric field lines between the trace and the ground plane are in both air (ϵ_0) and the PCB's dielectric ($\epsilon_r\epsilon_0$). It is estimated as:

$$\epsilon_{eff} = \frac{\epsilon_r + 1}{2} + \frac{\epsilon_r - 1}{2\sqrt{1 + 12\left(\frac{h}{w}\right)}}.$$ (2.65)

With Z_0 and ϵ_{eff}, the per-unit inductance and capacitance can be calculated since:

$$Z_0 = \sqrt{\frac{l}{c}},$$ (2.66)

$$v = \frac{v_0}{\sqrt{\epsilon_{eff}}} = \frac{1}{\sqrt{lc}},$$ (2.67)

where v_0 is the speed of light in a vacuum. Therefore,

$$l = \frac{Z_0}{v} = \frac{Z_0\sqrt{\epsilon_{eff}}}{v_0},$$ (2.68)

$$c = \frac{1}{Z_0 v} = \frac{\sqrt{\epsilon_{eff}}}{Z_0 v_0}.$$ (2.69)

2.3.2   Per-Unit Length Resistance, r

The resistance of transmission lines measured at dc is low, and can be calculated as:

$$r_{dc} = \frac{\rho}{wt},$$ (2.70)

where ρ is the dc resistivity of the metal conductor. Not only is r_{dc} small, a constant value of resistance does not produce the pulse-spreading phenomena seen in real transmission lines. Equation (2.70) assumes current is uniformly distributed across the entire cross-sectional area (wt), a situation depicted in Figure 2.16 (a). However, an alternating current is pushed outward in the conductor resulting in a much smaller effective cross-sectional area, shown by the shaded region in Figure 2.16 (b). Known

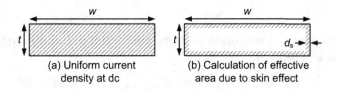

(a) Uniform current   (b) Calculation of effective
density at dc   area due to skin effect

Figure 2.16 Current density in a conductor: (a) uniform current density at dc; (b) current is
pushed to a layer of thickness, dS, at the outside of a conductor at high frequency.

as the skin effect, this results in an effective ac resistance specified in one of the two
following ways:

$$r_{ac}(f) = k_r(1+j)\sqrt{f} = r_s(1+j)\sqrt{\frac{f}{f_0}}, \qquad (2.71)$$

where k_r is a skin effect parameter. For the right-hand equation, r_s is the resistance
taken at $f = f_0$. In either case, (2.71) predicts two important phenomena:

1. The skin effect leads to an increase in *real* resistance as well as a reactance,
   $jX = jk_r\sqrt{f} = jr_s\sqrt{\frac{f}{f_0}}$.
2. The increase in impedance is with the square root of frequency rather than
   proportional to frequency as we are used to seeing in an inductor.

An approximate way to combine ac and dc resistance is:

$$r_{tot} = \sqrt{r_{dc}^2 + r_{ac}^2(f)}. \qquad (2.72)$$

Since $r_{ac}(f)$ is complex, r_{tot} will also include the skin effect-induced reactance.

2.3.3   Total Frequency-Dependent Series Reactance

As seen in Section 2.3.2, the skin effect contributes to a series reactance in addition to series resistance. The overall series reactance is the sum of that calculated
in Section 2.3.1 and Section 2.3.2, giving:

$$jX_{tot} = j2\pi f l + jk_r\sqrt{f}. \qquad (2.73)$$

If we equate (2.73) to $j2\pi f l_{eq}$, where l_{eq} is an equivalent inductance, we have:

$$l_{eq} = l + \frac{k_r}{2\pi\sqrt{f}}, \qquad (2.74)$$

giving the appearance that although the series reactance increases with frequency, the
equivalent inductance decreases with frequency.

2.3.4   Estimating k_r or r_s

Current density decreases with depth inside the conductor. The average current depth is given by the skin depth, d_s:

$$d_s = \sqrt{\frac{\rho}{\pi \mu f}}, \tag{2.75}$$

where μ is the material's magnetic permeability. The real part of the ac resistance in (2.71) can be found assuming a uniform current density in a cross-sectional area corresponding to a layer of thickness d_s around the perimeter of the conductor. The area of the shaded region in Figure 2.16 (b) is given by:

$$A = 2wd_s + 2(t - 2d_s)d_s = 2(w + t)d_s - 4d_s^2 \approx Pd_s, \tag{2.76}$$

where A is the cross-sectional area and P is the perimeter of the rectangle. The $4d_s^2$ term is neglected for small d_s. The result in (2.76) holds for other (i.e., non-rectangular) shapes of conductors as well.

Therefore, the real part of (2.71) is computed as:

$$k_r\sqrt{f} = \frac{\rho}{Pd_s} = \frac{\rho}{P\sqrt{\frac{\rho}{\pi \mu f}}} = \frac{\sqrt{\pi \mu \rho f}}{P}. \tag{2.77}$$

In the case of the conductors in Figure 2.15, $P = 2wt$. Thus:

$$k_r = \frac{\sqrt{\pi \mu \rho}}{2wt}. \tag{2.78}$$

2.3.5   Dielectric Loss and g

Dielectrics have $\epsilon_r > 1$ due to polarization of the material. Under an alternating electric field, the polarization also alternates. This process gives rise to dielectric loss each cycle. Therefore, as the frequency increases, we expect the dielectric losses *per-unit time* to increase. In an equivalent circuit, dielectric loss of a capacitor is modelled as a parallel conductance g. However, since the phenomenon corresponds to the same amount of energy conversion per cycle, the value of g must increase as a function of frequency.

Dielectric loss can be modelled by defining a complex dielectric constant:

$$\epsilon = \epsilon' - j\epsilon'', \tag{2.79}$$

where ϵ' is its real part, giving rise to conventional capacitance and ϵ'' is its imaginary part, giving rise to dielectric loss. The imaginary part is often specified relative to the real part as:

$$\epsilon = \epsilon' (1 - j\tan \delta), \tag{2.80}$$

where δ is the dielectric's loss angle and $\tan \delta$ is its "loss tangent." Since δ is usually small, for many practical cases, $\tan\delta \approx \delta$. The phasor diagram of capacitor currents is presented in Figure 2.17. We arrive at this diagram by observing that a dielectric constant in (2.80) leads to a capacitance given by:

$$C = C_0 (1 - j\tan\delta), \tag{2.81}$$

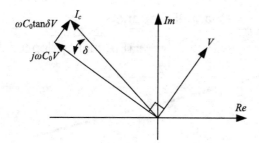

Figure 2.17 Phasor diagram of capacitor currents.

where C_0 is the nominal, real-valued capacitance of the element. In the phasor domain, the current through the overall element is:

$$I_c = j\omega CV = j\omega C_0 (1 - j\tan\delta) V = j\omega C_0 V + \omega C_0 \tan\delta V. \qquad (2.82)$$

Dielectric loss gives rise to a real current (in phase with the applied voltage) and a real conductance of:

$$G = \omega C_0 \tan\delta. \qquad (2.83)$$

Therefore, we model the per-unit conductance, g as:

$$g = \omega c \tan\delta. \qquad (2.84)$$

With r and g both increasing with frequency, not only is the line lossy, but its loss is frequency-dependent. This gives rise to a low-pass channel response and significant pulse widening (ISI) for longer channels.

Example 2.3   Consider a microstrip line similar to that in Example 2.2 but with the following parameters:

- $L = 1$ m
- $v_0 = 15$ cm/ns
- A lossless characteristic impedance of $\sqrt{\frac{l}{c}} = 50\ \Omega$
- $\tan\delta = 0.0017$
- $r_{dc} = 0.5\ \Omega/\text{m}$
- $r_s = 60\ \Omega/\text{m}$
- $f_0 = 10$ GHz

Solution
Low frequency values of l and c denoted as l_0 and c_0 are calculated as in Example 2.2 using (2.58) and (2.59). Considering the skin effect, ac resistance r_{ac} is calculated using (2.71) along with given values of r_s and f_0, giving a complex resistance as a function of f. This can be combined with r_{dc} using (2.72). The computed value of c_0 can be used for c. The dielectric loss tangent is used to calculate $g(f)$ using (2.84).

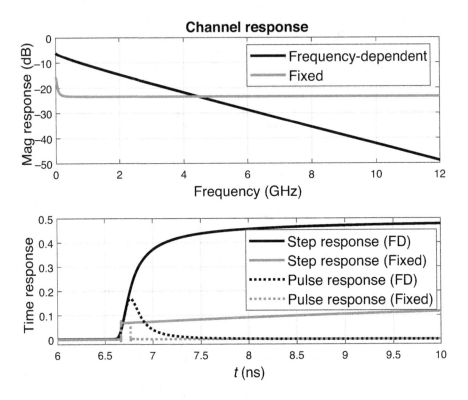

Figure 2.18 Behaviour of matched transmission line with fixed and frequency-dependent r, l, g, c parameters: (top) frequency response and (bottom) step and pulse responses (100 ps UI).

It may seem unintuitive to have a complex value of r since we think of resistances as being real. However, wherever we use r in the calculation of the propagation constant γ or the characteristic impedance Z_0, r appears in the term $r + j\omega l$. Thus it does not matter if we compute an imaginary part of r or convert it to an equivalent inductance $l_{eq} = \frac{\Im(r)}{\omega}$ that we add to the constant inductance l.

With frequency-dependent values of r, l, g and c, we compute γ, using (2.32), and Z_0, using (2.38), both frequency-dependent. Now, (2.46) can be evaluated as a function of frequency. This gives a transfer function for the entire channel. The inverse Fourier transform of the channel's frequency response gives the impulse response from which step and pulse responses can be computed.

The results of this are shown in Figures 2.18 and 2.19, generated using [17], denoted by "frequency-dependent" (FD), where the frequency, step and pulse responses of the channel in this example are plotted alongside those from Example 2.2 denoted by "Fixed." The parameters of Example 2.2 were selected so that both lines have the same loss near the Nyquist frequency, in this case, 5 GHz. This resulted in an unrealistically large r for the fixed-loss case. With less dc loss, the step response of the frequency-dependent case approaches 0.5 V, the voltage expected from a lossless line. The frequency-dependent pulse response is higher but wider, and lacks the sharp rising edge of the fixed-loss pulse.

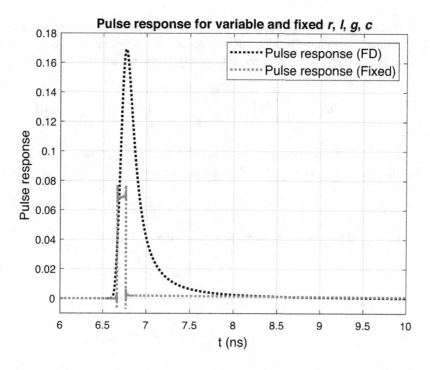

Figure 2.19 Expanded view of pulse response for 100 ps UI of a matched transmission line with fixed and frequency-dependent r, l, g, c parameters.

2.4   Packaging Parasitics

The electrical environment between the IC and the channel must be taken into account. This interface can include additional parallel capacitance in a package or series inductance from the package or from wirebonds.

2.4.1   Wirebonds

Though many high-performance chips use a flip-chip approach to packaging, wire-bonds are still commonplace for connecting laser diodes and photodetectors to electronic chips. Wirebonds present very little series resistance, but nontrivial series inductance L_{wb} given by:

$$L_{wb} = \frac{\mu h}{2\pi}\left(\ln\left(\frac{h}{r} + \sqrt{\left(\frac{h}{r}\right)^2 + 1}\right) + \frac{r}{h} + \frac{1}{4} - \sqrt{\left(\frac{r}{h}\right)^2 + 1}\right), \qquad (2.85)$$

where h is the length of the wire and r is its radius. Typical values are in the range of 1 nH/mm of length. At 10 GHz, a 1 mm wire presents a series impedance of $Z_L = j2\pi 10\ \Omega \approx j63\ \Omega$, which is non-trivial compared to a 50 Ω line. In some cases, multiple wires can be used to reduce the overall inductance. We might expect that two parallel wires each presenting 1 nH of inductance would combine to give an equivalent inductance of 0.5 nH. However, due to mutual inductance between the wirebonds, adding a second but nearby wire will reduce the inductance by less than the $\frac{1}{2}$ factor expected. The mutual inductance L_m can be computed as:

$$L_m = \frac{\mu h}{2\pi} \left(\ln \left(\frac{h}{s} + \sqrt{\left(\frac{h}{s}\right)^2 + 1} \right) + \frac{s}{h} - \sqrt{\left(\frac{s}{h}\right)^2 + 1} \right), \quad (2.86)$$

where s is the separation between parallel wirebonds. If two parallel wires conduct current in the same direction, the total per wirebond inductance (L_{tot}) is given by:

$$L_{tot} = L_{wb} + L_m, \quad (2.87)$$

whereas if they conduct current in opposite directions,

$$L_{tot} = L_{wb} - L_m. \quad (2.88)$$

For $h >> r$, (2.85) can be simplified to:

$$L_{wb} = \frac{\mu h}{2\pi} \left(\ln \left(\frac{2h}{r} \right) - \frac{3}{4} \right). \quad (2.89)$$

2.5   Simulating Channels with Active Devices

The *ABCD* parameter approach introduced in Section 2.2 works well for investigating a linear time-invariant channel consisting of a transmission line and passive (linear) terminations. However, once a transistor-level implementation of a transceiver front-end has been designed, simulations including the electronics and the channel must be performed. Circuit simulators such as Spectre support S-parameter blocks, for which S-parameters can be specified as a function of frequency. This block can be used in an ac, dc or transient simulation. Software such as MATLAB can convert *ABCD* parameters to S-parameters. The mapping from one two-port description to another is described in [18] for general cases, including complex reference terminations. For the case of a real-valued reference impedance Z_0, S-parameters can be computed as:

$$\begin{bmatrix} s_{11} & s_{12} \\ s_{21} & s_{22} \end{bmatrix} = \frac{1}{\Delta} \begin{bmatrix} B - Z_0 (D - A + Z_0 C) & 2Z_0 (AD - BC) \\ 2Z_0 & B - Z_0 (A - D + Z_0 C) \end{bmatrix}, \quad (2.90)$$

where:

$$\Delta = B + Z_0 (D + A + Z_0 C). \quad (2.91)$$

Typically, the lumped-elements that model pads, packages and terminations are left in a schematic while the transmission line is replaced by an S-parameter description.

2.6   Other Examples

Example 2.4   Consider the impact of a short section of an inadvertently mismatched transmission line between the output of an IC and a 50 Ω measurement environment as shown in Figure 2.20. In particular, the output of the analog front-end of a wireline receiver denoted as the device under test (DUT) is wirebonded to a PCB in order to connect it to 50 Ω equipment. A 5.5-mm-long section of transmission line is between the wirebond and the SMA connector's centre conductor. The important electrical characteristics of this channel are:

- The output resistance of the circuit is 100 Ω.
- There is an I/0 pad capacitance of 150 fF. Output resistance and capacitance are denoted by Z_S.
- 1.5 mm wirebond.
- Transmission line impedance of 16 Ω. Length is 5.5 mm.
- $\epsilon_r = 4.9$.
- The 50 Ω measurement environment is assumed to start at the tip of the edge-mounted SMA connector's centre conductor.

Solution

To simplify the investigation of the mismatched line, the SMA connector, cable and measurement equipment is assumed to present an ideal 50 Ω load. All frequency dependence of the circuit is assumed to be captured by the pad capacitance and the wirebond. A schematic of the equivalent circuit is shown in Figure 2.21. Using the *ABCD* parameter modelling approach, the frequency response of the channel is plotted in Figure 2.22 (generated using [17]) along with the case in which the line has $Z_0 = 50$ Ω. The grey curves show the frequency response and pulse response for the case in which the line is 50 Ω while the 16 Ω behaviour is shown in black. Both transfer

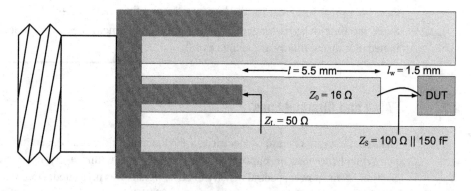

Figure 2.20 Example of an IC output connected to an edge-mounted SubMiniature version A (SMA) connector via a 1.5 mm bondwire and a 5.5 mm trace with $Z_0 = 16$ Ω.

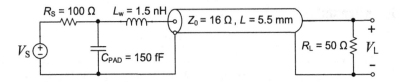

Figure 2.21 Equivalent circuit for the DUT, wirebond and transmission line.

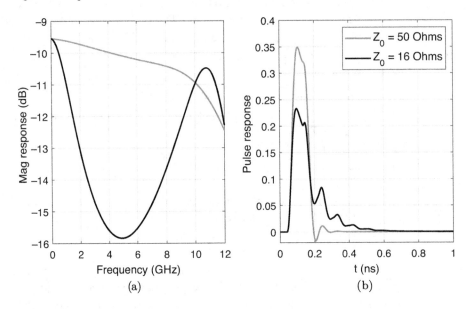

Figure 2.22 Effect of a short section of mismatched transmission line: (a) frequency response and (b) pulse response for 10 Gb/s data.

functions have low-frequency gain of $20\log\frac{50}{50+100} = 20\log\frac{1}{3} = -9.54\ dB$. The mismatched line significantly reduces the main cursor's height and introduces significant post-cursor ISI. Eye diagrams for the matched and mismatched lines are presented in Figure 2.23 showing that the mismatched line leads to significant eye closure. Given that $v = \frac{c}{\sqrt{\epsilon_r}} = 13.55$ cm/ns, the line introduces a time-delay of $\frac{0.55\ \text{cm}}{13.55\ \text{cmns}^{-1}} = 41$ ps. Since the time-delay is longer than the signal's rise time, reflections induced by the impedance discontinuity are visible as ISI.

2.7   PAM4 and Channel Loss

Given the low-pass nature of electrical-link channels (and circuits), sending more bits per symbol provides an opportunity to fit more data within a given bandwidth. In Section 1.9 the power spectral density of NRZ signals was presented. Here we consider a PAM4 signal as the sum of two streams of NRZ data in the time- and frequency-domains, thereby extending (1.5) and (1.22) to PAM4 signals. In the time-domain, (1.5) is modified to include two bits per symbol:

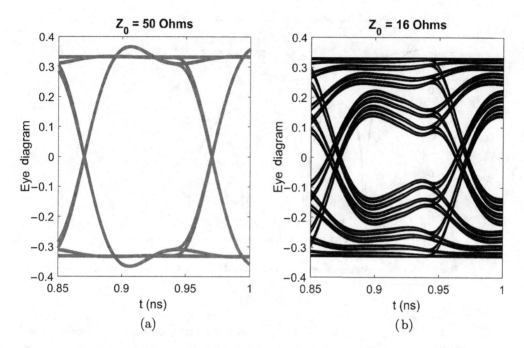

Figure 2.23 Eye diagrams for (a) matched ($Z_0 = 50\ \Omega$) line and (b) mismatched ($Z_0 = 16\ \Omega$) line.

$$x_{PAM}(t) = \frac{2}{3} \sum_{k=-\infty}^{\infty} d_{MSB_k} r(t - kT_b) + \frac{1}{3} \sum_{k=-\infty}^{\infty} d_{LSB_k} r(t - kT_b), \qquad (2.92)$$

where d_{MSB_k} refers to the most significant bit of the two bits encoded for the kth symbol and d_{LSB_k} is the least significant bit. The coefficients $\frac{2}{3}$ and $\frac{1}{3}$ give d_{MSB_k} twice as much weight in the signal as the LSB, required to generate four evenly spaced signal levels. At a given k, the $x_{PAM}(t)$ can take on one of four levels, namely: $-1, -\frac{1}{3}, +\frac{1}{3}$, or $+1$. This is the same peak-to-peak amplitude as the signal described in (1.5) which took on values of -1 and $+1$.

Each term in (2.92) gives rise to a power spectral density (PSD) related to $R(f)$. Since the LSB and MSB data streams are uncorrelated, the overall PSD of $x_{PAM}(t)$ is the sum of the PSDs of each term. Thus the PSD is given by:

$$
\begin{aligned}
S_{x_{PAM}}(f) &= \left(\frac{2}{3}\right)^2 \frac{1}{T_b} |R(f)|^2 + \left(\frac{1}{3}\right)^2 \frac{1}{T_b} |R(f)|^2 \\
&= \frac{5}{9} \frac{1}{T_b} |R(f)|^2 \\
&= \frac{5}{9} S_{x_{NRZ}}, \qquad (2.93)
\end{aligned}
$$

where the downscaling by $\frac{5}{9}$ accounts for the reduced average amplitude of the signal. Although the PAM4 signal was constructed to have the same peak-to-peak amplitude

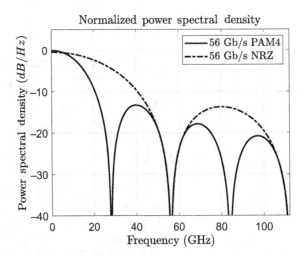

Figure 2.24 Power spectral density of 56 Gb/s PAM4 and 56 Gb/s NRZ, normalized to the PAM4 signal's dc PSD.

as the NRZ signal, for half of the symbols, the signal is only $\frac{1}{3}$ of the amplitude and, therefore, $\frac{1}{9}$ of the power of the NRZ signal. Thus, the PAM4 signal has the same spectral shape as an NRZ signal operating at the same *symbol* rate, but with lower power at all frequencies.

The spectra of random NRZ (PAM2) and PAM4 data at 56 Gb/s are shown in Figure 2.24. The PAM4 signal has the same symbol rate as 28 Gb/s NRZ, giving its PSD nulls at multiples of 28 GHz. On the other hand, the 56 Gb/s NRZ signal has its first spectral null at 56 GHz, giving considerably more power for frequencies above the Nyquist frequency of the PAM4 signal (14 GHz). Moving to PAM4 allows twice as many bits to be transmitted with the same channel loss. On the other hand, three decision thresholds must slice the data meaning that each eye is only 1/3 of the available eye opening.

Reducing the internal eye opening by $\frac{1}{3}$ corresponds to a reduction in SNR of $20 \log 3 = 9.5$ dB. When attempting to double the data rate from 28 to 56 Gb/s, PAM4 is often proposed if, in a given application, channel loss increases by more than 9.5 dB from 14 GHz to 28 GHz. Although this frequency-domain criterion is convenient, we must consider the time-domain consequences of moving to PAM4. Figure 2.25 (a) (generated using [17]) shows an example channel's attenuation with respect to frequency as well as its step response. Step response is a data rate agnostic time-domain signal. The lower portion of Figure 2.25 (a) also shows pulse responses for 56 Gb/s NRZ and PAM4, which are expanded in Figure 2.25 (b).

In looking at Figure 2.25 (b) a few observations can be made:

1. Both pulse responses are for an input pulse of unit amplitude.
2. The PAM4 pulse, having an input symbol duration of 36 ps, is wider than the NRZ pulse, having an input pulse duration of only 18 ps.

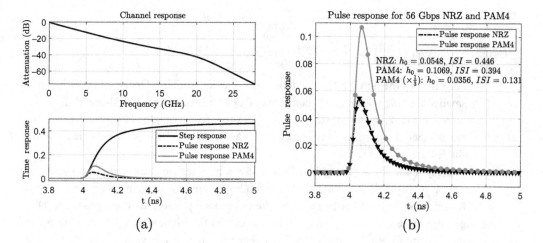

Figure 2.25 Channel characteristics: (a) top, attenuation relative to DC; bottom, channel step response and pulse responses for NRZ and PAM4 unit input pulses; (b) expanded view of pulse responses.

3. The PAM4 pulse response rises to a main cursor (h_0) of 0.1069 compared to the NRZ's main cursor of 0.0548, approximately 50%.

4. The NRZ pulse response, with ISI samples every 18 ps, has a total ISI of 0.446 while the PAM4 pulse response, with ISI samples 36 ps apart, has less total ISI (0.394). In both cases the total ISI is much larger than the respective h_0, meaning significant equalization is needed to open the eye diagram at the receiver. However, the ISI to h_0 ratio is about twice as large for the NRZ pulse response, due to the shorter symbol period.

5. Although the attenuation increases by much more than 9.5 dB from 14 GHz to 28 GHz, the pulse response decreases by only about a factor of 2 in height.

6. Finally, when the PAM4 pulse response is scaled by $\frac{1}{3}$ the scaled main cursor is 0.0356, which is smaller than the NRZ pulse response (0.0548) but much larger than $\frac{1}{3}$ of it. This is due to the increased channel loss that the NRZ signal sees. Since the PAM4's h_0 to ISI ratio is larger, less equalization is required, and equalization can take place at lower speeds. This is particularly beneficial for decision-feedback equalizers and other digital equalization techniques.

In looking at Figure 2.25 (b) it appears that the PAM4 pulse response is approximately equal to $2\times$ the NRZ pulse response. This makes sense when we consider h_{UI} as T_b decreases, since it is the system's response to $r(t)$, which can be written as:

$$r(t) = u(t) - u(t - T_b).$$

Consider $r^*(t)$, defined as the unit pulse for vanishing T_b:

$$r^*(t) = \lim_{T_b \to 0} r(t)$$

$$= \lim_{T_b \to 0} u(t) - u(t - T_b)$$

$$= \lim_{T_b \to 0} T_b \frac{u(t) - u(t - T_b)}{T_b}$$

$$= \lim_{T_b \to 0} T_b \delta(t).$$

For small T_b the unit input pulse looks like a scaled impulse. Hence, once T_b is much smaller than the relevant time constants of a channel or system, the shape of the pulse response stops depending strongly on T_b, except for its height. This was the case for the channel in Figure 2.25 (a). Hence the PAM4 and NRZ pulse responses are a similar shape but scaled in amplitude.

Despite the ISI advantages of PAM4, it has been resisted due to the complexity of setting multiple decision thresholds and other design requirements such as linearity. Recently, as multi-standard receivers are using analog-to-digital converters, PAM4 is gaining more attention. Now 56 Gb/s is seen as the highest speed with widespread deployment of NRZ, particularly for higher-loss channels, beyond which PAM4 will become dominant.

2.8   Summary

This chapter reviewed lossy and lossless transmissions lines, defining relevant terminology: characteristic impedance, phase velocity, propagation constant and reflection coefficient. Frequency-dependent loss mechanisms (skin effect and dielectric loss) were explained. The importance of the loss being frequency-dependent was emphasized by way of an example. The use of $ABCD$ parameters for combining transmission lines and lumped elements that model packages and chip terminations was explained. This chapter also presented various transmission line structures and gave equations for the characteristic impedance of microstrip lines.

Problems

2.1   Consider a 50 cm transmission line, assumed lossless with per-unit-length inductance $l = 0.2\ \mu\text{H/m}$ and per-unit-length capacitance $c = 0.2\ \text{nF/m}$, connected to $Z_S = 50\ \Omega$ and $Z_L = 200\ \Omega$ as shown in Figure 2.26. Assume V_1 is a voltage step at $t = 0$ of 1 V.

(a)   Find the phase velocity in the line and the line's characteristic impedance.
(b)   Determine and sketch V_2 and V_L for t from 0 until shortly after the wave travelling from source to load has reflected and arrived back at the source.

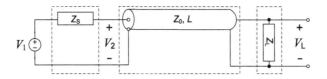

Figure 2.26 Terminated transmission line for Problem 2.1.

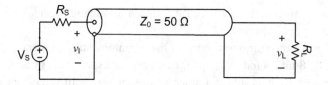

Figure 2.27 Schematic of lossless transmission line for Problem 2.7.

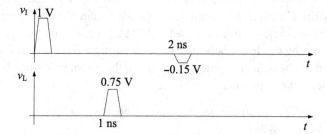

Figure 2.28 Source and load waveforms for Problem 2.7.

2.2 This problem uses a channel with the following parameters:

- $L = 0.25$ m
- $v_0 = 15$ cm/ns
- $\sqrt{\frac{l}{c}} = 50 \, \Omega$
- Loss tangent is 0.022
- $r_s = 87 \, \Omega/\text{m}$ at 10 GHz
- $Z_S = Z_L = 50 \, \Omega$

Using MATLAB code found at [17], determine the magnitude response of the channel above with matched source and load terminations. Modify the code to plot the pulse response for 10 Gb/s data and compute the VEO. At what data rate will the eye close completely?

2.3 Consider a modification to the channel in Problem 2.2 in which the 0.25 m line is split into two series lines of 0.125 m. At the interface between these two lines, a series LC network makes a path to ground with $L = 1$ nH and $C = 2$ pF. What is the highest data rate this channel can support before the eye is completely closed? At 10 Gb/s, what is the ratio between the pulse response's main cursor and total ISI?

2.4 Using the channel from Problem 2.2 investigate the importance of matching. Consider the following cases:

(a) $R_S = 50 \, \Omega, R_L = 50 \, \Omega$
(b) $R_S = 50 \, \Omega, R_L = 100 \, \Omega$
(c) $R_S = 20 \, \Omega, R_L = 50 \, \Omega$
(d) $R_S = 20 \, \Omega, R_L = 100 \, \Omega$
(e) $R_S = 50 \, \Omega, Z_L = 50 \, \Omega || 1 \, \text{pF}$

Compare the magnitude of h_0 and total ISI for each termination scenario at the data rate that closed the eye in Problem 2.2.

2.5 Design a 50 Ω microstrip transmission line assuming an $\epsilon_r = 4$ and a dielectric thickness h of 0.78 mm. Find w, ϵ_{eff}, v_0, l and c. Use expressions in Section 2.3.1. Note that these are not easily solved. Rather, you sweep w/h until you get the right Z_0, taking into account the corresponding change in ϵ_{eff}.

2.6 In Example 2.4 the mismatched line had a characteristic impedance approximately 1/3 that of the measurement environment. Consider a similar, relative mismatch but in the opposite direction by repeating the simulation but with the characteristic impedance of the mismatched line increased to 150 Ω. Comment on the eye opening degradation caused by this line relative to having a matched line or a 16 Ω line.

2.7 Consider the schematic in Figure 2.27. Assume the transmission line is lossless and has a characteristic impedance $Z_0 = 50$ Ω.

(a) To maximize the vertical eye opening at the load, what value should we choose for R_S and R_L? Why?

(b) From the waveforms in Figure 2.28, determine R_S and R_L.

2.8 Consider the channel from Example 2.3. Generate a plot of h_0/T_b vs T_b for T_b ranging from 10 ps to 1 ns. When $T_b = 10$ ps, h_0/T_b is effectively the height of the *impulse* response, $h(t)$. At what T_b is h_0/T_b within 10% of this value.

3   Decision Circuits

Figure 3.1 shows an expanded view of a generic wireline receiver. A small signal at the output of the channel is applied to the input of the analog front-end (AFE). It is amplified and equalized (EQ) and then applied to the clock and data recovery unit (CDR). The first block within the CDR is a decision circuit, sometimes known as a slicer or latch. It is an edge-sensitive circuit that compares an input signal to a voltage threshold and produces a full-rail output based on the polarity of this comparison. The arrows connecting the input and output of the decision circuit show how relatively small inputs are regenerated into full-rail outputs.

The operation of a high-speed latch is shown in greater detail in Figure 3.2, where an input signal and a clock are shown in the upper portion. The dashed line is the decision threshold, L_d. The rising edge of the clock samples the input signal, shown by the dots. Wherever the sampled input is larger than L_d, the output settles to a high voltage (1 V in this case). Likewise if the sampled input is less than L_d, the output will settle to a low voltage (0 V). There is a short delay ($\tau_{Clk \rightarrow Q}$) between the clock edge and the transition of the output. Regardless of the magnitude of the difference between the sampled input and L_d, the output settles quickly, based only on the polarity of the comparison. This is apparent at $t = 14$ UI where the input is only slightly below L_d but the output settles to 0 V. In this text, the term "latch" is assumed to be edge-triggered unless denoted specifically as a "level-sensitive" latch.

The operation by which an input level appears at the output shortly after a clock edge is similar to the D-flip-flop (DFF) we see in logic circuits. Its input range is presented in Figure 3.3 (a). Although all logic circuits exhibit noise margins, small transistor sizes and a lack of offset trimming mean that a considerable fraction of the input range will lead to ill-defined behaviour. For example, input voltages in the "undefined" range may lead to very slow settling of the output, variations in the DFF's decision from one instance of the circuit to another, or metastability. Metastability refers to very slow settling where the voltage to which the circuit settles is determined by noise in the circuit rather than the input signal. On the other hand, a high-speed latch used as a decision circuit has a much smaller input range for which the output is indeterminate. This is shown in Figure 3.3 (b) where a range of V_{MIN} on either side of L_d must be avoided, giving an overall undefined input range of $2V_{MIN}$.

A well-defined threshold is usually achieved using a differential circuit in which the desired L_d is applied to one input and the signal is applied to the other input, or

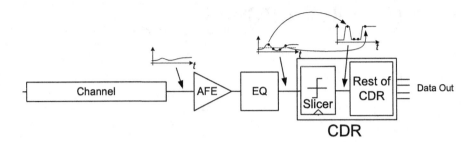

Figure 3.1 Generic block-diagram of a wireline receiver depicting the amplification/equalization in a receiver as well as the regeneration of the signal into full-rail digital values.

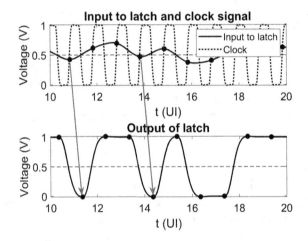

Figure 3.2 Input and output of a high-speed latch.

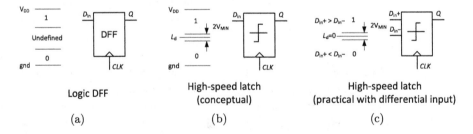

Figure 3.3 Difference between a high-speed latch and a DFF: (a) DFF input operating range with large undefined range; (b) single-ended high-speed latch with small undefined input range; and (c) differential high-speed latch with small undefined input range centred around 0 V.

the signal is differential with an implied threshold of $L_d = 0$. Differential operation is presented in Figure 3.3 (c).

Although only in the mV range, V_{MIN} imposes a requirement on minimum latch input signal amplitude and, in turn, the receiver's minimum gain. Also, decision circuits will have finite setup and hold times, which are usually a function of input amplitude. The response of a real latch is based on an average of the input signal over a time interval between the setup time and hold time. This interrelation between amplitude and timing parameters leads to a characterization technique based on an impulse sensitivity function (ISF).

After presenting key metrics such as voltage sensitivity and input timing constraints, the two main categories of decision circuits are presented. Offset cancellation schemes are also outlined.

3.1   Decision Circuit Metrics and Characterization

The main metrics of a decision circuit are the following:

- Power dissipation: lower power dissipation is preferred. Current-mode logic circuits (CML) dissipate static power, giving rise to power dissipation independent of clock period for a given design. However, in a given technology, a design optimized for high-speed operation will likely dissipate more power than one that can only operate at a lower data rate. On the other hand, sense-amplifier-based latches draw current only on clock edges. For these circuits it is necessary to specify power dissipation at a particular data rate.
- V_{MIN} or some other sensitivity measure: a smaller minimum input voltage swing required to give error-free operation will reduce the gain requirements of the rest of the receiver's front-end.
- Sampling window: the decision of the circuit should depend on the input voltage only in a small window of time around the clock edge. This is analogous to saying that we want the sum of the setup and hold times to be short. However, in decision circuits, timing parameters and voltage swing interact. One option to characterize a decision circuit is to use input voltages with some multiple of V_{MIN}, say $2V_{MIN}$, and then determine the sum of setup and hold times for this amplitude. This sum must be much less than the data's UI.
- Chip area: smaller chip area is preferred, particularly if several latches are used in a sub-rate clocking architecture where multiple latches are used.
- Offset voltage: due to asymmetry in the fabricated circuit a random input offset will be present. This is a shift in the latch's decision threshold. Calibration circuitry is usually included. It is preferred that the post-calibration offset is small.

3.2   Current-Mode Logic Latches

Current-mode logic decision circuits are constructed from two level-sensitive latches similar to how a DFF for logic applications can be built from two level-sensitive

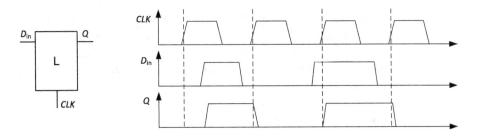

Figure 3.4 Operation of a level-sensitive latch.

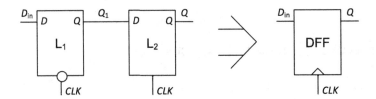

Figure 3.5 Construction of a DFF from two level-sensitive latches.

latches. Figure 3.4 shows the operation of a level-sensitive latch. When CLK is high, the latch is said to be "transparent," meaning that the signal applied at D_{in} propagates to Q, with a small delay. During the entire clock-high interval, changes in the signal polarity of D_{in} appear at Q. Once CLK goes low, the output of Q is held, maintaining its value from the end of the clock-high interval. Subsequent changes in D_{in} do not propagate to Q. Only when CLK goes high again will D_{in} transfer to the output.

When CLK is high, the latch is said to be "transparent" in a logic context, but in a wireline context this is "tracking mode." The output will track the input, ideally with a small gain. When CLK is low, it is said to be in "hold mode" for logic designers. However, in the wireline context, the clock going low usually is associated with additional amplification of the input D_{in} toward full-rail signal levels. Hence, in this context we use the term "regeneration mode" when CLK is low.

The formation of a DFF from level-sensitive latches is shown in Figure 3.5 where L_1 and L_2 are level-sensitive latches. Latch L_1 responds to the input in the opposite fashion as L_2, as implied by the bubble on its CLK input. Hence, when clock is low, L_1 tracks the overall input D_{in}, and L_2 is holding the state regenerated from the previous high-to-low transition of the clock. When CLK goes high, L_1 regenerates and L_2 tracks L_1's regenerated output. This operation corresponds to a positive-edge-triggered DFF with the ability to amplify small differences between D_{in} and a decision threshold.

A schematic of a resistively loaded level-sensitive CML latch is shown in Figure 3.6, using an ideal tail current source, I_B. Inputs CLK and D_{IN} are differential inputs, residing on common-mode signal levels, defined as $CLK = CLK_+ - CLK_-$ and $D_{IN} = D_{IN+} - D_{IN-}$. When CLK is high (meaning $CLK_+ \gg CLK_-$), the current I_B flows through the differential pair formed by M_3 and M_4. Along with the resistors, M_3 and M_4 form a buffer, amplifying the input signal $D_{IN+} - D_{IN-}$. Clearly, the latch

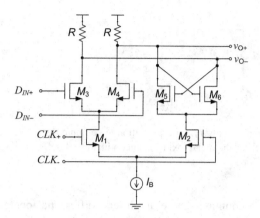

Figure 3.6 Level-sensitive CML latch, transparent when $CLK_+ > CLK_+$.

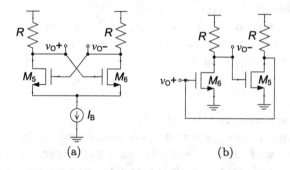

        (a)   (b)

Figure 3.7 Differential pair composed of $M_{5/6}$: (a) schematic and (b) differential operation shown with cross-coupling unrolled to show positive feedback.

is in the tracking mode. Changes in the polarity of D_{IN} appear as changes in the output voltage v_O, with a finite time delay due to the time constant at the output. When the polarity of the clock signal inverts, the current I_B flows through the differential pair formed by M_5 and M_6. Notice that the input of this differential pair is also the output of the pair. However, this output to input connection is made such that any difference in the output signal is reinforced by the amplification of the differential pair (i.e., positive feedback).

To better understand the operation of the cross-coupled pair formed by M_5 and M_6, consider the simplified schematic shown in Figure 3.7 (a). Assume that at the beginning of the hold/regeneration phase there is a small differential voltage between v_{O+} and v_{O-}, with $v_{O+} > v_{O-}$. With a small output voltage difference, approximately half of I_B flows through each of M_5 and M_6. Each of $M_{5/6}$ forms a common source amplifier with inverting gain. Thus, v_{O+}, applied at the gate of M_6, is amplified to its drain by an inverting gain, reducing v_{O-}. This in turn is applied to the gate of M_5 which will be amplified to v_{O+} through an inverting gain, further reinforcing $v_{O+} > v_{O-}$. Then, v_{O+} increases while v_{O-} decreases until all of I_B flows through M_6, the transistor with higher gate-source voltage at the start of the regeneration phase.

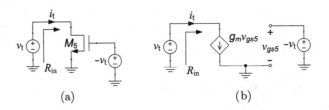

Figure 3.8 Differential pair composed of $M_{5/6}$: (a) excitation to compute resistance and (b) shown with small-signal model for the metal–oxide–semiconductor field-effect transistor (MOSFET).

Figure 3.7 (b) shows the common-source amplifiers' differential operation. The shared source terminal is viewed as an ac ground for small signals, allowing the removal of the current source. This figure is rearranged to show a loop of two common-source amplifiers giving positive feedback.

Regeneration from a small differential voltage to a fully-switched differential pair happens only if the gain of each common-source amplifier is greater than 1. Ignoring transistor output resistance, this occurs for

$$g_m R > 1, \tag{3.1}$$

where g_m is the transconductance of $M_{5/6}$ when $V_{O+} = V_{O-}$.

The cross-coupled pair can also be viewed as a negative resistance. The analysis of the circuit from this point of view is shown in Figures 3.8 (a) and (b). The resistance looking into the drain of each transistor (R_{in}) under differential excitation will be the ratio of v_t/i_t. Figure 3.8 (a) and (b) show the differential half circuit used for finding the drain resistance of the circuit. Bear in mind that under a differential input, terminals of the circuit see inputs of opposite polarity. Hence the drain of M_5, which is also the gate of M_6, sees an input of v_t, while the gate of M_5, which is also the drain of M_6, sees an input of $-v_t$. The ac small-signal equivalent circuit is shown in Figure 3.8 (b). From this it is clear that:

$$i_t = g_m(-v_t), \tag{3.2}$$
$$R_{in} = v_t/i_t = -1/g_m. \tag{3.3}$$

Capacitance at nodes V_{O+} and V_{O-} sees an equivalent resistance, R_{eq}, given by R in parallel with R_{in}. This is computed as:

$$1/R_{eq} = 1/R_{in} + 1/R,$$
$$1/R_{eq} = -g_m + 1/R,$$
$$R_{eq} = \frac{R}{1 - g_m R}. \tag{3.4}$$

A positive value of R_{eq} results in stable behaviour while a negative value of R_{eq} gives regenerative (unstable) behaviour. Notice that the condition for R_{eq} to be negative ($g_m R > 1$) is the same as that presented in (3.1).

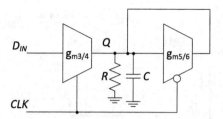

Figure 3.9 Model of a level-sensitive latch.

3.2.1 Model of a Current-Mode Level-Sensitive Latch

Figure 3.9 shows a linear model of the level-sensitive latch. It consists of two transconductors ($g_{m3/4}$ and $g_{m5/6}$) which model the two differential pairs. Resistance R is the load resistor while the capacitance C is the total single-ended capacitance at each output. The cross coupling is modelled by the positive feedback from the output of $g_{m5/6}$ to Q. The figures that follow show the simulation of this model. Initially, both D_{IN} and CLK have near zero rise/fall times. When $CLK = 1$, $g_{m3/4}$ is active and $g_{m5/6}$ is off. When $CLK \to 0$, $g_{m3/4}$ turns off and $g_{m5/6}$ is on, initiating regeneration.

When $CLK = 1$, if D_{in} is a step from v_1 to v_2 at $t = 0$, the voltage at Q follows a typical first-order step response. Recall that for a system with a time constant of τ, the step response, $v(t)$ is:

$$v(t) = v(\infty) - (v(\infty) - v(0)) \exp\left(-\frac{t}{\tau}\right). \tag{3.5}$$

where $v(0)$ and $v(\infty)$ are the output's initial and final voltages, respectively. These can be related to the inputs v_1 and v_2 through the dc gain, A_{dc} to give:

$$v(t) = A_{dc}\left(v_2 - (v_2 - v_1)\exp\left(-\frac{t}{\tau}\right)\right). \tag{3.6}$$

For the level-sensitive latch in Figure 3.9, $A_{dc} = g_{m3/4}R$ and $\tau = RC$. Therefore, for an input step at $t = 0$ from v_1 to v_2 the voltage at Q is given by:

$$v_Q(t) = g_{m3/4}R\left(v_2 - (v_2 - v_1)\exp\left(-\frac{t}{RC}\right)\right). \tag{3.7}$$

Consider the case in which $v_1 < 0$ and $v_2 > 0$. To "capture" the value of $D_{in} > 0$ we need $v_Q > 0$ before the clock edge, assumed to arrive t_{set} after the input data transition. Solving for $v_Q(t_{set}) = 0$, we find:

$$t_{set} = RC\ln\left(\frac{v_2 - v_1}{v_2}\right), \tag{3.8}$$

where t_{set} is the level-sensitive latch's setup time. If $v_2 = -v_1$, $t_{set} = RC\ln 2$, determined only by circuit element values. However, t_{set} increases if v_2 decreases or v_1 increases in the negative direction. Through (3.8) we see that latch timing parameters depend on element values, but also input signal conditions. It might seem reasonable

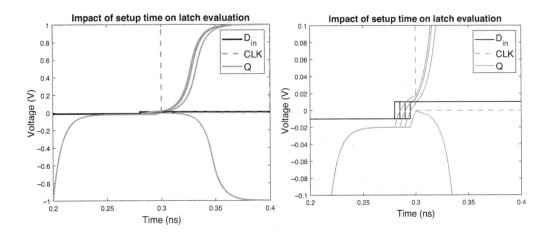

Figure 3.10 Impact of setup time on latch evaluation: $t_{set} = 5, 10, 15, 20$ ps. Axes expanded on the right-hand side.

to assume that a link would operate with a symmetrical differential signal, meaning $v_2 = -v_1$. That is, ones and zeros are encoded by equal and opposite voltages. However, due to noise or ISI, consecutive bits during bit transitions might not be equal and opposite. That is, v_2 could be much less than $-v_1$, prolonging t_{set}.

Figure 3.10 shows the response of the latch as the transition in D_{in} moves earlier relative to the clock edge. As seen in the expanded figure on the right, when D_{in} arrives only 5 ps before the clock edge, Q does not cross zero. In the regeneration phase it grows exponentially in the negative (incorrect) direction. With larger setup time Q crosses zero and the correct (high) value is regenerated. For this simulation, the calculated t_{set} is 5.5 ps.

Advancing D_{in} by 5 ps allows Q to cross zero and regeneration further amplifies Q toward ∞. During regeneration, $v_Q(t)$ follows:

$$v_Q(t) = v_Q(0) \exp \left(\frac{t}{\tau_{Regen}} \right), \tag{3.9}$$

where $v_Q(0)$ is the voltage at the clock edge and τ_{Regen} is derived below. Here, t is referenced to the clock edge.

As shown in (3.8) setup time depends on the relative amplitude of the current and previous inputs, modelled in this case as a step. If v_1 is larger (in the negative direction) setup time increases. For example, in Figure 3.11 when $v_2 = -v_1$ the latch regenerates in the positive (correct) direction with a $t_{set} = 10$ ps. If $v_1 < -3 \times v_2$, Q does not cross zero before the clock edge resulting in regeneration in the negative (incorrect) direction.

An ideal clock enables only one transconductor at a time. However, during a real clock transition, both transconductors are active. Therefore, a change to the input before the voltage at Q has significantly regenerated can reverse the direction of regeneration. Figure 3.12 shows how a slower falling clock edge allows a large change in

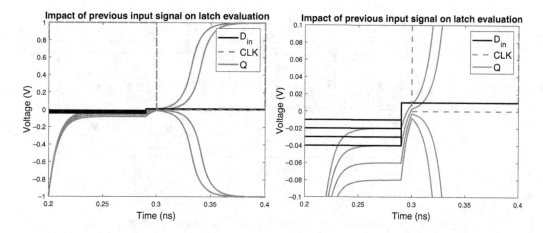

Figure 3.11 Impact of signal amplitude before transition on latch evaluation: $t_{set} = 10$ ps, $v_2 = 10$ mV; $v_1 = -10, -20, -30, -40$ mV. Axes expanded on the right-hand side.

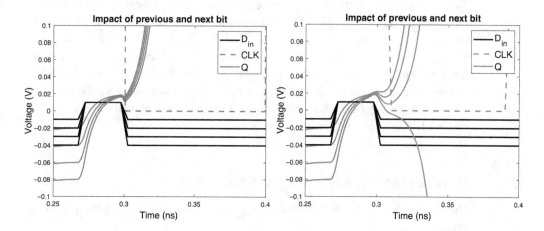

Figure 3.12 Impact of clock rise/fall time on latch evaluation: $t_{set} = 30$ ps, $t_h = 0$ ps; $v_2 = 10$ mV; $v_1 = -10, -20, -30, -40$ mV. Left-hand side: $t_{rclk} = 1$ ps. Right-hand side $t_{rclk} = 20$ ps.

input signal after the start of the clock edge to prevent the latch from evaluating in the correct direction. On the left, the clock edge has a near-zero fall time when the data changes. Regardless of the magnitude of the voltage transition, the latch evaluates correctly. However, on the right, with slower clock fall time and a larger negative v_1, the latch evaluates in the wrong direction.

In the plots in Figures 3.10 through to 3.12 the voltages do not regenerate to infinity, as predicted by (3.9). We know that the differential pair's output saturates when the current is fully switched to one side or the other. Similar saturation effects were added to the model used to generate these plots.

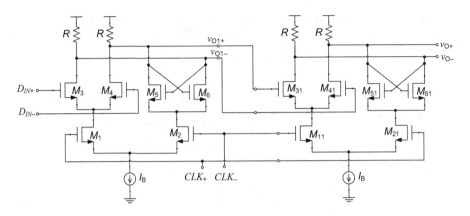

Figure 3.13 CML decision circuit formed from two level-sensitive latches.

3.2.2   Construction of a Decision Circuit from Level-Sensitive Latches

Two level-sensitive CML latches can be combined to implement an edge-triggered CML decision circuit. (Figure 3.13). In the case of CML circuits, the differential nature of the input gives the circuit a well-controlled decision threshold at $D_{IN+} = D_{IN-}$, making the circuit suitable as a wireline decision circuit. The output of the left-hand latch is connected to the D input of the right-hand latch. Notice that the CLK input is connected to the latches with opposite polarity. This ensures that when the left-hand latch tracks, the right-hand latch regenerates. The reader is encouraged to confirm that this is a falling-edge-triggered decision circuit.

3.2.3   Overview of Timing Analysis of CML Latches

During the tracking phase, the drains of $M_{3/4}$ see a time constant τ set by an RC load given by:

$$\tau_{Track} = RC_{OUT}, \qquad (3.10)$$

where C_{OUT} is the total capacitance connected to the drain of M_3 or M_4, along with any wiring capacitance present. During the regeneration phase, the relevant time constant, τ_{Regen} is computed as:

$$\tau_{Regen} = R_{eq}C_{OUT}, \qquad (3.11)$$

which for modest values of g_mR approaches:

$$\tau_{Regen} \approx -\frac{1}{g_m}C_{OUT}. \qquad (3.12)$$

Considerations for sizing are as follows:

- All MOSFETs, except those implementing the current sources, will have $L = L_{min}$.
- The product of $I_B R$ should be > 250 mV or so in order to have sufficient voltage swing.
- As an example, choose a value for R. The rest of the device sizes will be relative to this impedance level.
- Compute I_B to satisfy the swing guideline above.
- We assume that the clock input has a similar differential swing: i.e., $\approx$ 500 mVp–p. Therefore, $W_{1/2}$ must be large enough that most of I_B is switched for an input of the target swing. If $M_{1/2}$ are too small, the current will not fully switch the latch from tracking to regeneration. If $M_{1/2}$ are too wide, the latch will load the clock signal too much.
- Start with $W_{3/4} = W_{5/6}$. Choose a width so that the track phase has a gain of $\sim$2.

These sizes can be scaled by shrinking all values of W and I_B by a constant factor K and by increasing R by K. If the overall outputs' load capacitance can be correspondingly reduced, the latch's performance should remain relatively constant, but its power dissipation will reduce by K. In the event that there is a fixed load capacitance, eventually the speed of the latch will suffer and the sizes cannot be reduced as much.

3.3   Sense-Amplifier-Based Latches

One disadvantage of the CML decision circuit is that it dissipates static power. Sense-amplifier latches based on the sense-amplifier approach [19] only dissipate dynamic power. A schematic of a typical sense-amplifier latch is shown in Figure 3.14. To understand its operation, consider two phases of operation, namely the pre-charge phase and the evaluation phase. In the pre-charge phase, the clock is low and NMOS M_1 is off. PMOS transistors M_8 to M_{11} conduct, raising the output nodes $v_{O+/-}$ and internal nodes $v_{int+/-}$ to V_{DD}. Therefore the input signal $D_{IN+/-}$ has no effect on the circuit.

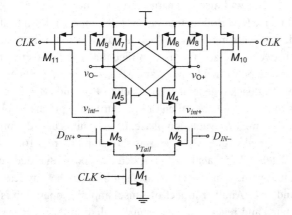

Figure 3.14 Sense-amplifier flip-flop used as a decision circuit.

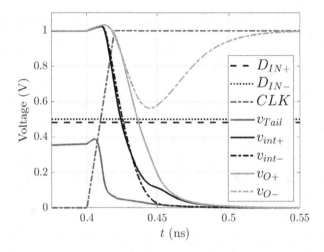

Figure 3.15 Sense-amplifier flip-flop waveforms.

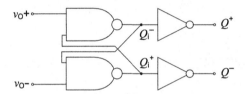

Figure 3.16 Set/reset latch implemented with NAND gates.

Waveforms during the evaluation phase are shown in Figure 3.15 for dc input. Here, D_{IN-} is 500 mV and D_{IN+} is 480 mV. On the rising edge of CLK, the circuit enters the evaluation phase; v_{Tail} is pulled from a mid-rail voltage, determined by leakage to GND, activating the differential pair of $M_{2/3}$. Conduction through $M_{2/3}$ discharges $v_{int+/-}$, which through $M_{4/5}$ will also discharge $v_{O+/-}$. Since $D_{IN+} < D_{IN-}$, $M_{2/4}$ conduct more current than $M_{3/5}$. Both output nodes will be pulled down near $V_{DD}/2$, but the righthand side conducts more current and is pulled down a little more. This difference in voltage between v_{O+} and v_{O-} is amplified through the positive feedback of the cross-coupled inverters formed by $M_{5/7}$ and $M_{4/6}$. As shown in the waveforms in Figure 3.15, positive feedback regenerates v_{O-} back to V_{DD} and v_{O+} close to GND.

Since both outputs of this circuit are pulled to V_{DD} during the precharge phase, and both drop momentarily during the evaluation phase, this circuit is usually followed by a set/reset latch, implemented with NAND gates as shown in Figure 3.16.

Unlike the CML latch whose analysis of speed has been distilled to simple equations based on small-signal parameters, sense-amplifier latches are less straightforward to analyze and size. Another aspect of sense-amplifier operation is that the evaluation phase captures and regenerates the signal, unlike in the CML case where the input signal has been captured at the end of the tracking phase. The line between

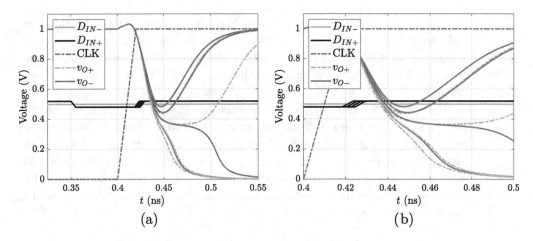

Figure 3.17 Sense-amplifier flip-flop waveforms showing hold time violation. D_{IN} flips polarity t_{Hold} after the clock edge. Outputs plotted for $t_{Hold} = 11, 13, 15, 17$ ps. For smaller hold time, direction of regeneration can be reversed: (a) waveforms and (b) waveforms with expanded time axis.

capture and regenerate is blurred in the sense-amplifier latch. To illustrate this, consider the situation where at the rising edge of *CLK* a small negative differential input is presented. Somewhere between 25 and 30 ps (Figure 3.15) the difference between the output voltages reaches a "point of no return," beyond which reversing the input polarity and even applying a large positive differential input will not reverse the direction of regeneration. However, before this point, such an input signal inversion may flip the direction of the regeneration. The time after which changes to the input polarity have no effect on how the latch resolves is the hold time of the circuit and will depend on how large the positive differential voltage is.

Figure 3.17 investigates the hold time. Before the clock edge, a small negative differential (-20 mV) is presented to the latch. Input polarity flips t_{Hold} after the clock edge. Taking the time at which the clock crossed $V_{DD}/2$ as the time of the clock edge, hold times of 11–17 ps are considered in 2 ps increments. For t_{Hold} of 15 and 17 ps, v_{O-} regenerates to V_{DD} as expected for a negative input polarity. However, for smaller t_{Hold}, the flipping of the input polarity reverses the direction of regeneration, violating the hold time of the circuit. In this specific example, D_{IN} changed from -20 mV to $+20$ mV. A larger amplitude after the input polarity change would require a larger hold time. Thus, the timing characteristics of sense-amplifier flip-flops are a complex function of input amplitude.

In general PMOS pre-charge transistors will be the minimum width so as to not load internal nodes. Even minimum width transistors can pre-charge internal nodes in half a UI. Consider a design with NMOS inverter transistors M_4 and M_5 also of minimum width. The choice of width for PMOS inverter transistors sets loading at output nodes and the equilibrium voltage of the cross-coupled inverters. If during evaluation phase both outputs drop low enough to put the SR latch into the "forbidden-input" state, the inverter threshold can be increased or the threshold of the SR latch lowered. With the cross-coupled inverter size ratio fixed, the input and clock transistors set how quickly the V_{int} and V_O nodes discharge. Increasing these transistors will lead to faster evaluation, but increase the loading on the circuits driving them.

The speed of a sense-amplifier latch depends on its input common-mode voltage. Unlike a CML latch for which the designer must only ensure that the input common-mode voltage is sufficiently high to keep the clock tail current source in saturation, the *CLK* transistor M_1 is not designed to be in saturation. Therefore, the amount of current that flows through $M_{1/2/3}$ depends on the input common-mode voltage. This requires simulations that consider the targeted common-mode voltage along with a suitable design margin.

3.4   Decision Circuit Simulation

A complete simulation-based characterization of a decision circuit involves simulations to estimate hysteresis, timing parameters, power dissipation, noise and offsets. Hysteresis, noise and offset characterization simulations are discussed in the following subsections.

3.4.1   Hysteresis

An ideal latch responds to the polarity of its input in a given UI, independent of past decisions. However, latches can show memory effects. Consider as an example a CML latch (Figure 3.13) that in one UI regenerated to have a +ve differential output voltage. In the next tracking phase, a small -ve input voltage must have sufficient time to invert the polarity of the output voltage before the subsequent regeneration phase. For a given UI, this imposes the minimum magnitude of input voltage that can flip the state of the latch even if the set-up time is increased to the entire UI. Thus, the hysteresis input voltage is UI-dependent.

In the case of a sense-amplifier-based latch, the pre-charge phase raises both output nodes to V_{DD}. However, due to the latch's regeneration in the previous UI, one output is already at V_{DD}. The other output is raised to V_{DD} through a PMOS transistor. The differential output voltage decays to zero. However, any remaining differential output voltage at the onset of the evaluation phase predisposes evaluation toward the previous bit decision. Therefore, if pre-charge transistors are too small, hysteresis increases. An additional PMOS transistor that shorts the two outputs together can help to reduce the hysteresis-inducing residual differential voltage.

The SR latch that loads the sense-amplifier-based comparator can also introduce hysteresis if each input's capacitance depends on the state stored in the latch. This phenomena, combined with the capacitive loading the SR latch presents, means that it is imperative to include the SR latch when simulating the sense-amplifier-based comparator.

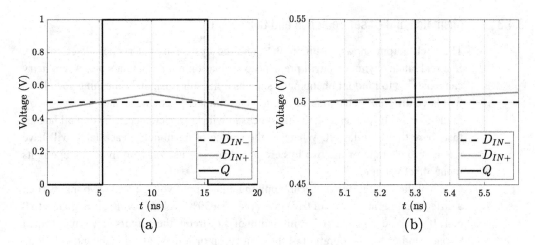

Figure 3.18 Decision-circuit simulation waveforms for estimating hysteresis: (a) overall view and (b) zoomed-in view around input that flips the decision circuit's state.

Figure 3.18 shows simulation waveforms for determining hysteresis. D_{IN-} is held at a constant voltage while a slow ramp signal is applied to D_{IN+}. The clock signal is not shown, although the behaviour of the circuit can be clock-period dependent. Up to 5 ns, $D_{IN+} < D_{IN-}$ and Q evaluates low. Ideally, Q would evaluate high at the first clock edge for which $D_{IN+} > D_{IN-}$ occurring at $t = 5$ ns, and then evaluate low as soon as $D_{IN+} < D_{IN-}$ (15 ns). However, as seen in Figure 3.18 (b), Q evaluates high at 5.3 ns. Since D_{IN+} increases from 0.5 V to 0.55 V as t increases from 5 to 10 ns, we estimate $D_{IN+} = 0.503$ V for $t = 5.3$ ns, giving a minimum input of 3 mV. Likewise, as D_{IN+} crosses D_{IN-} at 15 ns, an input of -3 mV is needed to flip the latch. The resolution of the estimate is limited by the clock period. For a finer estimate, the slope of the ramp input can be slowed.

In the example above, the latch shows 3 mV of hysteresis in both directions. The average of the required voltages is 0 and hence the latch has no net offset. A real latch will suffer from small deterministic asymmetry in layout and mismatch in transistor parameters, giving rise to asymmetry in behaviour, discussed in Section 3.4.3.

3.4.2   Noise

Thermal noise from transistors and resistors introduces uncertainty in how a decision circuit evaluates for inputs near its decision threshold. If we can model the real latch as a noiseless latch with an input-referred noise, σ_{Latch}, added to its input, we can combine its noise with that from the rest of the front-end. For example, consider Figure 3.1. If a simulation of the circuitry up to the output of the EQ gives a noise with standard deviation σ_{FE}, the overall noise σ_N to use in (1.30) can be computed as:

$$\sigma_N = \sqrt{\sigma_{FE}^2 + \sigma_{Latch}^2}. \tag{3.13}$$

3.4.3   Mismatch and Offset Detection and Correction

The latch output shown in Figure 3.18 showed hysteresis, but no net offset voltage. In a real latch, asymmetry will predispose evaluation to one output state. Asymmetry can exist due to random phenomena such as transistor mismatch, meaning each latch will have different behaviour, or due to deterministic phenomena such as asymmetry in the circuit's layout leading to differences in wiring capacitance, affecting all latches that share the same layout. Whatever the source of asymmetry, each latch will have a non-zero average of the two hysteresis voltages. This non-zero input voltage is its input offset voltage.

When the latch is symmetrical, applying equal input voltage to its two inputs (i.e., a differential input voltage of 0 V) gives rise to a 50% probability that it evaluates to 0 or 1. If no noise is added to the simulation of the circuit, the outputs may remain equal to one another for the duration of the simulation or it may take a comparatively long time to evaluate to 0 or 1. However, with the addition of transient noise, evaluation time is reduced.

In the design of wireline receivers, the deterministic input offset must be estimated along with the statistics of the random offset. Usually, latches are designed with offset compensation schemes to calibrate out both types of offset on a per-latch basis. Finally, the system must have the capability to detect offsets during run time to allow periodic zeroing of offsets.

The estimation of offsets can be done with a slowly increasing ramp applied to the latch's input. This is done without including random mismatch in transistors or transient noise. Initially, the latch will evaluate to a 0. The time at which the output evaluates to a 1 can be mapped to an input voltage corresponding to the minimum input required to flip the latch. If this simulation is repeated with a decreasing ramp, two input voltages leading to the latch flipping can be found. These voltages will be different from each other due to hysteresis of the latch and because the latch can only flip on clock edges, meaning that the input voltage at the clock edge will have moved past the minimum voltage for latch flipping. The average of these two voltages gives the deterministic offset.

To find the random offset, transistor mismatch is included and the ramp-based simulation is repeated. A common approach to include mismatch is through Monte Carlo simulation. This approach uses foundry-supplied standard deviations for threshold voltage mismatch and current-factor mismatch ($\mu_n C_{OX} \frac{W}{L}$) usually based on a Pelgrom model [20]. Each transistor has its parameters modified using samples of Gaussian distributions using these data. Then, a simulation based on this circuit with randomly generated parameters can be run. The offset of this particular instance of the circuit can be found using the approach in the previous paragraph. The challenge with this approach is that many iterations must be simulated to get a meaningful estimate of the mean and standard deviation of the latch's offset.

Another approach is to estimate a worst-case offset. As presented in [21], for low-voltage operation, threshold-voltage mismatch is more significant than current-factor mismatch. A mismatch in threshold voltage can be modelled by adding a voltage

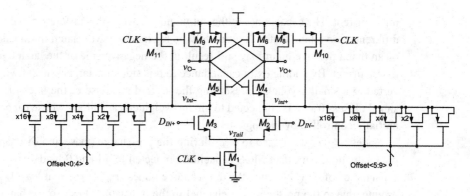

Figure 3.19 Programmable capacitor-based latch offset compensation.

source in series with the gate of the transistor. A designer can add a voltage source to each transistor with a value equal to its anticipated standard deviation of mismatch. The polarity of each source is set so as to give the worst-case combination of mismatch effects. Even if the designer is unsure of which polarity is needed for each of the N transistors in the circuit, this approach needs only $2N$ simulations, which will be fewer than the 1000, or more Monte Carlo simulations. To get an even more severe assessment of mismatch, the voltage sources can be set to the 3-σ mismatch for each transistor.

Each transistor's anticipated threshold voltage offset (δV_t) can be calculated using:

$$\delta V_t = \frac{A_{V_t}}{\sqrt{2WL}}, \tag{3.14}$$

where A_{V_t} is a foundry-supplied mismatch parameter and W and L are the width and length of the transistor, respectively. The reader may have seen this expression without the factor of $\sqrt{2}$ in the denominator. Equation (3.14) models the statistics of the difference between a given transistor's threshold voltage and the mean, whereas without the factor of two the equation models the difference between two transistors' threshold voltages, which for independent random variables of equal standard deviation will be $\sqrt{2}$ times larger.

With the latch offset characterized, the designer must then decide if it is too large, necessitating a compensation scheme. An estimate of offset can be added to V_{MIN} shown in Figure 1.21 and (1.30) and (1.31), thereby increasing the VEO needed for a given BER. Since offset compensation schemes are relatively straightforward, anything but the smallest increase in V_{MIN} would normally be reduced through a compensation scheme.

In the case of sense-amplifier-based latches, compensation schemes usually adjust the capacitive load of the two outputs as shown in Figure 3.19 [22]. At the start of the evaluation phase, the polarity of the input determines which branch should conduct more current. The capacitance at the output node coverts this difference in current to a difference in voltage that will ultimately be regenerated. Mismatch in transistors will mean that a positive input may still result in the negative side of the latch conducting more current. By adding extra capacitance to this side, this larger current will be converted to a smaller voltage, preserving the desired response of the latch. The digital signal $Offset<0:9>$ is connected to binary-weighted capacitors, whose capacitance is partially controlled by the logic level.

Nodes $v_{int+/-}$ are raised to V_{DD} during the pre-charge phase and discharged during evaluation. An offset MOS capacitor connected to a logic 0 loads $v_{int+/-}$ with a capacitance of $C_{OX}WL$ when the drain/source nodes ($v'_{int+/-}$) are at least $|V_{tp}|$ above ground, due to the presence of a channel in the transistor. On the other hand, a MOS capacitor connected to a logic 1 has no channel and presents only a smaller overlap capacitance between source/drain nodes and the gate.

The designer has to decide the total range of offset to be corrected and the step size of compensation. Notice that, regardless of the range selected, there is always a diminishing probability that the offset of a given latch is beyond the range of compensation. Therefore, the compensation range usually extends to at least the 3σ range. Even with this, a given latch has a 0.3% chance of being beyond this range, which given the number of latches per wireline channel and the number of wireline channels per chip is potentially too high of a percentage.

Once the range is selected, the step size is chosen, typically in the range of 4-bits per side, giving a total of 32 steps. This means that in the 3σ example, the step size is $\frac{3}{16}\sigma$, leading to a residual offset of not more than $\frac{3}{32}\sigma$.

Note that in addition to the capacitors and switches needed to implement the offset compensation, eight bits of memory are needed in which to store the state of the switches. Since sense-amplifier latches can be made with small transistors, the area of compensation/storage circuitry may exceed that of the core latch.

As a final note, the mismatch in threshold voltage predicted by (3.14) can be reduced by increasing the width of transistors. However, scaling up the width of every transistor in a latch will correspondingly increase power dissipation. The compensation scheme described above leads to lower power dissipation than upsizing transistors for a given residual offset.

3.4.4   Simulation of Dynamic Performance

DFFs are usually characterized with setup and hold times, assuming inputs are logic 0s and 1s. Wireline decision circuits must also regenerate small input signals to produce full-rail outputs. Consider a worst-case 1 applied to D_{IN+}, shown in Figure 3.20. An ideal data eye on D_{IN+} is also shown in light grey. However, an actual 1, due to noise, might start from a lower value (stronger 0) down at v_0 and only cross the decision threshold set by D_{IN-} by a small difference, reaching a weak one at v_1. For a given v_1 and v_0, setup and hold times can be reduced to find a minimum duration of the 1,

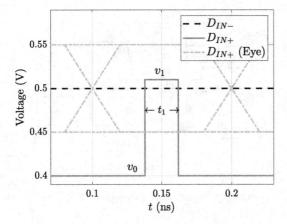

Figure 3.20 Worse-case input data for stressing latch timing parameters.

t_1 that is properly captured. Note that a larger v_1 may permit a shorter t_1. Regardless of the choice of v_1, t_1 must be less than the unit interval. To ensure sampling without additional bit errors we must ensure:

$$t_1 + t_{J,P-P} < T_b, \tag{3.15}$$

where $t_{J,P-P}$ is an estimate of peak-to-peak jitter. The voltage v_1 corresponds to V_{MIN} in (1.30).

The simulation shown in Figure 3.20 can be done to find a set of values for v_0, v_1 and t_1. A more involved approach presented in [23] gives a complete model for the latch showing the trade-off between timing and amplitude parameters.

3.5 Sub-Rate Operation and Latch Design

Figure 3.21 (a) shows a decision circuit activated by a clock operating at the data rate. The latch uses the rising edge of the clock signal to capture each bit. An alternative approach is shown in Figure 3.21 (b), where the data stream is connected to four latches, each activated by four phase-shifted clock signals each at $\frac{1}{4}$ of the data rate. Notice that each bit is still captured. The dashed lines show how latches capture bits in a round-robin fashion. As discussed in Chapter 1, wireline receivers usually demultiplex data into several lower-rate data streams. This happens automatically in (b) since each latch produces data at $\frac{1}{4}$ of the line rate. An advantage of this approach is that it relaxes some of the timing constraints on the latches. Although the sampling window must remain short such that each bit is sampled separately, the regeneration phase can be lengthened since each latch has 4 UI between consecutive input bits. Sub-rate clocking is also used to reduce the power dissipation associated with generating and distributing high-frequency clocks. As will be discussed in Chapter 13, ring oscillators can generate phase-shifted clock signals. Although it is possible to generate four signals each shifted by 90°, mismatch in circuit delay can introduce timing errors among the signals that must be kept to a small fraction of the UI.

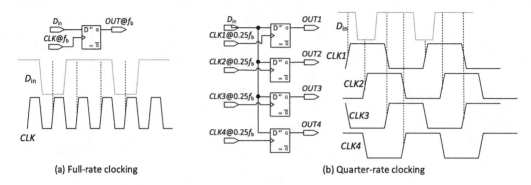

(a) Full-rate clocking   (b) Quarter-rate clocking

Figure 3.21 Different clocking strategies: (a) full-rate clocking and (b) quarter-rate clocking.

3.6   Summary

This chapter introduced decision circuits. It may seem out of place to discuss the last block in a receiver first. However, the operation and limitations of the decision circuit shape the requirements of the receiver circuitry between an electrical channel and the decision circuit. To this end, the ideal and typical operation of decision circuits was presented along with its behaviour relative to a digital DFF. CMOS and CML implementation was discussed, and CML circuit operation began with a reminder of how a DFF can be constructed from two transparent latches clocked with complementary clock phases. A model of a transparent latch was introduced, allowing an analysis of timing parameters and criteria for the negative resistance of the cross-coupled transistors. Sense-amplifier-based decision circuits were presented, including the need to follow them by SR latches.

The various types of decision circuit simulation were explained, including methods to determine the effects of hysteresis, noise, deterministic offset, mismatch and the minimum detectable pulse width. An example circuit for offset correction was also given. The chapter closed with an introduction to sub-rate clocking, a technique for operating decision circuits from several clock phases, allowing for demultiplexing and longer clock to Q delays.

Problems

3.1   This problem deals with the static characterization of the CML level-sensitive latch in Figure 3.6 using the parameters in Table 3.1. To get realistic performance, the ideal current sources should be replaced with suitably designed current mirrors.

(a)   Determine the small-signal voltage gain in the tracking phase. Is this an appropriate gain? Do you expect the circuit to properly regenerate in the regeneration phase? If the gain is too low, modify the transistor widths and/or R to increase the gain allowing correct operation.

(b)   Determine the hysteresis input voltage for a clock period of 200 ps. Confirm that the latch has no deterministic offset voltage.

Table 3.1 Recommended parameters for the CML
level-sensitive latch in Figure 3.6. All transistors have
$L = L_{min}$.

Parameter	Value
I_{BIAS}	100 μA
R	2 kΩ
V_{DD}	1 V
W_3 through W_6	1 μm
W_1 and W_2	2 μm
Load capacitance	Gate of a MOSFET with $W = 2$ μm

(c) Repeat for a clock period of 100 ps and comment on why this voltage is or is not different from that in 3.1.

(d) Investigate the input offset for the following asymmetries, each applied separately:

  (i) Increase of one load resistor by 5%.
  (ii) Addition of an extra 1 μm of transistor width to the load of M_5.
  (iii) Addition of a 5 mV voltage source between the drain of M_6 and the gate of M_5.

  Which one has the smallest effect? Which one has the largest?

3.2 For the CML level-sensitive latch in Figure 3.6 and with the parameters in Problem 3.1, determine the minimum pulse width for the following scenarios. See Figure 3.20 for plots that depict a minimum pulse.

(a) Assume v_1 is 10 mV above the reference level and v_0 is 20 mV below. These voltages refer to the level of D_{IN+} relative to D_{IN-}. Use a clock period of 200 ps. Is the pulse centred in time around the clock edge?

(b) Assume v_1 is 5 mV above the reference level and v_0 is 20 mV below. Use a clock period of 200 ps. Is the pulse centred in time around the clock edge?

(c) If an extra 2 μm of transistor width is added to each output, how does the minimum input pulse width change for the voltage scenario in Problem 3.2(a).

(d) If both resistors (R) are reduced down to 70% of the value in Table 3.1, how does the minimum input pulse width change for the voltage scenario in Problem 3.2(a).

3.3 Pat designed the level-sensitive latch shown in Figure 3.6. To reduce loading on the output node, Pat reduced the size of $M_{5/6}$ to only 2 μm compared to 15 μm for $M_{1/2}$. Figure 3.22 (a) shows the response that Pat wants. However, Figure 3.22 (b) shows the actual response. The input and clock signals for the two simulations are the same.

(a) Describe what the undesirable aspect of Figure 3.22 (b) is.

(b) Explain why it is occurring and what can solve the problem.

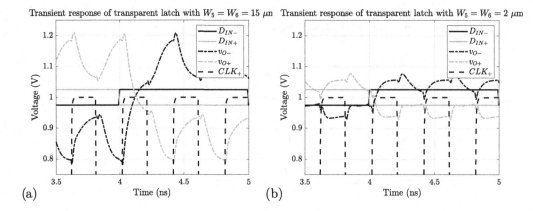

Figure 3.22 Level-sensitive latch responses: (a) desired response and (b) achieved response.

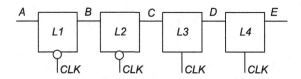

Figure 3.23 Four level-sensitive latches in cascade.

3.4 This question deals with the level-sensitive latch in Figure 3.6 and a possible connection of four of them as shown in Figure 3.23.

(a) Explain how the circuit in Figure 3.6 works. In particular explain the performance trade-offs of increasing and decreasing the resistance of the loads. Assume the starting point for the design is a circuit with $g_m R = 2$.

(b) In attempting to design an edge-triggered decision circuit, a designer connects up four level-sensitive latches as shown in Figure 3.23. Will this circuit work as an edge-triggered decision circuit? If not, why not? If yes, how will its behaviour differ from the conventional approach to building a decision circuit from two level-sensitive latches, shown in Figure 3.5?

3.5 This problem deals with the static characterization of the sense-amplifier based latch in Figure 3.14 using the parameters in Table 3.2. Use a common-mode input voltage of $V_{DD}/2$.

(a) Determine the hysteresis input voltage for a clock period of 200 ps. Confirm that the latch has no deterministic offset voltage.

(b) Repeat for a clock period of 100 ps and comment on why this voltage is or is not different from that in 3.5.

(b) Investigate the input offset for the following asymmetries, each applied separately:

   (i) Addition of an extra 1 μm of transistor width to node v_{O+}.

Table 3.2 Recommended parameters for the sense-amplifier based latch in Figure 3.14. All transistors have $L = L_{min}$.

Parameter	Value
V_{DD}	1 V
W_8 through W_{11}	W_{min}
W_1	2 μm
W_2 and W_3	2 μm
W_4 and W_5	2 μm
W_6 and W_7	3 μm
Load capacitance	Inverter with the same sizes as W_4/W_6

   (ii)   Connection of the gate of a 1 μm wide transistor to node v_{int+}.

   (iii)   Addition of a 5 mV voltage source between the node v_{0+} and the gates of M_5 and M_7.

Which one has the smallest effect? Which one has the largest effect?

3.6 For the sense-amplifier based latch in Figure 3.14 and with the parameters in Table 3.2, determine the minimum pulse width for the following scenarios. Unless otherwise indicated, use a common-mode input voltage of $V_{DD}/2$.

(a)   Assume v_I is 10 mV above the common-mode input and v_0 is 20 mV below. Use a clock period of 200 ps. Is the pulse centred around the clock edge?

(b)   Assume v_I is 5 mV above the common-mode input and v_0 is 20 mV below. Use a clock period of 200 ps. Is the pulse centred around the clock edge?

(c)   Repeat 3.6(a) for common-mode input voltages of $0.4V_{DD}$ and $0.6V_{DD}$.

(b)   If an extra 2 μm of transistor width is added to each output, how does the minimum input pulse width change for the voltage scenario in 3.6(a).

3.7 For the sense-amplifier based latch in Figure 3.14 and with the parameters in Table 3.2, explore the phenomenon of kick-back noise. To do this, look at the voltage at the inputs when they are driven through resistors of 1 kΩ rather than connected directly to ideal voltage sources.

4   Equalization

A central challenge in the design of electrical links is to compensate for frequency-dependent loss in the channel that introduces intersymbol interference (ISI). This chapter presents the overall objectives of joint Tx/Rx equalization. The system-level operation of transmitter-side feed-forward equalizers (FFEs) is discussed. Circuit details are presented in Chapter 5. Receiver-side continuous-time linear equalizers (CTLEs) and finite impulse response (FIR) filters will be discussed next, followed by decision feedback equalizers (DFEs). DFEs differ from FFEs, CTLEs and FIRs in that they only remove ISI rather than attempt to invert the low-pass channel characteristic. With the growing trend toward ADC-based receivers, the implementation of DFEs and Rx FFEs is discussed in the analog/mixed-signal domain in this chapter and in the digital domain in Chapter 11. The topics in this chapter are also a relevant background for the sections in Chapter 10 that discuss transimpedance amplifiers for reduced-bandwidth systems, where equalization is used to remove ISI from a high-gain but intentionally bandwidth-limited optical receiver front-end.

4.1   Overall Objectives

The data rates of most electrical link standards are high enough that the channel's pulse response shows significant ISI. If not addressed, the eye will be closed. In a conventional receiver where a decision circuit resolves the signal into 0s and 1s, a closed eye results in an unacceptably high BER. The general objective of transmitter and receiver equalization is to improve the BER. In conventional receivers, we equate this to reducing ISI and allowing an open eye to be sampled. However, if we view equalization as BER improvement we can see other approaches, such as maximum-likelihood sequence estimation (MLSE). Thus, we can view equalization as doing three possible things:

- Invert the channel.
- Remove pre- and post-cursor ISI from the pulse response.
- Figure out what bits were sent.

The first viewpoint originates from the observation that an electrical channel has a low-pass response. As presented in Figure 4.1, if the channel is cascaded with an equalizer that has a high-pass response, the overall response is flat over some range of frequency.

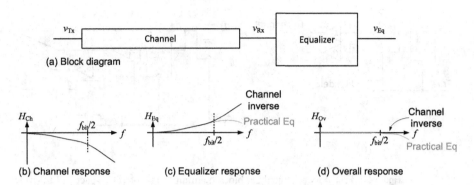

(a) Block diagram

(b) Channel response (c) Equalizer response (d) Overall response

Figure 4.1 Equalization as channel inversion: (a) block diagram of a channel followed by an equalizer; (b) frequency response of the channel; (c) frequency response of an equalizer; (d) overall frequency response.

Figure 4.1 (c) shows an equalizer response that is the exact inverse of the channel's response as well as a practical equalizer, whose response is close to being the channel inverse up to $f_{bit}/2$ but rolls off beyond. Part (d) shows the overall response of the cascade of the channel and each of the equalizers. The net low-pass response of the practical equalizer is preferred from a noise point of view.

We can also view equalization as a manipulation of the pulse response, subtracting ISI from the pulse response of previously received bits. This approach neither inverts the channel nor introduces high-frequency gain, giving it certain noise advantages. However, as we will see in Section 4.5.2, the timing requirements of the feedback loop can be challenging to meet.

The final viewpoint of equalization implies that we only need to estimate the transmitted bits, without necessarily removing ISI. This is the essence of MLSE (Section 11.4). In the sections that follow the broad classes of equalizer are introduced and explained.

4.2 Transmitter-Side Feed-Forward Equalizers

Knowing that a channel has a low-pass response, the transmitter can launch data with a pre-emphasis of the data stream's higher frequencies. Consider the signal waveform in Figure 1.9. Following two 0s, the signal at B' only rises part way to L_1 after the first 1 whereas at C', the signal has almost reached the full L_1 level. The amplitude of B' and C' can be made equal to one another by increasing the amplitude of the transmitted signal for a bit transition while decreasing the amplitude of the signal when consecutive identical digits are transmitted. Conceptually, the output voltage of the transmitter y at each bit n is formed by the values of the transmit data at the current bit (n) and the previous bit $(n-1)$:

$$y[n] = d_i[n] - \alpha d_i[n-1], \tag{4.1}$$

where d_i has values +1 and −1. In the previous example of a 1 following a long run of zeros ($d_i = -1$), the transmitter outputs a signal of $1 + \alpha$ for the first 1 and a signal of

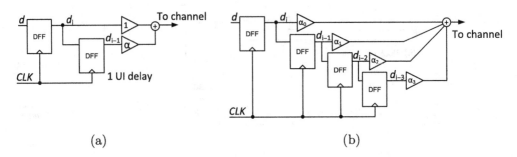

Figure 4.2 Feed-forward equalizer block diagram: (a) two-tap UI-spaced FFE; (b) four-tap FFE.

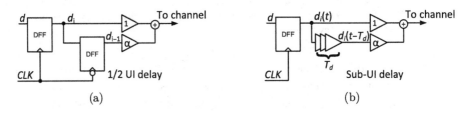

Figure 4.3 Feed-forward equalizer block diagram: (a) half-UI spaced; (b) continuous-time delay, sub-UI spaced.

$1 - \alpha$ for the second 1. Likewise, a bit transition downward is accentuated by the first 0 seeing an output of $-1 - \alpha$ followed by weaker 0s of $-1 + \alpha$.

The block diagram of such a system is shown in Figure 4.2 (a). A D flip-flop (DFF) provides a 1-UI delay allowing $d_i[n]$ and $d_i[n-1]$ to be combined with weights 1 and α, respectively. To be equivalent to (4.1) the α-input of the summing junction is inverted. Since the data see a delay of 1 UI, it is referred to as symbol-spaced equalization. The concept can be extended to generate an output signal based on more than two transmitted bits, as shown in Figure 4.2 (b).

To extend the effect of equalization to higher frequencies, the delay can be reduced to a fraction of the UI. In Figure 4.3 (a), the second DFF is falling-edge triggered, meaning that it implements a delay of only $\frac{1}{2}$-UI. To implement smaller delays, continuous-time (CT) delay elements can be used, such as in Figure 4.3 (b). However, these require a form of calibration or tuning to ensure that their implemented delay is accurately set. Although analog delays with good phase response can be difficult to implement, the CT-delay elements here implement delay less than $\frac{1}{2}$-UI and delay what is still a digital (two-valued) signal. Thus these delay elements are more straightforward to design compared to those implementing analog FFEs in the receiver.

To illustrate the operation of the block diagram in Figure 4.2 (a), consider the data stream d_i, as given in Table 4.1. It is assumed that bits before $i = 0$ are -1.

The input d_i and the output y_i are plotted in Figure 4.4. The output signal $y_i = \pm 1.25$ when there is a bit transition from $i - 1$ to i and is ± 0.75 when consecutive bits are identical.

Table 4.1 Tx FFE example for $\alpha = 0.25$.

i	d_i	d_{i-1}	$y_i = d_i - \alpha d_{i-1}$
0	-1	-1	-0.75
1	-1	-1	-0.75
2	1	-1	1.25
3	1	1	0.75
4	-1	1	-1.25
5	1	-1	1.25
6	-1	1	-1.25
7	-1	-1	-0.75
8	-1	-1	-0.75

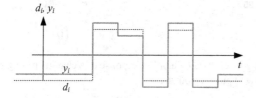

Figure 4.4 FFE example waveforms for $\alpha = 0.25$. Black dashed line is d_i. Thick grey line is the output y_i.

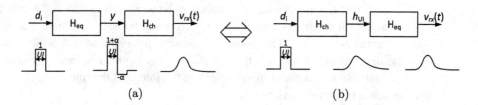

Figure 4.5 Block diagram of FFE and channel: (a) original block diagram; (b) channel and equalizer swapped.

To analyze an FFE in the frequency domain, we first consider the Z-transform of the equalizer. In this case, the equalizer gives:

$$H_{eq}(z) = 1 - \alpha z^{-1}. \tag{4.2}$$

Figure 4.5 (a) shows a block diagram of an equalizer preceding the channel. If the channel has a frequency response of $H_{ch}(s)$, we can combine the equalizer and channel by multiplying the frequency responses and substituting $z = e^{sT_d}$. In the case of symbol-spaced equalization, $T_d = T_b$. For our example with one delay, the overall transfer function H_{tot} is given by:

$$H_{tot}(s) = H_{eq}(s)H_{ch}(s) = \left(1 - \alpha e^{-sT_b}\right)H_{ch}(s). \tag{4.3}$$

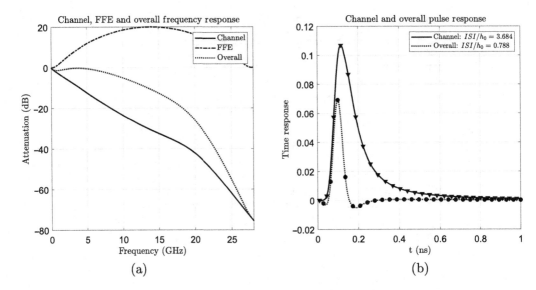

Figure 4.6 FFE example showing its effect on the pulse response of a channel having significant ISI: (a) frequency response and (b) pulse response.

Notice that H_{eq} is indeed high-pass, starting with a magnitude of $1 - \alpha$ for $s = 0$ and increasing to $1 + \alpha$ at $f_b/2$ where $s = j2\pi f_b/2$ and $e^{-sT_b} = e^{-j\pi} = -1$.

Note that H_{ch} is not usually a rational polynomial, given that in the case of electrical links it originates from distributed effects in transmission lines. Neither is H_{eq} a rational polynomial. Therefore, it is not straightforward to talk about zeros in the equalizer cancelling poles in the channel. By plotting the combined response we can qualitatively see that high-frequencies are boosted, and that the channel's frequency response is partially inverted.

Example 4.1   Investigate the ISI improvement provided by an FFE.

Solution

Figure 4.6 shows an example of an FFE's effect on the frequency and pulse response of a system. An example channel response is plotted in part (a) using a solid line and its pulse response for 28 Gb/s data is plotted in part (b). From part (a) we see that the channel has a little more than 30 dB of loss at the Nyquist frequency (14 GHz). The sampled pulse response is denoted by the triangle markers in part (b). There is significant ISI with total ISI $> 3.5\times$ the main cursor. Part (a) shows the frequency response of a candidate FFE with α_0 to α_3 equal to $[-0.1, 1, -0.8, 0.1]$.

In Figure 4.6 (a) the FFE's frequency response is plotted, boosting frequencies near Nyquist (14 GHz). The effect of equalization on the pulse is clear. The ratio of total ISI to main cursor dropped from 3.684 without equalization to 0.788 with an FFE. Usually, additional equalization would be carried out to further reduce ISI.

We can quantify the improvement in data transmission by considering the impact of equalization on the pulse response by applying the equalizer to the pulse response of the channel. To do this, consider Figure 4.5 (b) in which the order of the channel and the equalizer is swapped. This is mathematically equivalent because the system is a linear time invariant (LTI) system. Notice that if we apply d_i to the channel, we get the pulse response h_{UI} at the output. Then the equalizer is applied to the pulse response. For the example in (4.1),

$$v_{rx}(t) = h_{UI}(t) - \alpha h_{UI}(t - T_b).\tag{4.4}$$

From (4.4) we note that an FFE, even one with only one delay element, operates on the entire pulse response from $t = T_b$ onward. This means that for a well-behaved post-cursor ISI tail, a simple FFE can cancel the tail. To illustrate this mathematically, consider the following example.

Example 4.2 Use an FFE with one delay element to equalize a first-order pulse response shown in Figure 4.7.

Solution

As discussed in Example 1.1, a first-order system's pulse response is given by:

$$h_{UI}(t) = \begin{cases} 0, & \text{for } t < 0 \\ 1 - e^{-t/\tau}, & \text{for } 0 \leq t < T_b \\ \left(1 - e^{-T_b/\tau}\right)e^{-(t-T_b)/\tau} & t > T_b. \end{cases}\tag{4.5}$$

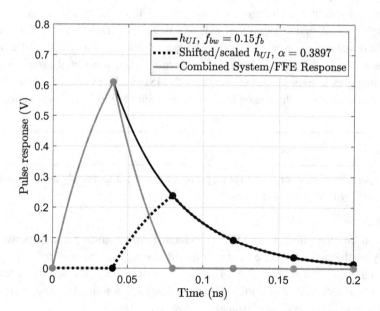

Figure 4.7 FFE operation on a first-order pulse response. Black line is the pulse response. Dashed line is the shifted, scaled pulse response. Grey line is the overall response.

If we scale the delayed pulse response by an appropriately selected value of α, we get the dotted black line in Figure 4.7, expressed mathematically as:

$$\alpha h_{UI}(t - T_b) = \begin{cases} 0, & \text{for } t < T_b \\ \alpha(1 - e^{-(t-T_b)/\tau}), & \text{for } T_b \le t < 2T_b \\ \alpha\left(1 - e^{-T_b/\tau}\right)e^{-(t-2T_b)/\tau} & t > 2T_b. \end{cases} \quad (4.6)$$

Subtracting the delayed, scaled pulse from the original pulse response, region by region gives:

$$h_{UI}(t) - h_{UI}(t - T_b) = \begin{cases} 0, & \text{for } t < 0 \\ 1 - e^{-t/\tau}, & \text{for } 0 \le t < T_b \\ \left(1 - e^{-T_b/\tau}\right)e^{-(t-T_b)/\tau} - \alpha(1 - e^{-(t-T_b)/\tau}), & T_b \le t < 2T_b \\ \left(1 - e^{-T_b/\tau}\right)e^{-(t-T_b)/\tau} \\ \quad -\alpha\left(1 - e^{-T_b/\tau}\right)e^{-(t-2T_b)/\tau} & t > 2T_b. \end{cases}$$

$$(4.7)$$

If we choose α so that the delayed/scaled pulse and the original pulse have the same height at $t = 2T_b$, we see that (4.7) is zero for the entire region of $t > 2T_b$. This is shown by the grey line in Figure 4.7. We start by substituting $t = 2T_b$ in to the last case of (4.7), setting it to 0 and solving for α:

$$0 = \left(1 - e^{-T_b/\tau}\right)e^{-(2T_b-T_b)/\tau} - \alpha\left(1 - e^{-T_b/\tau}\right)e^{-(2T_b-2T_b)/\tau}, \quad (4.8)$$

$$0 = \left(1 - e^{-T_b/\tau}\right)e^{-T_b/\tau} - \alpha\left(1 - e^{-T_b/\tau}\right), \quad (4.9)$$

$$\alpha = e^{-T_b/\tau}. \quad (4.10)$$

In the example shown in Figure 4.7, the bandwidth of the first-order system is 15% of the data rate, giving significant ISI. The first post-cursor (h_1) is 0.3897 times the height of the main cursor. Thus, selecting $\alpha = 0.3897$ gives a delayed pulse of height equal to the h_1 term and subsequent ISI terms. When subtracted using an FFE the resulting grey line shows no ISI over the entire ISI tail. Electrical link channels are not first-order. However, this example shows how a simple FFE can remove a long tail of ISI.

Example 4.3   Consider the pulse response in Table 1.2 and apply an FFE to equalize a channel with this pulse response.

Solution
We can find the effect of an FFE on a channel with a given pulse response (h_k) by applying the pulse response to the FFE and finding the resulting pulse response of the combined channel/equalizer. This is summarized in Table 4.2 where each case refers to a particular set of FFE coefficients. Each h_{rx_i} is found by convolving the FFE coefficients (h_{FFE}) with the channel pulse response (h_k):

$$h_{rx_i}[n] = h_k[n] * h_{FFE}[n]. \quad (4.11)$$

Table 4.2 Resulting pulse response after an FFE for a channel with $h_k = 0.02, 1, 0.5, 0.2$.

	Case 1	Case 2	Case 3	Case 4
FFE weights ($h_{FFE}[n]$)	0, 1, 0, 0	0, 1, −0.5, 0	0, 1, −0.5, 0.05	−0.02, 1, −0.5, 0.05
i	h_{rx_1}	h_{rx_2}	h_{rx_3}	h_{rx_4}
0	0	0	0	−0.0004
1	0.02	0.02	0.02	0
2	1	0.99	0.99	0.98
3	0.5	0	0.001	−0.003
4	0.2	−0.05	0	0
5	0	−0.1	−0.075	−0.075
6	0	0	0.01	0.01
ISI:	0.72	0.17	0.106	0.0884
VEO:	0.28	0.82	0.884	0.8916

The top row enumerates the various cases, while the second row specifies the FFE coefficients used in each case. The rest of the first column gives the index, i, of the input and output.

Case 1 has one non-zero FFE tap. Therefore, it has no equalization and the output pulse response h_{rx_1} is equal to h_k, but delayed by one UI. For each case, the resulting ISI and VEO are calculated. In Case 2, a second tap is added. By setting it to −0.5, $h_{rx_2} = h_k - 0.5h_{k-1}$. This removes the first post-cursor but has negligible effect on the main cursor. The VEO increases from 0.28 to 0.82. Cases 3 and 4 show further improvement by adding a third and fourth tap. In each case the tap was set such that the shifted main cursor nulled one post-cursor (Case 3) or one pre-cursor (Case 4). The four-tap FFE improves the VEO to 0.89.

More formally, the combined channel/FFE response is found by expanding (4.11) into summation form giving:

$$h_{rx_i}[n] = \Sigma_{i=0}^{\infty} h_k[n-i]h_{FFE}[i]. \tag{4.12}$$

For example, only writing non-zero terms:

$$h_{rx_4}[3] = \Sigma_{i=0}^{\infty} h_k[3-i]h_{FFE}[i], \tag{4.13}$$

$$= h_k[3]h_{FFE}[0] + h_k[2]h_{FFE}[1] + h_k[1]h_{FFE}[2] + h_k[0]h_{FFE}[3], \tag{4.14}$$

$$= (0.2)(-0.02) + (0.5)(1) + (1)(-0.5) + (0.02)(0.05) = -0.003. \tag{4.15}$$

Circuit implementation details of FFEs are presented in Section 5.4.

4.3 Receiver-Side Continuous-Time Linear Equalizers

Unlike a transmitter-side FFE which we often view as a discrete-time system, a continuous-time linear equalizer (CTLE) is a continuous-time amplifier designed with

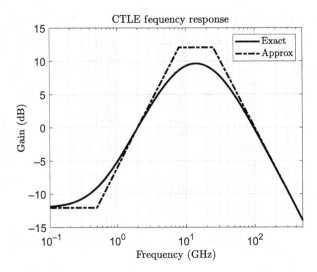

Figure 4.8 Frequency response of the CTLE given in Figure 6.7 showing both the exact transfer function and the straight-line approximation: $f_z = 500$ MHz, $f_{p1} = 8$ GHz and $f_{p2} = 25$ GHz.

a high-pass frequency response that partially inverts the channel's low-pass frequency response, meaning that the combined channel/CTLE frequency response has a higher bandwidth than the channel on its own.

A commonly used differential pair-based circuit has a transfer function given by:

$$A(s) = A_{dc}\frac{1 + s/\omega_z}{(1 + s/\omega_{p1})(1 + s/\omega_{p2})}. \tag{4.16}$$

The circuit has one zero at a lower frequency than its two poles. For the purpose of this discussion, assume $\omega_{p2} > \omega_{p1}$. If the poles are widely spaced, the gain increases by a factor of ω_{p1}/ω_z, reaching a midband gain of:

$$A_{mid} = A_{dc}\frac{\omega_{p1}}{\omega_z}. \tag{4.17}$$

An example of such an amplifier is shown in Figure 6.7. Section 6.5 relates element values to the poles, zero and gain.

Figure 4.8 shows an example of a CTLE frequency response for the case of $f_z = 500$ MHz, $f_{p1} = 8$ GHz and $f_{p2} = 25$ GHz. The dashed line is the straight-line approximation of the frequency response while the solid line is the exact transfer function. With the two poles only about a factor of three apart, the exact transfer function does not reach the midband value predicted by (4.17).

Example 4.4   Investigate the ISI improvement provided by a CTLE.

Solution
Figure 4.9 shows an example of a CTLE's effect on the frequency and pulse response of a system. An example channel response is plotted in part (a) using a solid line and

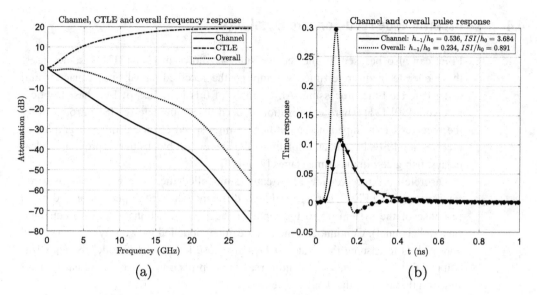

Figure 4.9 CTLE example showing its effect on the pulse response of a channel having significant ISI: (a) frequency response; (b) pulse response.

its pulse response for 28 Gb/s data is plotted in part (b). From part (a) we see that the channel has a little more than 30 dB of loss at the Nyquist frequency (14 GHz). The sampled pulse response is denoted by the triangle markers in part (b). There is significant ISI with the first pre-cursor sample (h_{-1}) more than 50% of the main cursor's height and total ISI $> 3.5\times$ the main cursor. Part (a) shows the frequency response of a candidate CTLE with the following parameters:

$$f_z = 1.5 \text{ GHz},$$
$$f_{p1} = f_{p2} = 28 \text{ GHz}.$$

If f_{p2} were much larger than f_{p1} we would expect the maximum gain of the CTLE to reach $A_{max} = \frac{f_{p1}}{f_z} = \frac{28}{1.5} = 18.7$. However, with f_{p2} also at 28 GHz, $A_{max} = 9.3$ or 19.3 dB. With only one pole/zero pair, matching the CTLE's response to the inverse of the channel is akin to fitting a square peg into a round hole. Nevertheless, the overall response is flattened such that it has a 3 dB bandwidth near 6 GHz. The improvement this provides in the pulse response is evident in part (b). The main cursor is much larger, although an amplifier with a flat frequency response would achieve this as well. What is important is that the ratio of ISI to the main cursor has reduced to less than 1, meaning that with no additional equalization, a very small VEO would be present. Typically, the VEO would be improved further either by adding transmitter-side FFE or additional receiver-side equalization in the form of a decision feedback equalizer. The reduced total ISI eases the DFE's design. Since a DFE cannot remove pre-cursor ISI, the effect of the CTLE on h_{-1} is of particular interest. With the CTLE, it was diminished to less than $\frac{1}{4}$ of the main cursor.

4.4   Receiver-Side Feed-Forward Equalizers

FFEs can also be used on the receive side, as shown in [24]–[31]. In some cases, these operate in discrete time by sampling the received signal on a capacitor and using discrete-time processing. An example of this is a class of FFE-based optical receivers [30] [31] where equalization is used to compensate for ISI introduced by the receiver's transimpedance amplifier. In high-speed electrical links, examples of receive-side FFEs have been implemented using circuit blocks that approximate ideal delays, using continuous-time stages [24]–[29].

Discrete-time receive-side FFEs can be modelled the same way we considered transmitter-side FFEs. That is, the FFE's difference equation operates on the pulse response of the system up to the input of the FFE (i.e., transmitter-side equalization and channel). Continuous-time FFEs can be modelled in two ways. An idealized approach is to assume the analog delays are close to ideal and model the equalizer using a difference equation. A more realistic approach is to use the actual transfer function that models the delay elements.

4.5   Decision Feedback Equalizers

Whereas CTLEs and FFEs invert the low-pass characteristic of the channel, decision feedback equalizers (DFEs) remove ISI without amplifying the high-frequency content of the signal. Thus, DFEs leverage knowledge of previously received bits to remove their post-cursor ISI. Broadly speaking, DFEs fall into two categories. The first, known as a finite impulse response (FIR) DFE, uses a clocked delay path to subtract ISI over a finite time duration, set by the number of delay taps. Infinite impulse response (IIR) DFEs, on the other hand, use an analog circuit that generates a replica of the ISI tail of the preceding channel/circuit that is subtracted over an infinite time duration, potentially cancelling the entire ISI tail. Both are discussed in the following sections.

4.5.1   Fundamentals

There are two conceptual points of view for understanding the operation of a DFE. They are:

1. The implementation of a variable-threshold decision circuit, whose threshold is varied according to previously received bits.
2. The cancellation of post-cursor ISI.

Both of these concepts will be presented, beginning with the first. Figure 4.10 shows the response of a first-order system to a 000010110100 unipolar data pattern at 10 Gb/s. Here, ISI is negligible since the system has a bandwidth of 7 GHz or 70% of the data rate. The signal nearly reaches its full-scale value, meaning that regardless of

Input and output of a first-order system with $f_{bw} = 0.7f_b$

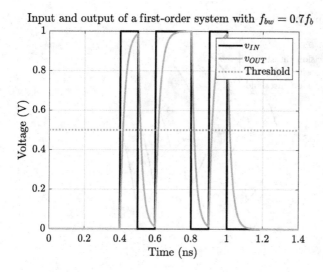

Figure 4.10 Response of a first-order system to a data pattern with sufficient bandwidth: circuit bandwidth is 70% of data rate.

Input and output of a first-order system with $f_{bw} = 0.3f_b$

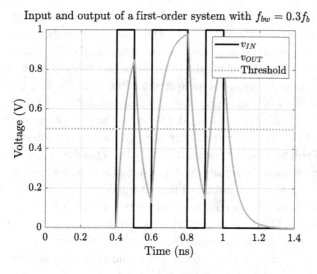

Figure 4.11 Response of a first-order system to a data pattern: circuit bandwidth is 30% of data rate.

the bit sequence the samples at the end of each UI reach 0 or 1 and are equally far from a decision threshold set at 0.5. Not only is a fixed threshold in the middle acceptable, it is the optimal location assuming 1s and 0s have equal noise associated with them. In the case of the 3 GHz bandwidth system in Figure 4.11, the signal does not reach the full-scale voltage after the first 1 at 0.5 ns. However, it does cross the fixed decision threshold regardless of the preceding bit(s). The BER will increase following (1.32) given the smaller VEO.

Input and output of a first-order system with $f_{bw} = 0.1f_b$

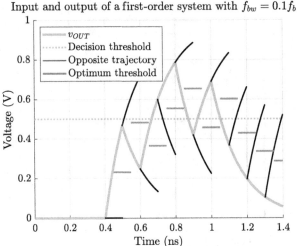

Figure 4.12 DFE example with 1 GHz bandwidth: circuit bandwidth is 10% of data rate.

When the bandwidth of the system is reduced to 1 GHz, as shown in Figure 4.12 the output signal resulting from the first 1 does not cross the fixed threshold located at 50% of the full-scale voltage. Therefore, a fixed decision threshold will not work. Consider an approach in which we lower the threshold when previous bits have lowered the output signal and raise the threshold when previous bits have raised the output signal. In the case of the first 1 following several 0s it is plausible that, since the output signal is low and the system is bandwidth limited, a lower threshold is appropriate. In fact the optimum threshold is halfway between where the signal will be if the next bit is a 0 and where the signal will be if the next bit is a 1. In the case of this first 1, the signal rises to 0.47 V. If that bit were a 0, the signal would have stayed at 0, as shown with the section of black horizontal line at 0 V extending from 0.4 to 0.5 ns. Therefore the optimum threshold for the signal when several 0s have been received is $(0.47 + 0)/2 = 0.23$ V. This optimum threshold is shown by a dark grey horizontal line at 0.23 V. In the case of the 0 following the first 1, the signal falls, but not all the way to zero, whereas if the next bit were a 1 instead of a 0, the signal would increase well above the fixed threshold at 0.5. For this bit, the ideal threshold is just below 0.5.

Table 4.3 summarizes signal dynamics. The first column gives the input sequence. The second column shows the channel's output at the end of each UI given the input sequence. The third column gives the complement of the input bits. The fourth column shows the output of the channel if the input sequence (from column 1) had been applied but the opposite bit were applied for the current bit. For example, the final row of the table shows that the output at the end of the entire input sequence is 0.06. Had the input sequence been the same, with the exception of the final input bit being replaced by a 1, the output would be 0.52. Therefore, the best threshold given an input sequence up to the last bit is $(0.52 + 0.06)/2 = 0.29$. The final column of the table gives the optimum threshold at each UI.

Table 4.3 Relevant analysis of optimum threshold for a first-order system with a bandwidth 10% of the data rate.

Input bit	Output	Opp. input bit	Output for opp. input	Optimum threshold
0	0			
0	0			
0	0			
1	0.47	0	0	0.23
0	0.25	1	0.72	0.48
1	0.60	0	0.13	0.37
1	0.79	0	0.32	0.55
0	0.42	1	0.89	0.65
1	0.69	0	0.22	0.46
0	0.37	1	0.83	0.60
0	0.20	1	0.66	0.43
0	0.10	1	0.57	0.34
0	0.06	1	0.52	0.29

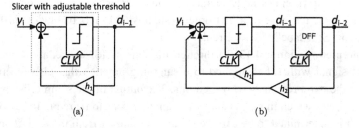

Figure 4.13 DFE block diagrams: (a) one-tap DFE; (b) two-tap DFE.

4.5.2 DFE Implementation

Figure 4.13 (a) shows a DFE. The summing block adds or subtracts a value of h_1 based on the previously received bit. In this block diagram, signals are assumed to be differential with a logic 0 represented by a differential signal of -1. As indicated by the dashed box we can view the summer and slicer as a slicer with an adjustable threshold of $\pm h_1$. In this case the threshold is adjusted based on one previously recovered bit. This is sufficient to remove the ISI of a channel with post-cursor ISI lasting only 1 UI and no pre-cursor ISI. Figure 4.13 (b) extends the DFE to include a second tap, h_2, that is adjusted to cancel the second post cursor. The block delaying d_{i-1} to give d_{i-2} is indicated as a DFF rather than a slicer, since d_{i-1} is a full-rail logic signal and need not have the ability to resolve small inputs. Extending a DFE to have a second tap allows four different slicer thresholds. The optimal thresholds computed in Figure 4.12 (a) require an infinite number of DFE taps, although depending on the length of the pulse response two taps may be sufficient.

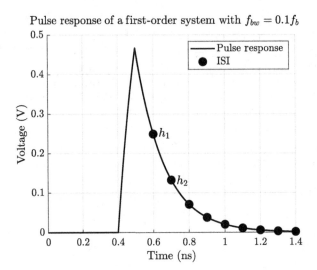

Figure 4.14 Pulse response for 100 ps UI input at 0.4 ns applied to a first-order system with 1 GHz bandwidth.

The other viewpoint of DFE operation is post-cursor ISI cancellation. Figure 4.14 shows the pulse response of a first-order system with 1 GHz bandwidth (100 ps UI). Post-cursor ISI is labelled with circular markers.

Reinterpreting Figure 4.13 (b) through the lens of ISI cancellation, we see that an input signal with ISI is applied to a summing junction that also receives input based on the two previously recovered bits. The output of the summer is applied to a decision circuit which in turn is applied to another decision circuit. To remove ISI, if the ith bit is determined to be a 1, we subtract h_1 and h_2 from the $(i+1)^{th}$ and $(i+2)^{th}$ bits, respectively. Similarly if the ith bit is determined to be a 0, we add h_1 and h_2 to the $(i+1)^{th}$ and $(i+2)^{th}$ bits. Therefore, we can view the summing junction as either removing ISI, or we can view it as being part of the decision circuit where it adjusts the decision threshold of the circuit.

The feedback paths produce inputs to the summer of magnitude h_1 and h_2 with polarity based on the previous bits. Note that if the input signal amplitude changes, the feedback taps must also change. One of the advantages of a DFE is that since the feedback taps are driven by full-rail outputs of decision circuits they can be designed to add very little noise to the signal. One limitation of a DFE is that it cannot remove pre-cursor ISI, since the fed-back signals only modify the input based on previous bits. The main design challenge is meeting the timing constraint around the h_1 loop. Within 1 UI the first decision circuit must produce a stable output whose value propagates through h_1 and the summer in time to satisfy the setup time of the decision circuit. This is summarized as:

$$t_{clk-Q} + t_{h1} + t_{sum} + t_{setup} < T_b, \qquad (4.18)$$

where t_{clk-Q}, t_{h1}, t_{sum} and t_{setup} are the clock-to-Q delay of the decision circuit, the delay through the h_1 tap, the settling time of the summer and the setup time of the

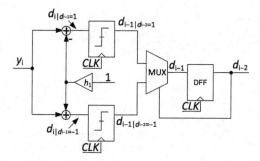

Figure 4.15 Block diagram of a speculative DFE [32], relaxing the constraint in (4.18).

decision circuit, respectively. The technique discussed in Section 4.5.3 is aimed at relaxing the constraint in (4.18).

4.5.3   Speculative DFEs (Loop Unrolling)

The block diagram in Figure 4.15 shows a widely used approach for relaxing the timing constraint in (4.18) in the context of a one-tap DFE [32]. In a conventional DFE the first feedback loop will either add or subtract h_1 from the input signal, depending on the previous bit. In Figure 4.15, the input signal is split into two paths. In the upper path, h_1 is subtracted from y_i giving the appropriate signal *if* d_{i-1} were a 1, labelled as $d_{i|d_{i-1}=1}$. In the lower path, h_1 is added to y_i, appropriate if $d_1 = -1$, labelled as $d_{i|d_{i-1}=-1}$. These two conditional signals are captured by decision circuits. The output of the DFF is the previous symbol, whose value controls the selection port of the MUX. If the previous bit were a 1, the upper path is selected, while if the previous bit were a -1, the lower path is selected. The resulting critical timing path is now:

$$t_{clk-Q} + t_{mux} + t_{setup} < T_b, \qquad (4.19)$$

where t_{mux} is the delay of the MUX. In comparing (4.18) and (4.19) we see that $t_{h1} + t_{sum}$ has been reduced to t_{mux}. Summers, particularly in DFEs with several taps, are notoriously slow. To eliminate these two terms with one relaxes timing. More subtle, though, is that the critical path in Figure 4.15 involves only full-rail signals, thus $t_{clk-Q} + t_{setup}$ for the DFF is smaller than that for the decision circuit in Figure 4.13. Speculation can be extended from one to N taps. However, the number of signal paths grows with 2^N.

Most wireline receivers also perform demultiplexing. Rather than add this after equalization has been done, the half-rate DFE architecture shown in Figure 4.16 produces two output data streams, referred to as d_{ei} and d_{oi}, denoting bits in d_i with even and odd indices, respectively. The clock signal $C2$ has a period of $2T_b$. Every input y_i is sampled because the decision circuit in the upper signal path samples on the rising edge of $C2$, whereas the decision circuit in the lower path samples on $C2$'s falling edge. The two paths are nevertheless coupled to one another. For example, the h_1 input to the even path is driven by the odd signal path. More generally, odd post-cursors are

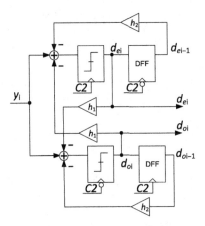

Figure 4.16 Two-tap half-rate DFE, implementing 1:2 demultiplexing.

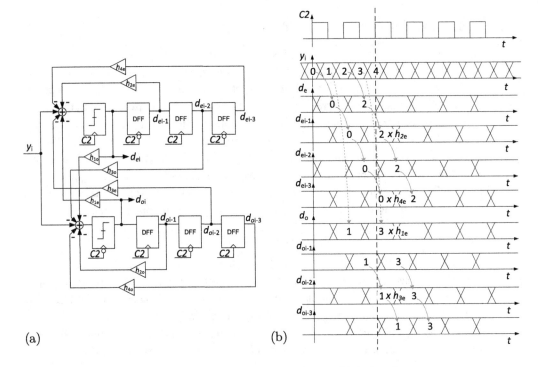

(a)   (b)

Figure 4.17 Four-tap half-rate DFE: (a) block diagram; (b) timing diagram.

cancelled using signals from the opposite path while even post-cursors are cancelled using signals from the given data path. This interaction between the even and odd data paths means that, although each path operates at half the data rate of the conventional DFE, the critical timing constraint is still (4.18).

Figure 4.17 (a) extends the half-rate DFE concept to four taps while Figure 4.17 (b) illustrates the timing of the relevant signals. The dashed line crosses the input data

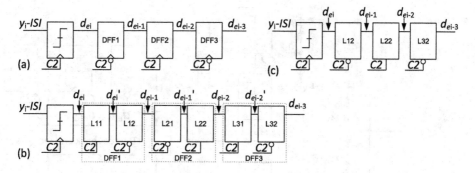

Figure 4.18 Converting DFFs to latches in a half-rate DFE.

stream y_i where $i = 4$. Numbers within a given interval of the other signals are the index i of y_i from which each interval is derived. For example, the interval of d_o where the dashed line intersects is a sample from y_3. This is the first post-cursor and is therefore scaled by h_1 in the equalization of y_4. The four-tap DFE processes bit decisions y_0 through to y_3 to equalize y_4. Although the DFFs are clocked with a half-rate clock, due to their alternating edge sensitivity, the even and odd chains of DFFs each produce delays of T_b. Usually, this architecture is simplified by using only level-sensitive latches.

Figure 4.18 shows the thought process of replacing DFFs with level-sensitive latches. Part (a) shows the DFF chain from the even path in Figure 4.17 (a). Each DFF is drawn as the cascade of two level-sensitive latches in Figure 4.18 (b). The first clocked circuit remains an edge-sensitive decision circuit. To see how part (b) can be simplified, consider that on the rising edge of $C2$, the input $y_i - ISI$ is sampled to give d_{ei}, which is held for $2T_b$ until the next sample is taken. However, on the rising edge of $C2$, latch $L11$ is transparent, meaning after a short delay, d_{ei} appears at d_{ei}'. Even when $C2$ is low and $L11$ is in the hold state, $d_{ei}' = d_{ei}$ since d_{ei} is does not change until the next rising edge of $C2$. Therefore, $L11$ is redundant. In fact, every second latch is unnecessary, allowing the use of the architecture in Figure 4.18 (c). Figure 4.19 shows a combination of speculation and half-rate operation using level-sensitive latches.

4.5.4 Infinite Impulse Response (IIR) Decision Feedback Equalizers

In the block diagram in Figure 4.13 (b), the signal through a given tap persists at the input to the summer for only 1 UI, at which point its value is updated based on the next received bit. The feedback path is said to have a finite impulse response. This means that one tap is needed for each non-trivial post-cursor ISI sample. In some cases, there can be a long, yet predictable tail of ISI that can be reproduced at the input of the summer by using an infinite impulse response tap.

The first-order pulse response presented in (1.19), an example of which is plotted in Figure 4.20 (a), can be rewritten for $t \geq T_b$ in terms of the main cursor as:

$$h_{UI}(t) = h_0 \exp\left(-2\pi f_{-3dB}(t - T_b)\right). \tag{4.20}$$

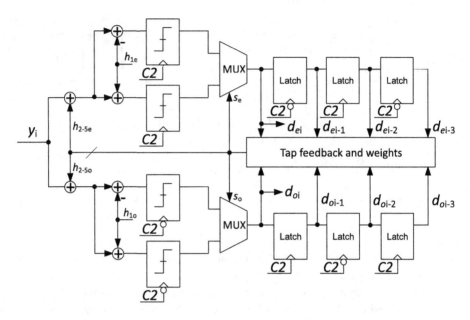

Figure 4.19 Combination of half-rate clocking and speculation in a four-tap DFE. Latches are level-sensitive circuits, not edge-triggered.

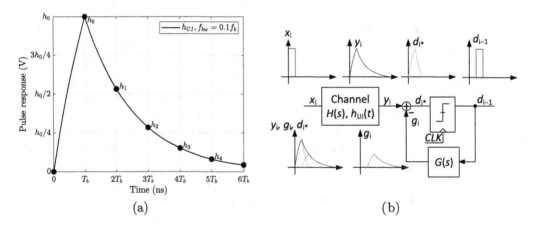

(a)   (b)

Figure 4.20 Motivation for an infinite impulse response DFE: (a) predictable post-cursor ISI in a first-order pulse response; (b) conceptual implementation.

The main cursor and post-cursor ISI can be expressed as:

$$h_k = h_0 \exp\left(-2\pi f_{-3dB}(kT_b)\right)$$

$$= h_0 \left(\exp\left(-\frac{T_b}{\tau}\right)\right)^k, \tag{4.21}$$

where $\tau = \frac{1}{2\pi f_{-3dB}}$, valid for $k \geq 0$. The ISI tail described by (4.21) for $k > 0$ decays predictably and, depending on the bandwidth of the system relative to the data rate, may persist for several UIs.

A block diagram of an IIR DFE is shown in Figure 4.20 (b) along with relevant signals when the input $x(t)$ is a unit pulse. The output of the decision circuit d_{i-1} persists for 1 UI at a time, although $G(s)$ has an infinite impulse response. To determine an appropriate signal for $G(s)$ to generate, consider the operation of the system under ideal conditions when $x(t) = r(t)$, where $r(t)$ is a unit pulse lasting T_b:

- The output of the channel, a first-order system in this example, is shown as y_i, which has significant ISI, lasting several UIs.
- The objective is to generate a signal at the input of the decision circuit (d_{i*}) that has no significant ISI. That is, we want $d_i \neq 0$ for only 1 UI.
- Therefore, $G(s)$ must generate the same post-cursor ISI tail introduced by $H(s)$ offset by 1 UI.
- For example, the main cursor of the fed-back signal g_0 cancels h_1, etc.
- In the case of a first-order channel as depicted in Figure 4.20 (a), $G(s)$ is an amplitude-tuned replica of $H(s)$.

In the example of the first-order channel, both the gain and time constant of $G(s)$ must be tuneable to match the $H(s)$. Circuit details are presented in Section 6.7. When $H(s)$ is due to a distributed electrical channel, $G(s)$ can only approximate its behaviour, since it is constructed from lumped resistors and capacitors. Some variants of the approach use an FIR tap to cancel h_1 and an IIR tap for h_2 onward [33].

4.6   Electrical Link Budgeting

With access to reasonable models for the overall channel, consisting of packages and PCB traces, overall link response can be predicted. A system-level designer can explore trade-offs among IC block parameters and the link's BER and power dissipation. For example, equalizers on the Tx and Rx might be needed to mitigate severe ISI due to 40 dB of channel loss at Nyquist. Ultra-short-reach links will get by with fewer equalizers. However, only with a study of the whole link can the best allocation be determined.

An example of a set of link design constraints is the following:

- The data rate is 25 Gb/s.
- Tx's output swing is limited to 1.0 V_{pp} differential due to power supply limitations.
- A particular channel and package are to be used with appropriate models.
- Rx latches need about 30 mV peak-to-peak differential to avoid contributing significantly to the BER.

The system engineer needs to determine the simplest combination of equalization required to present the latches with an eye sufficiently open and with low enough noise such that the target BER is achieved. Equalization can be implemented on the Tx side as a feed-forward equalizer. Rx-side equalization is typically implemented as a combination of continuous-time linear equalizers and decision feedback equalizers. The decisions that can result from this link budgeting/modelling exercise include:

- Number of taps in the Tx FFE.
- Pole/zero location of the CTLE.
- Gain of CTLE and other Rx amplifiers.
- Tolerable noise in the Rx front-end.
- Number of taps in the Rx DFE.

In addition to investigating the overall link's performance, link budgeting also must check for the compliance of each end of the link to its associated standard. It is not sufficient for the Tx and Rx from company X to function properly. The Tx from any company must function with the Rx from any company, giving rise to per-end compliance specifications, usually specified in terms of eye masks, signal swing, jitter, jitter tolerance, etc.

In the discussion above, the term target BER has been used, since it varies with the application. For example, links that do not use significant forward error correction (FEC) usually target 10^{-12}. On the other hand, PAM4 links frequently use FEC and need a raw BER in the range of 10^{-3} so that a post-FEC BER of 10^{-12} is achieved.

4.7   Summary

This chapter presented equalization: strategies for mitigating ISI introduced by channels or circuits. The focus was on transmitter-side FFE, receiver-side CTLE and DFE. Given a pulse response, we can compute the effect of the equalizer. For FFEs, we usually do this in the time domain using discrete-time convolution. The effect of CTLEs was explored using cascades of transfer functions while DFEs were considered in the time domain. All of the analysis we did assumed the channel and circuits are linear. Nonlinearity will give rise to signal-dependent ISI.

The stringent timing constraints of DFEs were introduced. Speculative DFEs and half-rate architectures were shown, and IIR DFEs were shown to efficiently cancel a well-behaved ISI tail using a smaller number of taps.

Problems

4.1   In this problem, the role of a CTLE in eliminating pre-cursor ISI is studied. Assume the channel has a frequency response approximated by:

$$H(s) = \frac{1}{(\frac{s}{\omega_0} + 1)(\frac{s}{\omega_1} + 1)(\frac{s}{\omega_2} + 1)(\frac{s}{\omega_3} + 1)(\frac{s}{\omega_4} + 1)}, \tag{4.22}$$

where $\omega_i = 2\pi f_i$ and f_i is given by:

$$f_0 = f_{bit}/80, \tag{4.23}$$
$$f_1 = f_{bit}/20, \tag{4.24}$$
$$f_2 = f_{bit}/10, \tag{4.25}$$
$$f_3 = f_{bit}/6.67, \tag{4.26}$$
$$f_4 = f_{bit}/5. \tag{4.27}$$

Table 4.4 Pulse response amplitude for Problem 4.2.

Sample	Value (mV)
$h_{k<-1}$	0
h_{-1}	100
h_0	1000
h_1	400
h_2	80
$h_{k>2}$	0

(a) Find the pulse response of the channel for 10 Gb/s NRZ data.

(b) Express the first pre-cursor as a percentage of the main cursor.

(c) Propose a CTLE transfer function of the form:

$$H_{EQ}(s) = \frac{\frac{s}{\omega_z} + 1}{\frac{s}{\omega_p} + 1}, \tag{4.28}$$

and consider its impact on the level of the first pre-cursor as well as the post-cursor ISI relative to the main cursor. Note that you select values for ω_z and ω_p.

(d) If a two-tap DFE is used instead of the CTLE, what should the tap weights be? What is the worst-case residual ISI as a percentage of the maincursor when the taps are optimally chosen?

(e) Propose parameters when both the CTLE and DFE are used. What is the worst-case residual ISI as a percentage of the maincursor using both the CTLE and the DFE?

4.2 Consider the sampled pulse response of a channel given in Table 4.4.

(a) What is the total ISI and resultant VEO?

(b) Propose an FFE with at most three taps (two 1-UI delay elements). Give tap weights and compute the combined FFE/channel pulse response.

(c) Determine the residual ISI and main cursor height.

4.3 Consider the use of a decision feedback equalizer (DFE) to equalize the pulse response in Table 4.4.

(a) Draw an FIR DFE with two feedback taps and explain how a DFE can be used to equalize the pulse response.

(b) Which ISI terms are cancelled and which remain? Be sure to explain/indicate the part of the DFE responsible for the cancellation of a given ISI term.

4.4 Consider the sampled pulse response of a channel given in Table 4.5.

(a) Propose an equalization strategy using an FFE with at most two taps (one 1-UI delay element) and a one-tap FIR DFE in order to minimize the residual ISI. Give tap weights and compute the combined FFE/channel pulse response.

Table 4.5 Pulse response amplitude.

Sample	Value (V)
$h_{k<-1}$	0
h_{-1}	0.4
h_0	1.0
h_1	0.2
h_2	0.1
$h_{k>2}$	0

(b)   Determine the residual ISI and main cursor height at the input of the DFE's summer.

4.5   Consider a first-order pulse response. Determine the gain and τ of an IIR DFE. Show graphically that this choice of parameters completely cancels post-cursor ISI. Separately consider a 25% error in gain and τ. How much post-cursor ISI is present? Which type of error is worse?

4.6   For the channel and UI specified in Problem 4.1 find the best tap weight for a two-tap (one 1-UI delay element), symbol-spaced FFE. If the goal is to eliminate pre-cursor ISI, determine if a half-UI spaced FFE is better.

4.7   Transmitter-side FFEs are constrained by a fixed supply voltage such that $\sum_{j=0}^{N} |\alpha_j|$ is a constant. Consult the equalization scenarios in Table 11.1 but scale FFE coefficients such that $\sum_{j=0}^{N} |\alpha_j| = 1$, rather than fixing the largest coefficient at 1 as was done in the example. Comment on the relative improvements among the cases when this new constraint is considered.

4.8   Compare the optimal thresholds shown in Figure 4.12 and given in Table 4.3 to what is achieved with a two-tap FIR DFE.

5   Electrical-Link Transmitter Circuits

This chapter starts with an overview of the requirements of an electrical-link transmitter. Electrostatic discharge protection is required in CMOS processes. The large
capacitance it adds severely loads the output driver. Two inductive peaking techniques
are presented to mitigate capacitive loading, with emphasis on T-coil-based compensation, a type of inductive peaking. In Section 1.12, two categories of circuits were
introduced, namely CML and CMOS. These give rise to current-mode and voltage-
mode transmitters, both of which are featured in this chapter. In both classes of drivers,
the input data are usually presented to the transmitter as parallel low-rate streams of
data that are serialized before transmission. However, some implementations perform
the last stage of multiplexing (serializing) in the output driver. Various approaches to
multiplexing are described toward the end of the chapter. A reader who is primarily
interested in optical transceiver design should read this chapter before moving on to
Chapter 8.

5.1   ESD Considerations

Inputs and outputs of chips must be protected against electrostatic discharge (ESD)
to avoid permanently damaging transistors during handling, packaging and assembly
onto PCBs. The thin oxide of scaled MOSFETs can be damaged by voltages as low as
a few volts. Even thick oxide devices frequently used for inputs and outputs will break
down under voltages of < 10 V. ESD events, often modelled as a 100 pF capacitor
charged to 2–8 kV discharged through a resistance of 1.5 kΩ, must not raise the inputs
of chips beyond safe operating ranges. Hence, circuit elements are added to chips to
ensure inputs/outputs remain approximately within the range of V_{DD} to ground. This
ESD circuitry is shown in Figure 5.1. Under normal operation, the inputs and outputs
see only reversed-bias diodes and a small series resistor in the range of 100 Ω, or
much less in the case of inputs that must be matched to 50 Ω. During an ESD event
that pulls the off-chip side high, diodes D1 and D3 will forward bias, clamping the
input voltage to V_{DD} plus the diodes' forward-bias voltage. The impedance of the on-
chip supply network can be large enough that an ESD event could pull the on-chip
V_{DD} up during an ESD event. To combat this, on-chip clamping circuits can provide
a short-circuit path from the on-chip V_{DD} to ground, ensuring that transistors are not

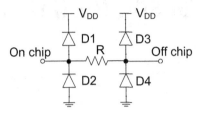

Figure 5.1 Basic ESD structure.

exposed to undue voltage stress. Similarly, under a negative ESD event diodes D2 and D4 will forward bias, preventing the input voltage from dropping much below ground.

Reverse-biased ESD diodes load the inputs and outputs with depletion capacitance. If the area of the diodes is increased to offer more reliable protection, capacitance, which is proportional to area, loads the inputs and outputs and limits operation speed.

Example 5.1   If a 400 fF total capacitance consisting of an I/O pad and ESD capacitance sits in parallel with a 50 Ω channel, find the resulting -3 dB bandwidth and maximum data rate for ISI-free operation.

Solution
For a first-order circuit, the pole and -3 dB bandwidth is given by:

$$f_{-3dB} = \frac{1}{2\pi RC} = \frac{1}{2\pi \, (50\Omega) \, (400 \text{ fF})} = 8 \text{ GHz.} \qquad (5.1)$$

In Chapter 1 a guideline for the bandwidth of a system being 50 to 70% of the data rate was given to avoid significant ISI. Therefore, we need to limit the data rate to 1.4 to 2$\times$ the bandwidth. Taking 1.5$\times$ as an intermediate value, the ESD capacitance will add non-trivial ISI to data rates above approximately $1.5 \times f_{-3dB}$ or about 12 Gb/s.

The most common approach to mitigate the loading effect of ESD diodes is to use a form of inductive peaking, using a T-coil. In the next section, inductive peaking is introduced, starting with a simpler technique known as shunt-peaking. Following this, the bridged T-coil is presented.

5.2   Inductive-Peaking Techniques

In this section, two techniques to extend the bandwidth of the parallel RC load from Example 5.1 are presented. In the first scheme, presented in Section 5.2.1, a single inductor is added to the load, whereas in Section 5.2.2 two coupled inductors are used for a more dramatic extension of circuit bandwidth.

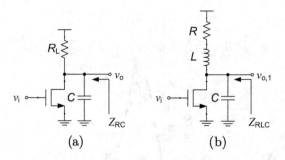

Figure 5.2 Common-source amplifier (a) with RC-load and (b) using shunt-peaking.

5.2.1 Inductive Shunt-Peaking

Figure 5.2 (a) shows a resistively loaded common-source amplifier. Though shown as a single-ended circuit, this can represent the differential half circuit of a CML driver. The output v_o is loaded by a capacitor C. Although in this chapter we are introducing inductive peaking to mitigate ESD capacitance in electrical-link transmitters, these techniques are applicable wherever we have long time constants due to large capacitors. Ignoring other parasitic capacitances, this amplifier has a gain of:

$$A_{RC}(s) = \frac{v_o}{v_i} = -g_m Z_{RC}, \tag{5.2}$$

$$= -\frac{g_m R_L}{s R_L C + 1}. \tag{5.3}$$

The output impedance this circuit presents to the channel is also Z_{RC}. If an inductor is added in series with the resistor, the gain of the amplifier in Figure 5.2 (b) is given by:

$$A_{RLC}(s) = \frac{v_o}{v_i} = -g_m Z_{LRC}, \tag{5.4}$$

$$= -g_m (R + sL) \parallel \left(\frac{1}{sC} \right), \tag{5.5}$$

$$= -g_m \frac{sL + R}{s^2 LC + sRC + 1}. \tag{5.6}$$

As with the RC load, the shunt-peaked load has a dc gain of $-g_m R$. However, the inductor converts the first-order transfer function to a second-order transfer function with a zero (ω_z) and poles (described by ω_n and ζ) as:

$$\omega_z = -\frac{R}{L}, \tag{5.7}$$

$$\omega_n = \frac{1}{\sqrt{LC}}, \tag{5.8}$$

$$\zeta = \frac{R}{2} \sqrt{\frac{C}{L}}, \tag{5.9}$$

where ω_n and ζ are the natural frequency and damping factor of the poles. With C determined by the circuit and R determined by gain requirements, the designer is left

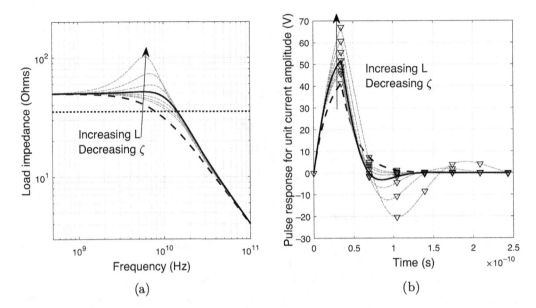

Figure 5.3 Effect of shunt-peaking on an *RC* load. Thick dashed lines are for $L = 0$. Thick solid lines are for $\zeta = 0.7$. Thin dashed lines are for various values of ζ. (a) Frequency response of load impedance; (b) pulse response of the load for 1 A input pulse with $T_b = 35$ ps.

to choose L. Smaller L leads to higher ω_n but ζ can increase such that the transfer function will have real and distinct poles, with the lower frequency pole setting the bandwidth. Larger L leads to a lower damping factor that can lead to too much peaking in the frequency response or overshoot in the pulse response. The transfer function also has a zero which decreases as L increases.

Example 5.2   For the parameters in Example 5.1, apply shunt-peaking to maximize the VEO when the UI is 35 ps ($f_{bit} = 28.57$ Gb/s).

Solution
Without the inductor, the bandwidth of the load is only 8 GHz. At 28.57 Gb/s we expect considerable ISI. To reduce ISI, we can consider the value of the inductor that maximizes f_{-3dB}. However, we should look at the amount of peaking in the frequency response, limiting it to some upper bound. Alternatively we can analyze the pulse response to find the VEO for different values of L.

   Figure 5.3 (a) shows the load impedance as a function of frequency for various values of ζ and L. As L increases, ζ decreases, initially widening the bandwidth but also increasing peaking. The thick dashed line shows the impedance of the load with $L = 0$. The thick solid line shows the frequency response for the second-order load when $\zeta = \frac{1}{\sqrt{2}}$. Although the impedance is second-order, at high frequency it follows the *RC* load due to the zero in the impedance function.

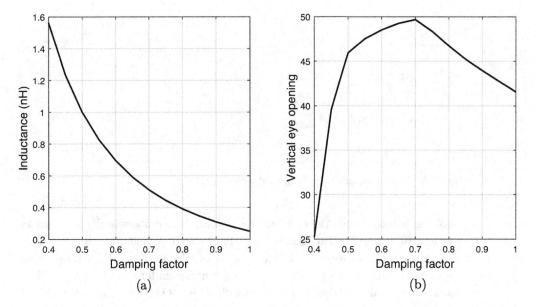

Figure 5.4 Shunt-peaking: (a) required inductance for a given ζ for $R = 50\ \Omega$, $C = 400$ fF; (b) vertical eye opening for a given ζ.

The pulse response of the load when a unit pulse (1 A) of current is applied is shown in Figure 5.3 (b) for a UI of 35 ps. Triangle markers are the UI-spaced samples of the pulse response. By subtracting total ISI from the main cursor, the VEO is computed and plotted against ζ in Figure 5.4 (b), while part (a) gives the inductor value as a function of ζ. Part (b) is the vertical eye opening (in V) when a unit current pulse is applied. The best VEO occurs for $\zeta = \frac{1}{\sqrt{2}}$. Rearranging (5.9) we can write:

$$L = C\left(\frac{R}{2\zeta}\right)^2,$$

(5.10)

giving $L = 500$ pH when $\zeta = \frac{1}{\sqrt{2}}$.

The zero significantly improves the response of the shunt-peaked load. For $\zeta = \frac{1}{\sqrt{2}}$ the bandwidth extension ratio (BWER) is approximately 1.8, meaning that the 8 GHz bandwidth of the RC load is extended to more than 14 GHz. An all-pole transfer function (i.e., no zeros) with the same ζ would have a BWER of only $\sqrt{2}$. In the time domain, the first-order load gave a VEO gain of 32.4 V. The shunt-peaked load increased it to 49.7 V (eliminating nearly all ISI), whereas the hypothetical all-pole transfer function has a VEO gain of only 34.7 V.

The quality factor of an inductor is the ratio of the imaginary part of its impedance (reactance) to the real part (resistance). On-chip inductors tend to be of lower quality factor than off-chip inductors or inductors implemented through bondwires.

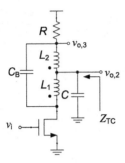

Figure 5.5 Common-source amplifier loaded with a T-coil.

Unlike an inductor used in an oscillator, the inductor here is intentionally put in series with a resistor, meaning that parasitic series resistance of the inductor will not significantly degrade the performance of the load. At 14 GHz the selected inductor has an impedance of $j44$ Ω. Its equivalent series resistance, even for a modest Q, would be small relative to the 50 Ω resistor in series with it. High-Q inductors are often laid out as a single turn, which occupies more area than a multi-turn, lower-Q inductor. The inductor in a shunt-peaked load can therefore be made smaller in area than the high-Q inductor in an oscillator. Nevertheless, inductive peaking will increase chip area.

5.2.2   Inductive Peaking Using a Bridged T-Coil

A second and more powerful inductive peaking technique that uses a "T-coil" is shown in Figure 5.5. Two, coupled inductors are added, along with a "bridging" capacitor C_B. Whereas the gain of the amplifiers in Figure 5.2 is simply the transconductance of the MOSFET multiplied by the load impedance, the gain of the T-coil-loaded amplifier requires the current-to-voltage conversion of the load when the current is injected at the drain of the MOSFET but the output voltage is taken between the coils. An excellent overview of the design considerations of this circuit can be found in [34].

Two different analysis approaches [35], [36] show an equivalent circuit with the coupled coils replaced by three uncoupled coils, as depicted in Figure 5.8 (a). The transfer function is fourth-order with a pair of complex zeros. For the symmetrical case of $L = L_1 = L_2$, by assuming:

$$C_B = \frac{1}{4}\frac{1-k}{1+k}C,\tag{5.11}$$

$$\frac{k}{1+k}C = \frac{(1+k)L}{R^2} - 2C_B,\tag{5.12}$$

where k is the coupling coefficient between L_1 and L_2 ($k = \frac{M}{\sqrt{L_1L_2}}$, where M is the mutual inductance of the coils), the overall transfer function reduces to second-order, becoming:

$$\frac{v_{o,2}}{v_i} = -g_mR\frac{\omega_n^2}{s^2 + 2\zeta\omega_ns + \omega_n^2},\tag{5.13}$$

where

$$\omega_n^2 = \frac{2}{(1-k)LC},$$ (5.14)

$$\zeta = \frac{RC - (1-k)\frac{L}{R}}{\sqrt{2(1-k)LC}}.$$ (5.15)

These equations can be used as follows: given R and C as dictated by the channel impedance and ESD capacitor impedance, select the damping factor ζ. The maximally flat (i.e., Butterworth) response of $\zeta = \frac{1}{\sqrt{2}}$ is a common starting point. From this, the element values can be calculated as follows:

$$L_1 = L_2 = \frac{R^2 C}{4}(1 + \frac{1}{4\zeta^2}),$$ (5.16)

$$k = \frac{4\zeta^2 - 1}{4\zeta^2 + 1},$$ (5.17)

$$C_B = \frac{C}{16\zeta^2}.$$ (5.18)

In the case of $\zeta = \frac{1}{\sqrt{2}}$,

$$L_1 = L_2 = \frac{3R}{8\omega_0},$$ (5.19)

$$k = \frac{1}{3},$$ (5.20)

$$C_B = \frac{C}{8},$$ (5.21)

where ω_0 is the pole in the original RC-loaded circuit. With this assumption for ζ, the natural frequency becomes:

$$\omega_n^2 = \frac{2}{(1-k)LC} = \frac{2}{\frac{2}{3}\frac{3R}{8\omega_0}C} = 8\omega_0^2.$$ (5.22)

Therefore, $\omega_n = \sqrt{8}\omega_0$. The bandwidth-extension ratio for the maximally flat (Butterworth) response is given by:

$$BWER = \frac{\omega_n}{\omega_0},$$ (5.23)

$$= \sqrt{8}.$$ (5.24)

As mentioned above and shown in Figure 5.8 (a), the two coupled coils can be replaced by three uncoupled coils for both analysis purposes or in the design of the circuit. The advantage of using coupled coils in the implemented circuit is a reduction in the number and the inductance of the required coils, saving chip area.

Example 5.3 For the ESD capacitance and channel impedance of 400 fF and 50 Ω choose element values to implement the bridged T-coil load of Figure 5.5.

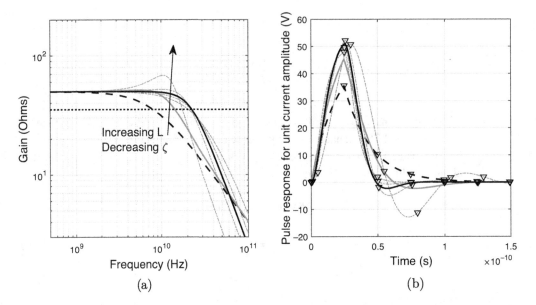

Figure 5.6 Current-to-voltage gain of the T-coil load for $C = 400$ fF, $R = 50\ \Omega$. Dashed thick lines are for the load without a T-coil. Thick solid lines are for $\zeta = 0.707$. Thin dashed lines are for various values of ζ. The thick grey line shows the behaviour of the shunt-peaked load with $\zeta = 0.707$. (a) Frequency response of gain; (b) pulse response of the load for $UI = 25$ ps.

Solution

Starting with a damping factor of $\zeta = \frac{1}{\sqrt{2}}$, (5.19)-(5.21) can be evaluated to give:

$$L_1 = L_2 = \frac{50^2 \times 400\ \text{fF}}{4}\left(1 + \frac{1}{4\left(\frac{1}{2}\right)}\right) = 0.375\ \text{nH}, \tag{5.25}$$

$$k = \frac{4\frac{1}{2} - 1}{4\frac{1}{2} + 1} = \frac{1}{3}, \tag{5.26}$$

$$C_B = \frac{400\ \text{fF}}{16\frac{1}{2}} = 50\ \text{fF}. \tag{5.27}$$

The RC load had a bandwidth of 8 GHz. The T-coil loaded amplifier has a bandwidth of $\sqrt{8} \times 8$ GHz or 22.5 GHz. The behaviour of the T-coil load is presented in Figures 5.6 and 5.7: $\zeta = \frac{1}{\sqrt{2}}$ gives the best performance. Figure 5.7 (b) shows negligible ISI for a UI reduced to 25 ps from the 35 ps used in the shunt-peaking exercise in Example 5.2.

Example 5.3 illustrates how the bandwidth can be extended significantly when the output is taken across the capacitor. Work on distributed amplifiers used T-coils to mitigate the capacitive loading of the next stage [37]. However, in a wireline transmitter, the large ESD capacitance need not be the output of the circuit. Instead, the load

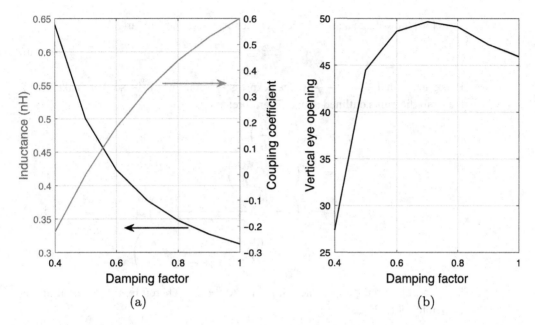

Figure 5.7 T-coil parameters for $R = 50\ \Omega$, $C = 400$ fF : (a) Required inductance ($L_1 = L_2$, black line) and coupling coefficient (k, grey line) for a given damping factor (ζ). (b) Vertical-eye-opening gain for a given ζ.

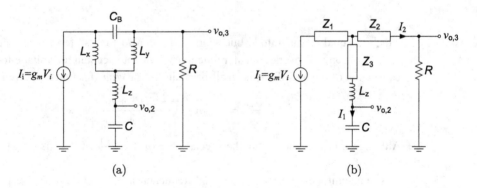

Figure 5.8 Equivalent circuit of the T-coil load: (a) transformation to uncoupled coils; and (b) use of $\Delta - Y$ transformation to allow current division [36].

resistance in Figure 5.5 represents the equivalent impedance of the channel in parallel with the transmitter's termination resistance. Hence, the relevant gain is $\frac{v_{o,3}}{v_i}$.

In [36] the transfer function was derived using a $\Delta - Y$ transformation to yield the circuit in Figure 5.8 (b). Current division was used to find the voltage across the capacitor (C). Here, we take a similar approach to find the voltage across the load resistance, considering the current division between I_i and I_2, given by:

$$\frac{v_{o,3}}{I_i} = -R\frac{I_2}{I_i} = -R\frac{Z_3 + sL_3 + \frac{1}{sC}}{Z_3 + sL_3 + \frac{1}{sC} + Z_2 + R}. \tag{5.28}$$

In [36] the voltage across the load capacitor was found similarly as:

$$\frac{v_{o,2}}{I_i} = -Z_C \frac{I_1}{I_i}.$$

(5.29)

Recognizing that $\frac{I_2}{I_i} = 1 - \frac{I_1}{I_i}$, we can express $\frac{v_{o,3}}{I_i}$ in terms of $\frac{v_{o,2}}{I_i}$ so as to make use of the simplifications outlined above. Therefore:

$$\frac{v_{o,3}}{I_i} = -R \left(1 - \frac{I_1}{I_i} \right),$$

(5.30)

$$= -R \left(1 - sC \frac{v_{o,2}}{I_i} \right),$$

$$= -R \left(1 - sCR \frac{\omega_n^2}{s^2 + 2\zeta \omega_n s + \omega_n^2} \right),$$

(5.31)

$$= -R \left(\frac{s^2 + \left(2\zeta \omega_n - CR\omega_n^2 \right) s + \omega_n^2}{s^2 + 2\zeta \omega_n s + \omega_n^2} \right).$$

(5.32)

From (5.22), at the $\zeta = \frac{1}{\sqrt{2}}$ design point, $RC = \frac{\sqrt{8}}{\omega_n}$. Therefore, the "s"-term of the numerator of (5.32) is:

$$2\zeta \omega_n - CR\omega_n^2 = \omega_n \left(\sqrt{2} - \sqrt{8} \right),$$

(5.33)

$$= \omega_n \left(\sqrt{2} - 2\sqrt{2} \right),$$

(5.34)

$$= -\sqrt{2}\omega_n,$$

(5.35)

which is equal in magnitude but opposite in sign to the "s" term in the denominator. Therefore, the overall transfer function, taking into account the voltage-to-current conversion of the MOSFET, is an all-pass filter (APF) transfer function given by:

$$\frac{v_{o,3}}{v_i} = -g_m R \frac{s^2 - \sqrt{2}\omega_n s + \omega_n^2}{s^2 + \sqrt{2}\omega_n s + \omega_n^2}.$$

(5.36)

Although this was shown for the special case of $\zeta = \frac{1}{\sqrt{2}}$, (5.32) is all-pass for all values of ζ.

The magnitude of (5.36) is $g_m R$ at all frequencies, although the phase response can introduce ISI. To see how constant gain magnitude can still introduce ISI, consider the transfer function in (5.31). The "1" gives an ideal pulse response. The second term, which is a subtraction of a band-pass response, introduces ISI.

An all-pass response is non-physical as all circuits will have high-frequency attenuation. For example, extra parasitic capacitance would be present at $v_{o,3}$. The T-coil response to the load resistor ($v_{o,3}$) is explored in Figure 5.9. Here, it is assumed that the APF transfer function has the same damping factor ($\zeta = \frac{1}{\sqrt{2}}$) and natural frequency as the low-pass filter (LPF) transfer function considered previously. Figure 5.9 (a) shows the frequency response of the APF (grey). The thick dashed line shows the response of the T-coil from an input current to the voltage across the capacitor ($v_{o,2}$). In order to show a physically realistic scenario, the overall frequency response of the APF cascaded by a LPF of different cutoff frequencies is shown with thin dashed lines. The LPF has $\zeta = \frac{1}{\sqrt{2}}$ and a bandwidth relative to the APF poles ranging from

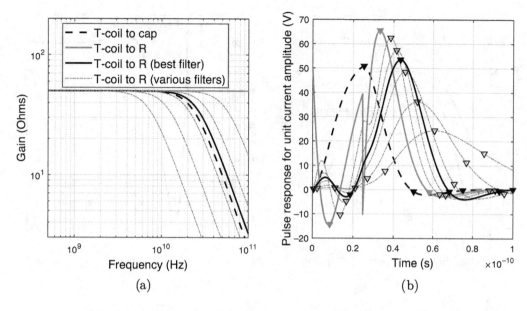

Figure 5.9 Investigation of T-coil response from input current to $v_{o,3}$ (voltage across the resistor) for $R = 50\ \Omega$, $C = 400$ fF: (a) frequency response; (b) pulse response for 25 ps UI.

$\frac{1}{\sqrt{10}}$ and $\sqrt{10}$. The filter cutoff that gives the best overall response is plotted with a dark solid line.

The pulse responses are shown in Figure 5.9 (b) following the same line types. The pulse response of the APF (grey) is clearly non-physical. It rises abruptly to 50 V at $t = 0$, swings down and up toward its peak. However, the upswing is interrupted by a sharp "glitch" caused by the input pulse's falling edge. This behaviour is smoothed out by the follow-on LPF. All responses have some pre-cursor ISI. The main cursor increases as the LPF bandwidth increases, while ISI however increases.

Figure 5.10 (a) shows main cursor, ISI and VEO as a function of the ratio between the LPF and APF filter poles. Figure 5.10 (b) shows the step response of the APF for reference.

An important takeaway is that even though the transfer function from input to the voltage across the resistor is an APF, it is not useful for an arbitrarily high data rate since the phase distortion is significant. Figure 5.11 (b) shows the pulse response when the UI is shrunk from 25 ps to 15 ps. The main cursor has dropped slightly, but the ISI has increased considerably, reducing the VEO from close to 50 V down to less than 30 V when considering a 1 A input pulse.

5.3 Current-Mode-Logic Drivers

The electrical-link transmitter circuits presented in this section are based on differential pairs, biased by a current source, similar to that shown in Figure 1.27. Section 5.3.1 presents a terminated CML driver that can be designed to be matched to a channel.

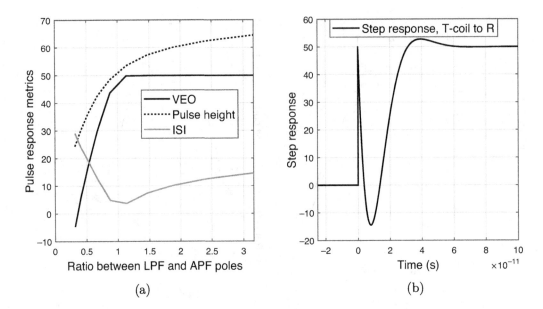

(a)   (b)

Figure 5.10 Investigation of T-coil response from input current to $v_{o,3}$ (voltage across the resistor) for $R = 50\ \Omega$, $C = 400$ fF: (a) height of pulse response, ISI and VEO as a function of the ratio between the poles of a cascaded LPF and the poles of the T-coil circuit; (b) step response from input current to $v_{o,3}$ (APF).

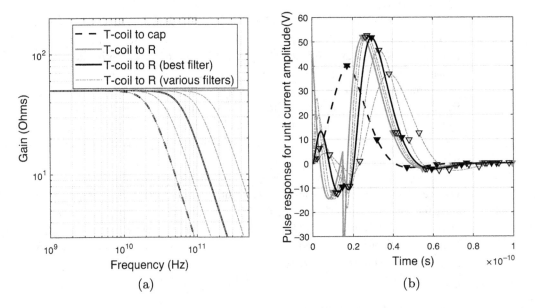

(a)   (b)

Figure 5.11 Investigation of T-coil response from input current to $v_{o,3}$ (voltage across the resistor) for $R = 50\ \Omega$, $C = 400$ fF: (a) frequency response; (b) pulse response for 15 ps UI.

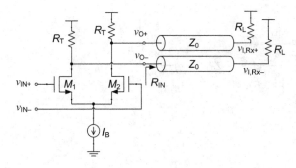

Figure 5.12 Matched CML driver.

Doing so minimizes signal reflections at the cost of increased power dissipation. The trade-offs of mismatching the driver to the channel are considered. Section 5.3.2 presents an open-drain driver that is mismatched to the channel, but operates with lower power dissipation.

5.3.1 Matched Current-Mode Drivers

Figure 5.12 shows a differential CML driver. Load resistors (R_L) model the input resistance of the receiver, usually matched (equal) to the transmission lines' characteristic impedance, Z_0. Therefore, the input impedance of the line R_{IN} is equal to R_L. Due to Tx and Rx package parasitics, Rx input capacitance and the channel itself, the input impedance of the channel will be frequency-dependent. However, to understand the fundamentals of this circuit, an idealized scenario is considered.

Matching R_L to Z_0 eliminates Rx-side reflections, in principle. Should there be reflections, matching R_T to Z_0 sees these reflections absorbed in the transmitter. The circuit operates as follows:

- Differential input $v_{IN+} - v_{IN-}$ is assumed to be large enough to fully switch the differential pair formed by transistors M_1 and M_2.
- When the input is > 0, no current flows through M_2 and I_B flows entirely through M_1 with half of this current flowing from the load. Then, $v_{I,Rx+} = V_{DD}$ and $v_{I,Rx-} = V_{DD} - \frac{I_B R_L}{2}$ leading to a differential signal at the receiver of $\frac{I_B R_L}{2}$. This assumes the line is lossless.
- When the input is < 0, I_B flows through M_2 with half of this current flowing from the load. Then, $v_{I,Rx+} = V_{DD} - \frac{I_B R_L}{2}$ and $v_{I,Rx-} = V_{DD}$ leading to a differential signal at the receiver of $-\frac{I_B R_L}{2}$.
- The peak-to-peak differential input at the receiver is $V_{p-p,diff} = I_B R_L$.
- The power dissipation of the driver is $V_{DD} I_B$.

In the configuration shown in Figure 5.12 the channel is differential. With either polarity of input voltage, a current of $I_B/2$ flows from the on-chip supply V_{DD} and a current $I_B/2$ flows from the receiver. This means that half of the bias current is wasted in the

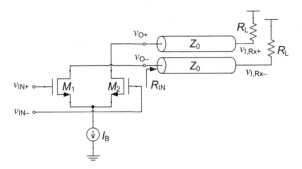

Figure 5.13 Open-drain differential CML driver.

on-chip loads. If R_T is increased, a larger percentage of I_B flows from the receiver, but transmitter-side matching is degraded. This is summarized as:

$$\alpha = \frac{R_T}{R_T + R_L}, \tag{5.37}$$

$$\Gamma_{Tx} = \frac{R_T - Z_L}{R_T + Z_L}, \tag{5.38}$$

where αI_B is the current that flows from the receiver when the differential pair is fully switched and Γ_{Tx} is the transmitter-side reflection coefficient. For example, suppose $R_L = Z_L = 50\ \Omega$ and $R_T = 150\ \Omega$. With these values, $\alpha = \frac{150}{150+50} = 0.75$ and $\Gamma = \frac{150-50}{150+50} = 0.5$. This means that only $1 - \alpha = 1/4$ of the bias current is wasted in the termination resistors. However, the transmit-side reflection coefficient has increased from 0 in the case when $R_T = Z_L$ to 0.5. This means that any signals travelling from the receiver back to the transmitter due to imperfect receiver-side matching are reflected at the transmitter, attenuated by a factor of 0.5. Whether this is acceptable depends on the loss in the line and the realizable matching at the receiver side. This approach can be extended to the case in which $R_T \rightarrow \infty$, known as an open-drain driver.

5.3.2   Open-Drain Current-Mode-Logic Drivers

Figure 5.13 shows an open-drain driver. In this case, $\alpha = \Gamma = 1$. The peak-to-peak differential input at the receiver is $V_{p-p,diff} = 2I_B R_L$. The power dissipation of the driver is $V_{DD} I_B$ but, for a given specification of receiver voltage swing, the bias current is half as large as the case of the matched driver in Figure 5.12. Although the entire bias current, I_B, flows from the load, it develops a signal voltage by flowing through only one side of the receiver termination at a time. Recall from Section 5.3 that for a given receiver termination resistor, current is either pulled through it, or set to zero. Greater efficiency can be realized in a differential channel if current is pushed and pulled. This is shown in Section 5.6.

5.3.3   Transistor Sizing

There are three design choices for a CML driver, namely R_T, I_B and W, the width of the MOSFETs. The design considerations are summarized as follows:

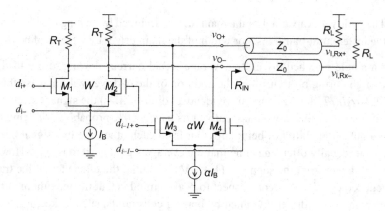

Figure 5.14 CML-based transmitter-side FFE.

- R_T is chosen based on the quality of matching required. If the channel is such that sufficient receiver-side matching is not possible, R_T will be set close to R_L. If the driver is used for a co-packaged laser, connected by a short wirebond, channel matching may not be a priority and an open-drain configuration can be used. On the other hand, having a small output impedance can help damp ringing introduced by the wirebond.
- Once R_T is chosen I_B is dictated by the required voltage swing.
- The width of the MOSFETs is chosen such that the typical voltage swings in CML drivers fully switch the output driver. Larger transistors will load preceding stages while smaller transistors will lead to reduced output amplitude by not fully switching the differential pair. Smaller W also leads to higher V_{GS} and less headroom for the current source.

In the analysis of differential pairs based on a square-law model of the MOSFET, a differential input voltage of $\sqrt{2}V_{OV}$ is needed to fully switch the differential pair where V_{OV} is the overdrive voltage of the differential-pair transistors when the current is divided equally between them. Since modern MOSFETs are highly scaled and do not follow square-law behaviour, a parametric sweep can be conducted to find a width that is just large enough to adequately switch the differential pair.

5.4 CML Implementation of a Tx FFE

The circuit implementation of the symbol-spaced FFE in Figure 4.2 (a) is shown in Figure 5.14. Its is explained as follows:

- The differential input $d_{i-1+/-}$ is generated from $d_{i+/-}$ using a DFF.
- Two CML drivers are connected together using a single set of load resistors. The addition of their outputs takes place by adding the signal currents of the drivers. Notice that the polarity of the output connection for the driver driven by $d_{i-1+/-}$ (referred to below as the α-driver) is reversed compared to the main driver, assuming α is negative.

- The α-driver is a variant of the main driver, designed by scaling the bias current and the differential-pair transistor width of the main driver by a factor of α.

For a single matched driver, the average signal current flowing to the load is $I_B/2$. Although during bit transitions the addition of the α-driver will increase this current to $(1 + \alpha)I_B/2$, during consecutive identical digits (CIDs) the signal current will drop to $(1 - \alpha)I_B/2$. Since two consecutive bits have a 50% probability of being the same and a 50% probability of being different from each other, the two signal current levels occur equally often, meaning that the average signal current is $I_B/2$. However, the current drawn from the supply is $(1 + \alpha)I_B$, meaning the efficiency of the transmitter is reduced by $\frac{1}{1+\alpha}$. A second aspect to bear in mind is that the maximum output voltage swing of the driver is limited by biasing constraints, such as keeping the current source in saturation. For example, suppose without equalization, $I_B R_T = 600$ mV; with equalization the bias current will be reduced such that $(1 + \alpha)I_B R_T = 600$ mV, reducing the average signal swing.

In the design presented in Figure 5.14 the α-driver is shown to have a bias current of αI_B. A transmitter with fixed equalization capability could implement this using a single current source transistor of width αW_B where W_B is the width of the transistor implementing I_B in the main driver. However, the amount of equalization is usually tuneable, meaning that the αI_B current source is usually implemented using a digitally controlled current source.

Since the output impedance of the differential pairs is assumed to be high, the matching of the driver to the channel is done by the load resistors. This makes the conversion of a single output driver to an FFE relatively straight forward. Source-series terminated (SST) drivers, discussed in Section 5.6, on the other hand present additional challenges in order to have good matching.

5.5   Low-Voltage Current-Mode Drivers

One way to increase the output swing of the current-mode driver from Figure 5.12 relative to the supply voltage is to remove the current source. This is beneficial because the current source requires a minimum V_{DS} to remain in saturation. If it leaves saturation, the circuit's output current decreases. An alternative approach is shown in Figure 5.15. Important aspects of this circuit include:

- This is a matched design, where R_T is tuned to match the channel's characteristic impedance.
- The driver has three identical branches, each consisting of M_{CX}, M_{NX} and M_{PX}.
- Each output sees a saturation-region cascode transistor $M_{C1/2}$.
- NMOS rail-transistors $M_{N0/1/2}$ are switched into triode, acting as degeneration resistors for the cascode transistors.
- The target output swing is set using the replica bias feedback loop consisting of the operational amplifier (opamp), M_{N0} and M_{C0}. This feedback loop settles so that the drain of M_{C0} is equal to V_{REF} when the gate of M_{N0} is V_{DD}.

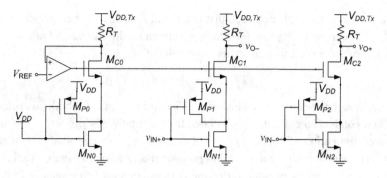

Figure 5.15 Low-voltage current-mode driver adapted from [3].

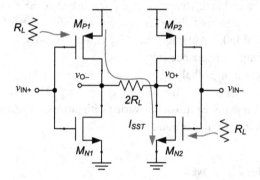

Figure 5.16 Source-series terminated voltage-mode driver.

- Since the other branches are sized identically, outputs swing from $V_{DD,Tx}$ down to V_{REF} for a peak-to-peak differential output swing of $2\left(V_{DD,Tx} - V_{REF}\right)$
- PMOS transistors M_{PX} act as common-source amplifiers that pull the intermediate nodes high when $v_{IN+/-}$ fall. This helps the circuit shut off more quickly. Without them, the intermediate nodes rise slowly as $M_{C1/2}$ shut off.

In the CML driver of Figure 5.12 gate and drain voltages of a M_1 and M_2 move in opposite direction, increasing voltage stress. On the other hand, in Figure 5.15, cascode devices have fixed gate voltage, reducing voltage stress considerably.

5.6 Voltage-Mode Source-Series Terminated Drivers

Figure 5.16 shows a simplified schematic of a voltage-mode driver. As with the current-mode driver, data are input differentially and the channel is driven differentially. In this case, the two receiver termination resistors are shown as a single, differential resistor of $2R_L$, where R_L is chosen to match the channel's characteristic impedance. The inputs switch between 0 and V_{DD}. Consider the case highlighted in Figure 5.16 where v_{IN+} is low and v_{IN-} is high. Transistors M_{P1} and M_{N2} conduct in triode whereas M_{N1} and M_{P2} are off. The sizes of the transistors and the supply voltage have been selected such that conducting transistors each present a resistance of R_L. A current of $I_{SST} = \frac{V_{DD}}{4R_L}$ flows developing a differential voltage of $-2I_{SST}R_L$, giving rise to a peak-to-peak differential voltage of:

$$V_{p-p,diff} = 4I_{SST}R_L = V_{DD}. \tag{5.39}$$

Even though the circuit is not biased with a current source, for both polarities of input data it conducts a current of I_{SST} drawn from a supply of V_{DD}, leading to an average power dissipation of $P_D = V_{DD}I_{SST}$. In comparing voltage-mode and current-mode drivers, assume both are designed to produce the same output swing. In this case, the SST driver requires one-quarter of the current compared to a matched CML driver and one-half of the current compared to an open-drain CML driver. Note that the SST driver is still matched to the channel. Hence we see a trend toward SST drivers for electrical links where efficiency and matching are desired. We also see SST drivers for driving laser diodes in optical links. Although matching is not a priority, having good efficiency along with low output impedance is beneficial.

SST drivers present a few design challenges. In particular, since V_{DD} determines the output swing and output impedance, it must be set properly. Also, the width of transistor needed to produce a matched resistance varies with process. The approaches to tuning SST drivers are discussed in Section 5.6.3.

5.6.1   Linearity Considerations for SST Drivers

Although the transistors in Figure 5.16 are assumed to act like resistors, MOSFETs are nonlinear devices, whose incremental resistance will vary with operating point but behave most like linear resistors only in deep triode. If we assume that the MOSFETs do indeed present large-signal and small-signal resistance of R_L, for the scenario depicted in Figure 5.16, $v_{IN-} = V_{DD}$ and $v_{O+} = \frac{V_{DD}}{4}$. Since $V_{th} \ll 0.75V_{DD}$, M_{N2} is indeed in triode. However, as the MOSFETs switch, the transmitter's output resistance changes. Also, as we discuss in Section 5.6.4, to implement equalization, several voltage-mode drivers can be connected together that may operate in opposition to one another during runs of CIDs.

To improve the linearity of the driver's output resistance, the approach shown in Figure 5.17 is used. In this case, the overall output resistance designed to match R_L is given by:

$$R_O = R_L = R_P + R_{sw}, \tag{5.40}$$

where R_P is a polysilicon resistor in series with the driver. The subscript sw denotes that the MOSFETs are used as switches, presenting an on-resistance (deep triode) of R_{sw}. To reduce R_{sw} to only a fraction of R_L dictates that the transistors in Figure 5.17 are larger than those in Figure 5.16 by approximately $\frac{R_L}{R_{sw}}$.

This configuration has the advantage that fluctuations in R_{sw} translate to smaller fluctuations in the overall driver output resistance. However, this approach introduces two disadvantages. The first is that larger switch transistors present larger capacitance to the preceding stage and therefore increase the dynamic power dissipation of the

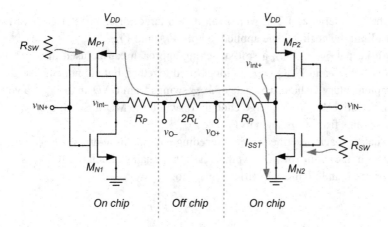

Figure 5.17 Source-series terminated voltage-mode driver with reduced switch resistance.

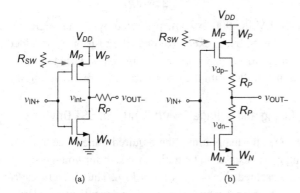

Figure 5.18 Possible locations of polysilicon resistors in SST drivers: (a) one resistor; (b) two resistors.

driver. The second disadvantage is that the voltage stress on the switching MOSFETs is increased, as discussed in the next sub-section.

Figure 5.18 shows two topologies that place a polysilicon resistor (R_P) in series with the switch's resistor (R_{SW}). In Figure 5.18, (a) and (b) are equivalent from the point of view of the dc current path when the input is fully switched. Circuit (b) will add more capacitance, since resistors introduce parasitic capacitance, but reduces voltage stress on the off transistor compared to circuit (a). Circuit (b) also introduces the chance for greater pull-up/pull-down impedance mismatch since the two R_P resistors have mismatch.

5.6.2   Voltage Stress for SST Drivers

The peak-to-peak differential output swing of the drivers in Figure 5.16, Figure 5.17 and Figure 5.18 is V_{DD}. To maintain large swing, the supply voltage of the output stage might be as large as 1.2 V, which can be too large for nanometer CMOS. Designers must limit the peak voltage applied to both V_{GS} and V_{DS}.

To keep $V_{GS} < V_{DD}$, a soft-switching approach can be used in which the gates of the NMOS and PMOS are no longer tied together, but are nonetheless driven in a complementary fashion. The gate of M_{N2} swings from GND up to V_{Max}, a voltage less than the driver's V_{DD} but large enough to switch the MOSFET. Similarly, the gate of M_{P2} swings from V_{DD} down to $V_{DD} - V_{Max}$.

The approach described in the preceding paragraphs addresses V_{GS} voltage stress, although the issue of V_{DS} voltage stress remains. Each NMOS drain node in Figure 5.16 and Figure 5.18 (b) is limited to:

$$v_{DS_N} < \frac{3}{4} V_{DD}. \tag{5.41}$$

For the circuit in Figure 5.17 and Figure 5.18 (a), this is increased to:

$$v_{DS_N} < V_{DD} \left(\frac{3}{4} + \frac{R_P}{4R_L} \right) \tag{5.42}$$

With a driver V_{DD} of 1.2 V, the 900 mV v_{DS} predicted by (5.41) might be just tolerable by highly scaled CMOS, but is excessive once the resistor R_P is added in (5.42) or Figure 5.18 (a). An approach that maintains a push-pull structure but moves the termination resistance in parallel with the output driver is presented in [38].

5.6.3   Output Impedance Tuning and Voltage Swing Control of SST Drivers

In this section, techniques for maintaining the required impedance matching and the desired voltage swing are discussed. The main approach for voltage-swing tuning is supply regulation. As presented above, $V_{p-p,diff} = V_{DD}$. Thus, a regulator can be used to raise and lower V_{DD}. In general, a higher-loss channel would use the largest V_{DD} the transistors can tolerate, and V_{DD} would only be reduced if a low-loss channel already allowed the removal of some equalization circuits.

The output impedance of SST drivers is tuned by controlling the effective width of transistors and equivalent series resistance. Figures 5.19 and 5.20 show two implementation approaches. Figure 5.19 shows an SST driver with a polysilicon series resistor R_P to improve linearity. The impedance of the switch transistors will increase at high temperature, low supply voltage (i.e., low output swing settings) and when the CMOS technology's process parameters drift toward those giving slower digital circuit operation, known as the slow process corner. Assume simulations give values for W_P, W_N and R_P, allowing suitable matching to the channel impedance for these worst-case conditions. For lower temperature, higher supply voltage and faster process corners, the driver in Figure 5.19 (a) sets output impedance too low. The impedance can be controlled using the approaches in part (b) and Figure 5.20. In Figure 5.19 (b) the driver has been divided into N segments, connected in parallel to the input v_{IN+} and output v_{OUT-}. Each segment (or slice) is controlled by complementary enable signals $EN_{n/p}$. Elements in (b) are sized as follows:

$$R_{Pi} = NR_P, \tag{5.43}$$

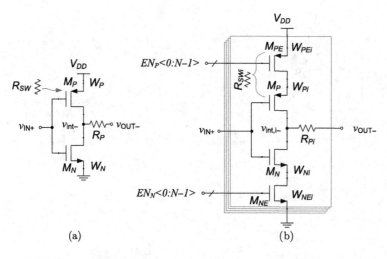

(a)   (b)

Figure 5.19 Approaches to segmentation of a source-series terminated voltage-mode driver for impedance control: (a) reference circuit; (b) segmentation into slices where each slice is enabled/disabled through footer transistors M_{PE} and M_{NE}.

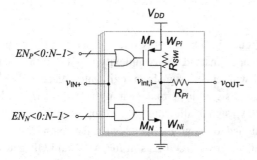

Figure 5.20 Segmentation into slices where each slice's input is gated via logic gates.

$$\frac{1}{W_N} = \frac{1}{N}\left(\frac{1}{W_{Ni}} + \frac{1}{W_{NEi}}\right),$$ (5.44)

$$\frac{1}{W_P} = \frac{1}{N}\left(\frac{1}{W_{Pi}} + \frac{1}{W_{PEi}}\right),$$ (5.45)

assuming square-law operation. This sizing leads to a per-segment output resistance that is N-times larger than that in Figure 5.19 (a). The sizing guideline in (5.44) and (5.45) has no unique solution. One approach would be to set $W_{Ni} = W_{NEi}$ and $W_{Pi} = W_{PEi}$, leading to $W_{Ni} = \frac{2}{N}W_N$ and $W_{Pi} = \frac{2}{N}W_P$. With N slices always connected to the input terminals, this sizing approach minimizes chip area but doubles the output driver's loading on the preceding stage, thereby increasing pre-driver power dissipation and chip area. Since the enable signals are not switched at the bit rate, $W_{P/NEi}$ can be increased and $W_{P/Ni}$ can be reduced. This leads to $W_{Ni} \approx \frac{1}{N}W_N$ and $W_{Pi} \approx \frac{1}{N}W_P$, reducing loading on the data path toward what it is in Figure 5.19 (a).

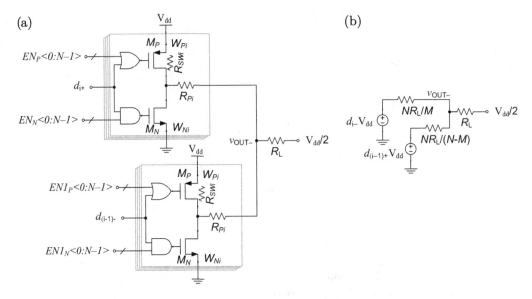

Figure 5.21 Implementation of an FFE in an SST Tx: (a) single-ended schematic; (b) equivalent circuit.

The slice approach in Figure 5.20 keeps only one transistor in the output driver's pull-up and pull-down networks. Instead, the complexity is put into the data path, by combining the input signal with enable signals in logic gates. Each EN_N signal that is a logic high allows the input signal v_{IN+} to appear at the gate of M_N. A high EN_N signal is paired with a low EN_P signal, allowing v_{IN+} to appear at the gate of M_P. The logic gates may also be designed to favour the speed of the data path by increasing the width of the enable-controlled series MOSFETs in the AND and OR gates. Finally, the gates will usually be transformed to use NAND and NOR gates.

5.6.4   SST Implementation of a Tx FFE

Figure 5.21 (a) shows a schematic of one side of an SST Tx that implements a one-tap FFE. It is assumed that signals $d_{i+/-}$ and $d_{(i-1)+/-}$ are generated by DFFs. This would be extended to a clocked delay line of DFFs if the FFE had more taps. Both the main-cursor tap, controlled by $EN_{P/N}$, and the post-cursor tap, controlled by $EN1_{P/N}$, have been divided into N slices. For the purposes of this example, it is assumed that when N slices are enabled, an impedance of R_L is presented. Enabling N slices in the main tap and 0 slices in the post-cursor tap reduces this circuit to the conventional SST driver. We generate equalization by enabling M (where $M < N$) slices in the main tap and then $N - M$ slices in the post-cursor tap. To maintain matching, a total of N slices must be enabled.

An equivalent circuit of the driver is shown in Figure 5.21 (b). For the circuits in (a) and (b) to both generate v_{OUT-} the polarity of the input signals must be reversed since the voltage sources in (b) do not invert as the driver in (a) does. The voltage sources in (b) toggle between 0 and V_{DD}. The output resistance of each tap as a function of M is shown. The reader can confirm that the overall output resistance is R_L. Through superposition, the single-ended output voltage resulting from the implied $V_{DD}/2$ source (there because of the circuit's symmetry) along with $d_{i+/-}$ and $d_{(i-1)+/-}$ can be found as:

$$v_{OUT-} = V_{DD}\left(\frac{1}{4} + \frac{M}{2N}d_{i-} + \frac{N-M}{2N}d_{(i-1)+}\right). \tag{5.46}$$

Note that in Figure 5.21 (b) and in (5.46), the current bit, d_{i-}, appears with negative polarity while the previous bit, $d_{(i-1)+}$, appears with positive polarity. When consecutive bits are different ($d_{i+} \neq d_{(i-1)+}$) the signals applied to the driver are equal. That is, $d_{i-} = d_{(i-1)+}$, and the driver produces a voltage of 1/4 or 3/4 V_{DD}. For consecutive identical digitals ($d_{i+} = d_{(i-1)+}$) or ($d_{i-} \neq d_{(i-1)+}$), the driver's output is

$$v_{OUT-} = \begin{cases} V_{DD}\left(\frac{1}{4} + \frac{M}{2N}\right) & \text{for } d_{i-} = 1,\ d_{(i-1)+} = 0, \\ V_{DD}\left(\frac{1}{4} + \frac{N-M}{2N}\right) & \text{for } d_{i-} = 0,\ d_{(i-1)+} = 1. \end{cases} \tag{5.47}$$

In Section 4.2 we saw that when the current bit is a 1, the output is $1 \pm \alpha$, and when the current bit is a 0, the output is $-1 \pm \alpha$. The single-ended SST driver generates signals centred around $V_{DD}/2$. We can subtract one half from the two possible signal levels for $d_{i-} = 1$ to give an expression for α:

$$\frac{1-\alpha}{1+\alpha} = \frac{\frac{1}{4} + \frac{M}{2N} - \frac{1}{2}}{\frac{3}{4} - \frac{1}{2}} \tag{5.48}$$

$$\alpha = \frac{N-M}{M}. \tag{5.49}$$

Example 5.4   If a given SST implementation, at a particular process corner, temperature and supply voltage, needs 16 slices to be matched to the channel, determine the two smallest non-zero values of α that are realizable and the resulting four possible output voltages as fractions of V_{DD} for each α.

Solution
If $N = 16$, we can consider $M = 15$ and $M = 14$. That is, all but one or two slices are allocated to the main tap and one or two are allocated to the post-cursor tap. Table 5.1 summarizes the four possible single-ended output voltage levels for each value of M. The input signals are listed using the notation in Figure 5.21 (b) and (5.46).

When matched to the channel, an SST FFE reduces the output swing since the maximum levels are limited to 1/4 and 3/4 V_{DD}.

Table 5.1 SST FFE with $N = 16$ and $M = 15$ or $M = 14$. Output voltages are fractions of V_{DD}.

M	α	$d_{i-} = 0,$ $d_{(i-1)+} = 0$	v_{OUT-} $d_{i-} = 0,$ $d_{(i-1)+} = 1$	$d_{i-} = 1,$ $d_{(i-1)+} = 0$	$d_{i-} = 1,$ $d_{(i-1)+} = 1$
15	1/15	1/4	9/32	23/32	3/4
14	2/14 = 1/7	1/4	5/16	11/16	3/4

5.7 Overview of Serialization Approaches

High-speed wireline links send data at a rate much higher than the clock speeds of CPUs or GPUs. Therefore, multiple data streams are multiplexed together before transmission. The transmitter in Figure 1.4 (c) takes 80 parallel inputs at 800 Mb/s and ultimately generates a 64 Gb/s PAM4 output data stream. Before discussing this implementation in more detail, general considerations are listed:

- Do many circuits process the full-rate signal? Since inductive peaking and other bandwidth extension techniques might be needed at these data rates, we want to minimize the full-rate signal path.
- How is the final stage of multiplexing implemented? For example, is there a 2:1 multiplexer at the end? In some cases a 4:1 or even 8:1 multiplexer is used. When contrasting a 4:1 to a 2:1 multiplexer, it is implicit that the 4:1 is not simply the combination of three 2:1 multiplexers, but directly implements 4:1 multiplexing.
- How are the clocks generated? And does IQ mismatch or duty-cycle distortion introduce jitter? These clocking issues are discussed more in Sections 12.3.4 and 12.3.5.

Returning to Figure 1.4 (c), the Tx generates eight 32 Gb/s data streams using 8:1 multiplexers that are built around 4:1 multiplexers. Four full-rate data streams are combined in an FFE to generate the LSB of a PAM4 data stream while the other four are combined to generate the MSB. All of the FFE outputs are connected together to sum their output current, possibly through inductive networks. The output node always operates with full-rate signals. In this design the inputs to the LSB and MSB FFE taps (labelled as [3, 12, 6, 3] and [6, 24, 12, 6], respectively) are also full-rate signals.

An alternative to having full-rate internal nodes is to perform multiplexing in the final driver stage, as argued in [39]. An earlier discussion of multiplexing trade-offs can be found in [40]. Although multiplexing can be discussed in terms of a truth table that combines select lines and data inputs, a switch-based or tri-state gate-based discussion is more frequently taken in wireline applications.

Figure 5.22 (a) shows the schematic of a 4:1 multiplexer. Input signals $v_i < 0:3>$ each last for four UIs. One-hot encoded select signals are applied to transmission

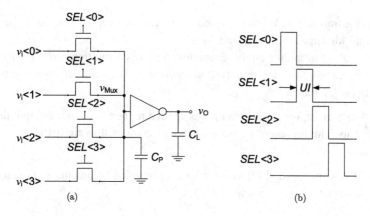

(a)   (b)

Figure 5.22 Implementation of a 4:1 pass-transistor multiplexer: (a) schematic where NMOS transistors represent transmission gates, composed of suitably driven PMOS and NMOS transistors; (b) 1 UI *SEL* signals.

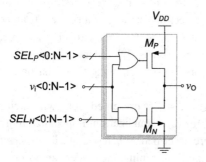

Figure 5.23 Implementation of a 4:1 tristate multiplexer.

gates, drawn as NMOS only in the schematic. Since the select signals are each high for only one UI, they operate with 25% duty cycle. Quadrature clock signals can be combined to generate the select signals. Challenges of this approach include:

- Generating equal width and properly aligned select signs.
- Driving node v_{Mux} through transmission gates due to the loading C_P.

An alternative to the pass-transistor multiplexer is shown in Figure 5.23, but using tri-stated inverters. A similar structure was used in Figure 5.20 to enable/disable slices for impedance control. In that context all slices receive the same input signal and enable signals are not changing frequently. In the multiplexer application, a different input signal is applied to each slice, where each slice is activated for one UI, every four UIs. Thus, some of the sizing approaches discussed in Section 5.6.3 would not apply here as both the input and the select signals are high-speed signals.

    The output (v_O) of the circuit in Figure 5.23 can be applied to an electrical or optical link driver. Alternatively, this approach to multiplexing can be built into a transmitter's final driver stage. Doing so limits full-rate operation to the output signal, lowering data rates in the transmitter. When built into the driver, series resistance would be added at the output for impedance matching.

The above multiplexer circuit examples showed 4:1 multiplexers. Other ratios are common, leading to different overall architectures. Two scenarios were compared in detail in [40], namely:

- An 8-to-1 multiplexer whose output is applied to the transmitter's output driver.
- Two 4-to-1 multiplexers whose outputs are combined in the transmitter's output driver.

The reader is directed to [40] to consider the presented arguments and to apply them to the particular design scenario.

5.8   Summary

Whereas Chapter 2 introduced the bandwidth limitations of electrical channels, this chapter presented how, due to ESD protection, the chip boundary introduces another low-pass response formed by ESD capacitance and the channel resistance. Conventional inductive-peaking techniques were discussed. Readers can consider more elaborate schemes in which ESD capacitors are split allowing transmission-line-inspired structures.

Current-mode and voltage-mode drivers were presented along with their respective advantages and disadvantages. Feed-forward equalization, impedance tuning topologies and multiplexing approaches were described. When seeing a large driver split into parallel sections to allow FFEs, or impedance tuning, the reader can observe that if we increase the number of individually controlled slices of the driver, we are building a DAC, a technique that is explored further in Chapter 11.

Problems

5.1 Consider the current-mode logic driver shown in Figure 5.12. Assume the transmission line has a characteristic impedance of 50 Ω.

(a) For a matched design choose the resistors and bias current for a peak-to-peak differential voltage at the receiver input of 1000 mV$_{pp}$.

(b) How would you select the transistor width for the differential pair? That is, explain the trade-offs of increasing and decreasing the transistors' width.

(c) Sketch single-ended voltages v_{O+}, v_{O-}, $v_{I,Rx+}$ and $v_{I,Rx-}$ for a few bits of random data applied at the input of the driver. Assume the line is lossless.

5.2 Design a matched CML driver for a load and channel of 50 Ω. Select parameters for an output swing of $V_{p-p,diff} = 800$ mV. Assume input data originate from a CML circuit with an output swing that can be as low as 500 mV$_{p-p,diff}$. Make sure the circuit has enough gain to produce the required output swing when the input swing is at its minimum.

5.3 Use a real channel to investigate the Tx from Problem 5.2's performance at 10 Gb/s assuming an ESD capacitance of 400 fF.

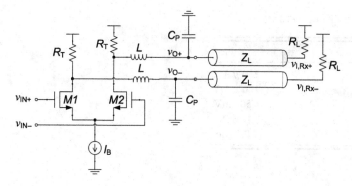

Figure 5.24 CML Tx with series inductors.

(a) Show the eye at the output of the transmitter.
(b) Show the eye at the input of the receiver.

5.4 For your design in Problem 5.1 assume a capacitor of 500 fF is added at the drain of each transistor in Figure 5.12.

(a) What is the bandwidth of the circuit? That is, what is the frequency response of the impedance seen by the drains of the transistors and what is its bandwidth?
(b) Apply shunt inductive-peaking to the Tx to extend the bandwidth. Draw your circuit with inductive-peaking. Be sure to compute element values. Specify the poles and zeros of the new circuit. Please bear in mind that due to the loading from the transmission line and receiver input resistance, the circuit topology is different from what is presented in Figure 5.2 and Section 5.2.1.

5.5 Consider the transmitter in Figure 5.24, designed with $R_T = R_L = Z_L = 50\ \Omega$, $I_B = 10$ mA, $C_P = 500$ fF.

(a) Sketch the voltages at $v_{I,Rx+}$ and $v_{I,Rx-}$ as well as the differential voltage at the receiver for a few bits of data, assuming no ISI and a lossless line.
(b) A designer considers series inductive-peaking shown in the figure. Is adding an inductor as shown useful? Compare the bandwidth without an inductor ($L = 0$) to the case of an inductor sized for $\zeta = 0.707$.

5.6 This problem deals with an improvement to the conventional inductive shunt-peaking presented in Section 5.2.1, shown in Figure 5.25, designed with $R = 200\ \Omega$, $C = 500$ fF.

(a) Determine the bandwidth when $L = 0$ H and $C_B = 0$ F.
(b) Read Section II of [41] where analysis of Figure 5.25 is presented.
(c) Design the network for BWER = 1.83 and what you decide is a good response. Give values for L and C_B. Justify your choice.
(d) For the elements you chose, determine the location of the zeros in the transfer function (Equation (3)) from [41].

Table 5.2 MOSFET parameters.

Parameter	Value
$\mu_n C_{OX}$	$250\ \mu A/V^2$
$\mu_p C_{OX}$	$100\ \mu A/V^2$
$V_{tn} = |V_{tp}|$	$500\ mV$
$\lambda_n = \lambda_p$	$0.1\ \mu mV^{-1}$
L_{min}	$0.1\ \mu m$

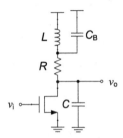

Figure 5.25 Shunt-peaking with a bridging capacitor (C_B) added, adapted from [41].

(e) Compare the time-domain behaviour of this circuit to the conventional shunt peaking (without C_B) at the same BWER at a data rate of at least twice the achieved bandwidth. Which one has larger vertical eye opening?

5.7 Consider the SST Tx shown in Figure 5.19 (a); $V_{DD} = 1.0$ V.

(a) For a transmitter matched to 50 Ω, choose the W and L of M_P and M_N such that their ON-resistance is 16.7 Ω using MOSFET parameters from Table 5.2.
(b) By what % does the total output resistance of the driver change when V_{DD} is increased to 1.2 V assuming the input signal now steps between 0 and 1.2 V?
(c) For $V_{DD} = 1.0$ V, what is the pull-up and pull-down resistance of the driver if the threshold voltage of the NMOS transistors decreases to 400 mV and $\mu_p C_{OX} = 80\ \mu A/V^2$?

5.8 Design a matched SST driver for the impedance and swing specified in Problem 5.2.

5.9 Compare the topologies in Figure 5.18 in a technology of your choice. Assume a matched channel and $R_{sw} = R_L/4$. Start with a single slice design and then compare the topologies when the driver is subdivided in 12 slices. Make sure you respect current density limits for the resistors.

5.10 The schematic in Figure 5.19 (c) is a conceptual schematic that should be modified in practice to use inverting gates. Modify it, according to the following steps:

(a) Replace the AND gate with a NOR gate and the OR gate with a NAND gate, making the necessary input polarity changes.

(b) Take advantage of the fact that the enable signals will not be switching quickly and can therefore see larger loading capacitance than the data signals. Consider making the gates asymmetrical so that they are fast, but do not load the data path significantly.

5.11 Revisit the FFE coefficients presented in Table 4.2 in the context of an SST Tx using 64 slices and a supply voltage of 1.2 V.

(a) Propose an allocation of slices that implements Case 4 in Table 4.2. How many slices are allocated to each tap? What are the resulting coefficients?

(b) Compute the residual ISI and VEO resulting from the realizable FFE in part (a).

(c) Compute the output voltages at the driver's outputs for runs of CIDs and compare this to $V_{DD}/2$.

6   Electrical-Link Receiver Front-Ends

Beginning with a discussion of receiver metrics, this chapter discusses electrical-link receiver implementation by first considering input termination and electrostatic discharge (ESD) capacitance mitigation. The receiver front-end has two main jobs. The first is to amplify the signal to levels that can be captured by a decision circuit and the second is to remove enough ISI such that low BER detection is possible. With the types of receiver-side equalization already described at the system-level in Chapter 4, this chapter focuses on their transistor-level implementation. Both CML and inverter-based circuits are presented.

6.1   Metrics

The receiver front-end's specifications are tightly coupled to the intended channel's characteristics and the amount of transmitter-side equalization. Once these are specified, the receiver must achieve a given BER for a given transmitter signal amplitude.

Although an electrical link receiver front-end must provide enough gain and equalization for a worst-case channel as dictated by the standard for which it is designed, not every use-case presents the worst-case channel. Also, circuits designed to provide a certain gain or frequency response will deviate from their target performance due to variations in CMOS process parameters, voltage and temperature, collectively referred to as PVT variation. To accommodate diverse channels and PVT variation, circuit blocks must be tuneable. This manifests itself in the following ways:

- Parameters such as CTLE frequency response and DFE tap weights must be adjustable. Also, the *location* of non-zero DFE taps is also sometimes configurable.
- The gain of amplifiers must be variable over a wide range and with relatively fine control.

The signal at the decision circuit should be as large as possible. This minimizes the effect of latch offsets. In the event that an ADC is used to capture the analog signal, having a large signal reduces quantization error due to the finite ADC resolution. However, the signal cannot be too large so as to be clipped. Therefore, a variable-gain amplifier is used to adjust signal swing along the receiver.

6.1.1   Linearity Considerations

All of the equalization strategies presented in Chapter 4 rely on the assumption that the channel and receiver are linear. For example, the CTLE inverts a portion of the channel's low-pass frequency response. To determine the output of the overall system by multiplying frequency responses requires a linear response from the CTLE. As a second example, the DFE feeds back an estimate of ISI for a given bit, regardless of the bits on either side. If the receiver were nonlinear, the required ISI mitigation would be dependent on the signal level and not only the previously received bit(s).

Intuitively, the larger the worst-case ISI at a given point in the signal path is relative to the main cursor, the more stringent the linearity requirement is. Nonlinearity will manifest itself as residual ISI that cannot be removed, even if equalization parameters are fine-tuned.

Linearity is quantified by metrics such as total harmonic distortion (THD) at a given input amplitude and frequency. To measure THD, a sinusoidal input is applied to the circuit and the output spectrum is measured. The amplitude of the fundamental is a_1 and the nth harmonic is a_n. The definition of THD is:

$$THD = \frac{\sqrt{\sum_{n=2}^{\infty} a_n^2}}{a_1}. \tag{6.1}$$

Values of a few percent are usually targeted where the input amplitude would be set by $\sum_{-\infty}^{\infty} |h_n|$ for the pulse response up to the input of the circuit being characterized. One could also consider an "equivalent number of bits" (ENOB) metric. Less common in wireline is the use of a second-order or third-order intercept point.

6.2   Input Termination and Matching

The circuitry at the input of the receiver sets the input impedance and, in conjunction with the transmitter, sets the dc operating point of the receiver's first stage. In the case of a dc-coupled link, the transmitter's output will be at the same dc voltage as the receiver's input. Although dc current may flow through the channel, resistance at dc will be negligible. Links can also be ac coupled, meaning that the transmitter and receiver are separated by a dc-blocking series capacitor. In this case, the input dc common-mode of the receiver is set by the receiver exclusively.

Exploring further the dc-coupling example, consider the case of a CML output driver shown in Figure 5.12. The Rx can achieve 50 Ω matching by tying its inputs to V_{DD} through 50 Ω resistors. In the case of a driver also terminated in 50 Ω, each input of the receiver sits at a dc voltage of:

$$V_{IN} = V_{DD} - \frac{I_B}{2} R_{EQ}, \tag{6.2}$$

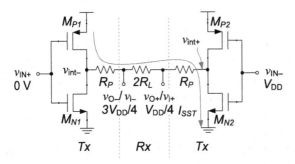

Figure 6.1 SST driver connected to a receiver input termination consisting of a floating resistance of $2R_L$. Node voltages given assume matched output impedance.

where R_{EQ} is the parallel combination of the driver's termination resistance and the input termination of the receiver. In this case we have two 50 Ω resistors in parallel. Therefore, the dc input voltage is:

$$V_{IN} = V_{DD} - \frac{I_B}{4}R_T,$$ (6.3)

where R_T is the termination resistance.

The case of an SST driver is a little different in that the power efficiency of this driver topology was argued with the receiver termination shown in Figure 6.1. Recalling the analysis in Section 5.6, when each side of the driver presents an output resistance of R_L, the current drawn is:

$$I_{SST} = \frac{V_{DD}}{4R_L}.$$ (6.4)

There are two potential problems with using a "floating" resistor of $2R_L$ as the input termination. The first is that this scheme, while presenting perfect differential imped-ance matching, presents a common-mode open circuit. One of the arguments for using a symmetrical circuit is that it can cancel the inevitable common-mode components of the signal, introduced through mismatch, power-supply noise or crosstalk. However, when a floating resistor is used, any signal of a common-mode nature is reflected rather than absorbed in the Rx termination. The common-mode open circuit also implies that the common-mode input voltage is controlled entirely by the transmitter, which if there are residual errors in its impedance tuning, will lead to a shift in the input common-mode of the receiver.

Figure 6.2 shows a scenario that preserves differential matching while also match-ing the receiver to the channel for common-mode signals, provided a voltage source is connected to $v_{RX,CM}$. There are various schemes to control $v_{RX,CM}$ using a feedback loop such that the input common-mode of the receiver can be controlled.

A single-ended input termination resistance of 50 Ω can be set by placing two 100 Ω resistors between V_{DD} and GND as shown in Figure 6.3. This sets an input dc voltage of $V_{DD}/2$ as well as a differential input resistance of 100 Ω. The down-side of this approach is that a large current flows through the termination resistors.

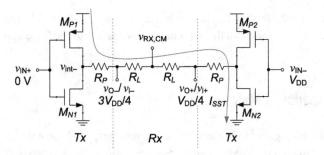

Figure 6.2 SST driver connected to a receiver input termination consisting of common-mode biased resistors of R_L. Node voltages given assume matched output impedance.

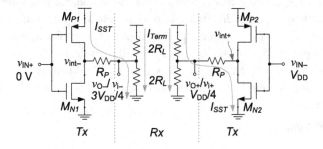

Figure 6.3 SST driver connected to a receiver input termination consisting of a voltage-division generation of common-mode input voltage using resistors of $2R_L$.

In Figure 6.1 the current that is sourced by the left side of the driver is sunk by the right side. No additional current paths are present. However, in Figure 6.3 the current sunk by the right side is also drawn from V_{DD}. This alone doubles its power dissipation compared to Figure 6.1. The additional current in the termination resistors can be calculated by considering the particular voltages shown in Figure 6.3. These voltages correspond to the case of a transmitted 0, since v_{O-} is raised to $3V_{DD}/4$ while v_{O+} is lowered to $V_{DD}/4$. The current I_{SST} is sourced by M_{P1} and flows through the lower $2R_L$ termination resistor in the Rx along with a current I_{Term}. Since only I_{Term} flows through the upper $2R_L$ termination resistance on the negative (left) side of the Rx, its current is found using Ohm's law as:

$$I_{Term} = \frac{V_{DD}/4}{2R_L}. \tag{6.5}$$

The total current draw by the driver and resistive terminations is:

$$I_{Tot} = 2I_{SST} + 2I_{Term} = 2\left(\frac{V_{DD}}{4R_L}\right) + 2\left(\frac{V_{DD}/4}{2R_L}\right),$$

$$I_{Tot} = 3\left(\frac{V_{DD}}{4R_L}\right) = 3I_{SST}. \tag{6.6}$$

Thus, if not carefully designed, the power dissipated in the Rx termination can be much larger than that required to generate a signal of a given peak-to-peak voltage swing.

This analysis motivates feedback schemes to set $v_{RX,CM}$ in Figure 6.2, that preserve common-mode matching without the extra power dissipation of the large current paths in Figure 6.3.

6.3   Wideband Impedance Matching

Regardless of the configuration of the termination resistors, ESD capacitance can limit the bandwidth of the input impedance, increasing reflections at the receiver and attenuating the high-frequency components of the received signal. In Figure 6.4 (a) the ESD capacitance appears in parallel with the receiver's input termination, degrading the input match above $\frac{1}{RC}$. The techniques presented in Section 5.2 that were used to extend the bandwidth of output drivers are also applicable to input terminations. Figure 6.4 (b) shows the use of a T-coil for improved input matching. The transfer function from the Rx's input current to the voltage across the termination was analyzed in Section 5.2 and shown to be all-pass. To show that the T-coil load is also useful for impedance matching, we must determine the impedance seen from the receiver's input.

6.3.1   T-Coil Input Impedance

Referring to Figure 6.5, the input impedance can be found by writing a KVL equation relating V_i to the voltages across Z_1, Z_2 and R:

$$V_i = (Z_2 + R)I_2 + Z_1 I_i, \qquad (6.7)$$

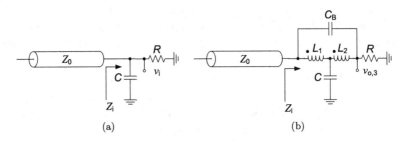

Figure 6.4 ESD capacitance (C) and input impedance (Z_i) for (a) parallel connection of ESD and termination resistance R and (b) use of a T-coil.

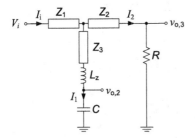

Figure 6.5 Modified version of Figure 5.8 adapted to find input impedance in Figure 6.4.

$$Z_i = \frac{V_i}{I_i} = (Z_2 + R)\frac{I_2}{I_i} + Z_1. \tag{6.8}$$

From [36] the impedance terms above are given by:

$$Z_1 = \frac{sL_x}{D(s)}, \quad Z_2 = \frac{sL_y}{D(s)}, \quad Z_3 = \frac{s^2 L_x L_y C_B}{D(s)}, \tag{6.9}$$

$$\text{where } D(s) = s^2(L_x + L_y)C_B + 1. \tag{6.10}$$

These expressions can be related back to the original circuit as:

$$L_x = L_1 + M \quad L_y = L_2 + M \quad L_z = -M. \tag{6.11}$$

For the symmetrical case $L_1 = L_2 = L$ and $M = kL$. Using these simplifications Z_1 and Z_2 become:

$$Z_{1/2} = Z_1 = Z_2 = \frac{s(1+k)L}{2s^2(1+k)LC_B + 1}. \tag{6.12}$$

Therefore, Z_i becomes:

$$Z_i = (Z_{1/2} + R)\frac{I_2}{I_i} + Z_{1/2}, \tag{6.13}$$

$$Z_i = R_T \frac{s^2 - \sqrt{2}\omega_n s + \omega_n^2}{s^2 + \sqrt{2}\omega_n s + \omega_n^2} + Z_{1/2}\left(\frac{s^2 - \sqrt{2}\omega_n s + \omega_n^2}{s^2 + \sqrt{2}\omega_n s + \omega_n^2} + 1\right). \tag{6.14}$$

The input impedance is the sum of two terms (i.e., $Z_i = Z_x + Z_y$). Considering only the second term (Z_y):

$$Z_y = Z_{1/2}\left(\frac{s^2 - \sqrt{2}\omega_n s + \omega_n^2}{s^2 + \sqrt{2}\omega_n s + \omega_n^2} + 1\right) =$$

$$Z_{1/2}\left(\frac{s^2 - \sqrt{2}\omega_n s + \omega_n^2}{s^2 + \sqrt{2}\omega_n s + \omega_n^2} + \frac{s^2 + \sqrt{2}\omega_n s + \omega_n^2}{s^2 + \sqrt{2}\omega_n s + \omega_n^2}\right). \tag{6.15}$$

Adding the two fractions and substituting for $Z_{1/2}$, we have:

$$Z_y = \frac{s(1+k)L}{2s^2(1+k)LC_B + 1}\left(2\frac{s^2 + \omega_n^2}{s^2 + \sqrt{2}\omega_n s + \omega_n^2}\right). \tag{6.16}$$

Substituting (5.11) and (5.14),

$$Z_y = \frac{s(1+k)L}{2s^2(1+k)LC\frac{1-k}{4(1+k)} + 1}\left(2\frac{s^2 + \omega_n^2}{s^2 + \sqrt{2}\omega_n s + \omega_n^2}\right) \tag{6.17}$$

$$= \frac{s(1+k)L}{s^2 LC\frac{1-k}{2} + 1}\left(2\frac{s^2 + \omega_n^2}{s^2 + \sqrt{2}\omega_n s + \omega_n^2}\right) \tag{6.18}$$

$$= \frac{s(1+k)L}{\frac{s^2}{\omega_n^2} + 1}\left(2\frac{s^2 + \omega_n^2}{s^2 + \sqrt{2}\omega_n s + \omega_n^2}\right) \tag{6.19}$$

$$= 2\frac{s(1+k)L\omega_n^2}{s^2 + \sqrt{2}\omega_n s + \omega_n^2}. \tag{6.20}$$

The numerator can be simplified using (5.19) and (5.20) to give $2(1+k)L = \frac{R}{\omega_o}$, which in turn equals $\frac{\sqrt{8}R}{\omega_n}$. With these substitutions, the overall input impedance is:

$$Z_i = R\frac{s^2 - \sqrt{2}\omega_n s + \omega_n^2}{s^2 + \sqrt{2}\omega_n s + \omega_n^2} + R\frac{\sqrt{8}\omega_n s}{s^2 + \sqrt{2}\omega_n s + \omega_n^2} \tag{6.21}$$

$$= R\frac{s^2 + \sqrt{2}\omega_n s + \omega_n^2}{s^2 + \sqrt{2}\omega_n s + \omega_n^2} = R. \tag{6.22}$$

This result is significant because it means that the input impedance of the circuit is constant with respect to frequency. If the bridged T-coil is used on the input of an IC, regardless of how large the ESD capacitance is, the circuit is still matched to the transmission-line impedance. Although the transmission line can be matched to R, we must also consider the transfer function from the input to the voltage across R, which will be all-pass. Due to its phase variation we must still limit the bandwidth of operation. However, the useful bandwidth is much higher than the case without the T-coil.

6.4   Basic Gain Stages

Electrical-link receivers use amplifiers to increase the amplitude of the received signal and compensate for high-frequency channel loss. Since the amount of loss introduced by the channel varies with channel length and other characteristics, the gain of amplifiers must be tuneable. Likewise the frequency response of equalizers must also be tuneable in order to adapt to differences in channel loss but also to compensate for variations in circuit behaviour introduced by process, voltage and temperature variation.

The tuning mechanisms will be discussed where variable-gain amplifiers (VGAs) and the circuit implementation of CTLEs are presented. Here, we discuss the basics of operation of gain cells. The two main types of gain cells are:

- Differential-pair based amplifiers; and
- CMOS inverters.

Each type is discussed in more detail in the next two sections.

6.4.1   Differential-Pair-Based Amplifiers

While the basic differential pair has been discussed in the context of transmitter circuits and decision circuits, its analog characteristics are presented in more detail here. Figure 6.6 (a) shows a differential pair consisting of transistors M_1 and M_2, biased by a current source I_B. Complex load (Z_L) and degeneration (Z_S) impedances are added, allowing us to derive a generalized gain expression applicable to the design of a CTLE and a variable-gain amplifier (VGA).

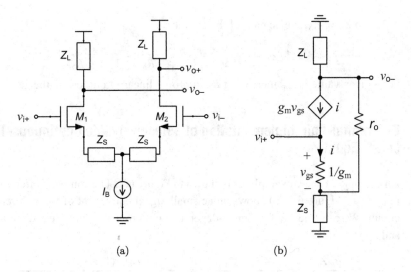

Figure 6.6 Generic schematic of a differential pair with complex degeneration and load impedances: (a) complete circuit; (b) differential half circuit.

Figure 6.6 (b) shows an ac small-signal model of the differential half circuit using the T-model of the MOSFET that will be used to derive the differential gain, $A_d = \frac{v_o}{v_i}$. The MOSFET's dependent source can either be a voltage-controlled current source (VCCS) parameterized by v_{gs} or a current-controlled current source (CCCS) parameterized by i, where i is the current through $1/g_m$. Although MOSFET capacitors have been ignored, their effects can be approximately accounted for by adding the capacitive loading (C_{gs} and C_{gd}) of the next stage to Z_L. If we ignore r_o, the differential gain is easily found by taking the ratio of the impedance of the elements in the drain to the impedance of the elements in the source:

$$A_d = \frac{v_o}{v_i} = \frac{Z_L}{1/g_m + Z_S} = \frac{g_m Z_L}{1 + g_m Z_S}, \tag{6.23}$$

where g_m is the MOSFET's transconductance when biased with $I_B/2$. Meanwhile, Z_L is assumed to be the parallel combination of R_L and C_L. When $Z_S = 0$, the gain is:

$$A_{d,Z_S=0} = g_m Z_L = \frac{g_m R_L}{1 + sR_L C_L}, \tag{6.24}$$

which simplifies to $g_m R_L$ at dc. The two main purposes of using $Z_S \neq 0$ are as follows:

1 $Z_S = R_S$: this can improve the linearity of the differential pair. R_S can also serve as a knob for adjusting the gain of a VGA to what a particular channel's loss requires.
2 Z_S is a parallel combination of R_S and C_S: this gives a frequency-dependent gain, lower at dc and increasing with frequency. This can be used to build a CTLE.

Due to parasitic capacitance, a circuit designed to be a VGA may nevertheless exhibit frequency-dependent peaking as discussed in Section 6.6.

When $Z_S = R_S$, the dc gain of the circuit is:

$$A_{d,Z_S=R_S} = \frac{v_o}{v_i} = \frac{R_L}{1/g_m + R_S} = \frac{g_m R_L}{1 + g_m R_S}. \tag{6.25}$$

For a given value of g_m increasing R_S improves linearity but lowers the gain.

6.5 Differential-Pair Implementation of a Receiver-Side Continuous-Time Linear Equalizer

An example of such an amplifier is shown in Figure 6.7. The complete circuit is shown in part (a). Figure 6.7 (b) shows an ac small-signal equivalent of the differential half circuit. We can find the frequency-dependent gain by substituting expressions for Z_L and Z_S into 6.23:

$$A_d = \frac{g_m Z_L}{1 + g_m Z_S} = \frac{g_m \frac{R_L}{1+sR_L C_L}}{1 + g_m \frac{R_S}{1+sR_S C_S}} = \frac{g_m R_L(1 + sR_S C_S)}{(1 + sR_L C_L)(1 + g_m R_S + sR_S C_S)}. \tag{6.26}$$

An example of (6.26) is plotted in Figure 4.8, showing both the straight-line approximation and the exact magnitude response.

The zero, f_z, and poles (f_{p1}/f_{p2}) are related to the circuit's parameters as follows:

$$f_z = \frac{1}{2\pi R_S C_S}, \tag{6.27}$$

$$f_{p1} = \frac{1}{2\pi \left(R_S || \frac{1}{g_m}\right) C_S}, \tag{6.28}$$

$$f_{p2} = \frac{1}{2\pi R_L C_L}. \tag{6.29}$$

The circuit has a low-frequency gain of:

$$A(0) = \frac{R_L}{\frac{1}{g_m} + R_S}. \tag{6.30}$$

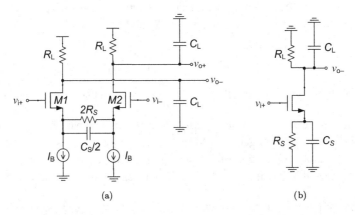

(a)   (b)

Figure 6.7 CML continuous-time linear equalizer: (a) complete circuit; (b) differential half circuit.

Here, $A(f)$ increases starting at f_z. Assuming the poles are widely spaced with $f_{p1} < f_{p2}$, $A(f)$ reaches a constant, mid-band value of:

$$A_{mid} = g_m R_L. \tag{6.31}$$

Beyond the second pole, the gain decreases. As with any pole/zero system the ratio between the low-frequency and midband gains is the same as the ratio between the pole and zero. That is:

$$\frac{A_{mid}}{A(0)} = \frac{f_{p1}}{f_z} = 1 + g_m R_S. \tag{6.32}$$

Notice that although the midband gain is higher than the dc gain, it is no higher than the gain of a resistively loaded differential pair without R_S/C_S. Also, f_{p2} is the same as the pole of a simple differential pair. Indeed, CTLEs are often designed with the gain ratio in (6.32) around 10 dB and a dc gain near 0 dB. In some cases, $A(0) < 1$.

The expressions for the zero and poles are more complicated when r_o is taken into account, but can be easily found through simulation. The effects are summarized as:

1. The zero, f_z, is unchanged.
2. The midband gain drops to $A_{mid} = g_m (R_L \parallel r_o)$.
3. The output pole, f_{p2}, increases because the resistance seen by C_L at high frequency drops from R_L to $R_L \parallel r_o$.
4. The pole set by C_S, f_{p1} moves to lower frequency because the resistance looking into the source of the differential-pair transistors increases, due to r_o.

Suppose a designer anticipates these effects due to r_o and selects a new, larger value of the load resistance, denoted as $R_{L,new}$, such that $R_{L,new} \parallel r_o = R_L$. This restores the midband gain and f_{p2} but the first pole (f_{p1}) remains lower than it was estimated to be ignoring r_o.

The frequency response of the CTLE is usually tuneable by implementing C_S as a bank of switch-selectable capacitors. Additionally, R_S can also be tuneable by similar methods or by using a triode-region MOSFET. A challenge in the design of CTLEs is matching the rational polynomial of the CTLE to the frequency response of the distributed channel. In some cases, additional series RC branches are added that span the sources of M_1 and M_2 to allow a more elaborate frequency response. Regardless of their complexity, CTLEs provide equalization by boosting the gain at high frequency, amplifying both the signal and the noise. The high-frequency boosting can invert the channel over a range of frequency, partially mitigating ISI.

6.6 Variable-Gain Amplifiers

Figure 6.8 (a) shows one of the simplest approaches to VGA design where the differential output is taken at the drains of the MOSFETs. Here, the degeneration resistor R_S is variable, either by switching in and out parallel resistors or by controlling a

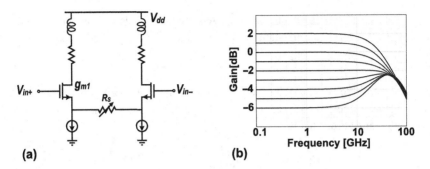

Figure 6.8 Source-degeneration-based variable-gain amplifier. (a) Schematic. The differential output is taken at the drains of the transistors. (b) Frequency response. © 2019 IEEE. Reprinted, with permission, from [3].

triode-region MOSFET's gate voltage in an analog sense. The digital approach is more common. The low-frequency gain of the circuit is:

$$A(0) = \frac{g_{m1}R_L}{1 + g_{m1}R_S/2}. \tag{6.33}$$

The gain of a (differential) common-source amplifier can also be changed by varying the load resistors. The source-degeneration approach is preferred by considering the scenario in which lower, even sub-unity gain is required. Lower required gain indicates a larger input signal, and hence a greater linearity requirement. The linear range of a common-source amplifier without degeneration will be some fraction of the V_{OV} of the differential pair. With degeneration, the linear input range increases as the gain decreases. This is one advantage of this approach. A second advantage is that degeneration reduces the equivalent gate-source capacitance from C_{GS} to $\frac{C_{GS}}{1+g_m R_S/2}$.

On the other hand, adding switched resistors increases the capacitance already present in the source of the circuit. At low gain settings where the equivalent R_S is increased, the zero in the transfer function decreases, leading to peaking in the frequency response, visible in the lower traces in Figure 6.8 (b). Rather than behaving like an amplifier with a low-pass transfer function, the circuit operates like a continuous-time linear equalizer with the transfer function presented in (6.26) with a zero given by:

$$f_z = \frac{1}{2\pi C_{EQ}R_S/2}, \tag{6.34}$$

where C_{EQ} is the equivalent capacitance to ground from the transistor's source. When R_S is increased to lower the gain, f_z also decreases, giving rise to more peaking in the frequency response.

Figure 6.9 (a) shows an alternative approach in which the gain is varied by adjusting the number of slices that connect to the output with positive and negative polarity. This reduces the gain from the maximum gain, achieved when all slices operate with positive polarity down to zero gain when an equal number of slices are connected with positive and negative polarity. The circuit works as follows:

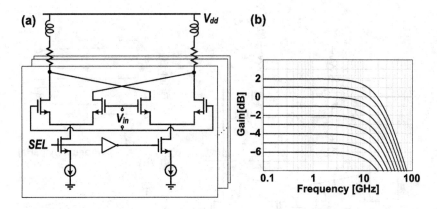

Figure 6.9 Cross-coupled pair-based variable-gain amplifier. (a) Schematic. Differential output is taken at the drains of the differential-pair transistors. (b) Frequency response. © 2019 IEEE. Reprinted, with permission, from [3].

- Assume N identical slices, each controlled by its own select bit, SEL.
- When $SEL = 1$, the left-hand differential pair is connected to the output whereas when $SEL = 0$, the right-hand differential pair is connected to the output, but with opposite polarity.
- If each slice has a transconductance g_m, the maximum gain is Ng_mR_L, achieved when all SEL-bits are 1.
- If $N/2$ SEL-bits are high and $N/2$ are low, the gain reduces to 0, because no net current flows to the loads.
- Making all SEL-bits low gives a gain of $-Ng_mR_L$.

In general, if n of N SEL-bits are high, the gain is:

$$A_d = (2n - N)g_mR_L. \tag{6.35}$$

The frequency response is shown in Figure 6.9 (b). The capacitance at the output node is constant, since the same total number of slices is always connected. The effect of gate-to-drain capacitance gets amplified due to the Miller effect. However, each slice presents two gate-to-drain capacitances per input, which see opposite gain. Since C_{GD} is the result of overlap capacitance, we assume each side of the slice has the same capacitance whether it is on or off. Therefore, for a given overall gain A, each slice presents an equivalent input capacitance of:

$$C_{GD,Eq} = C_{GD}(1 + A) + C_{GD}(1 - A) = 2C_{GD}. \tag{6.36}$$

Therefore the Miller effect is eliminated. Although each input sees two transistors, each having a C_{GD} and G_{GS}, and each output sees two transistors each having a C_{DB} and C_{GD}, the shape of the frequency response is independent of the gain setting. However, the increased capacitance due to having two differential pairs per slice reduces the achievable bandwidth of the circuit. Also, the linear input range does not increase at low gain settings as it does for the degeneration-based design of Figure 6.8.

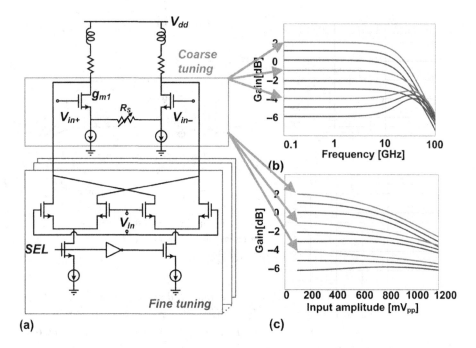

Figure 6.10 Hybrid VGA. (a) Schematic. Differential output is taken at the drains of the differential-pair transistors. (b) Frequency response. (c) Gain as a function of input amplitude. © 2019 IEEE. Reprinted, with permission, from [3].

A hybrid solution can be used, as presented in Figure 6.10 (a). Again, the output is taken at the drains of the differential-pair transistors. Here, the degeneration approach is used for coarse-tuning. Three different coarse gain settings are plotted with grey traces at in Figure 6.10 (b), indicated by grey arrows from (a). Having only a few steps of gain control in the degeneration scheme reduces the complexity of the network in the source of the differential pair, helping to keep the zero at higher frequency, reducing the amount of peaking. The cross-coupled pair approach is used for fine-tuning, which can adjust the gain down from the grey plots in (b). Since most electrical link receivers use a CTLE, the peaking introduced by the degeneration approach can be absorbed into the peaking provided by the CTLE. However, calibration of the link is simplified if fine adjustments to the VGA setting do not require CTLE recalibration. Therefore, using the cross-coupled approach for fine control reduces the impact of peaking variation of the degeneration-based VGA.

Figure 6.10 (c) shows the small-signal gain of the VGA as a function of the input amplitude. A reduction in gain with increasing amplitude indicates nonlinearity. This reduction is most pronounced when the gain is highest, a setting that is used when the input amplitude is smallest, and thus linearity is less important. On the other hand, the lowest gain setting (largest degeneration resistor) leads to the best linearity, which is required when the input amplitude is largest.

6.7   DFE Circuit Implementation

This section presents an overview of the implementation of DFEs. Section 6.7.1 shows a conventional summing junction implementation. The need for wide bandwidth at the summing node is eliminated in the integrating (latching) summer presented in Section 6.7.2.

6.7.1   Conventional Summer Implementation

A schematic of a conventional summer for a DFE is shown in Figure 6.11. The current from the input differential pair and the feedback taps is summed (through KCL) and applied to the load resistors. The input differential pair must operate linearly. Recall that we assume that the received signal is a linear combination of pulse responses. Thus, ISI cancellation relies on the premise that the ISI contributed by a given 1 (or 0) is independent of the signal magnitude resulting from previous or subsequent bits. For this premise to be valid, the signal path up to the latch must be linear. This is achieved either by designing the input differential pair's MOSFETs to have higher V_{OV} or to use source degeneration as shown in Figure 6.11. On the other hand, in an FIR DFE, the differential pairs connected to feedback taps are switched. Their inputs ($d_{i-1+/-}$ and $d_{i-2+/-}$) are full-rail signals (i.e, $0 \rightarrow V_{DD}$ in the case of sense-amplifier based latches or $V_{DD} - I_B R_L \rightarrow V_{DD}$ in the case of CML latches. The strength of each tap is set by its current source ($I_{h1/2}$).

Fully switching the differential pair of FIR DFE taps has two consequences:

1. There is no linearity concern.
2. Since the conducting differential pair transistor sits in series with a current source, it contributes no thermal noise to the summed signal. Noise analysis of circuits is presented in Appendix B.

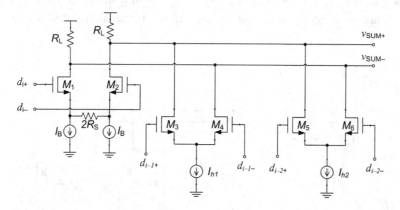

Figure 6.11 Conventional DFE summer circuit. Each tap adds an additional differential pair to the output nodes.

The noise from the current source is added to the summed signal. However, current sources can be designed to have low g_m and thus contribute less noise than the higher g_m input transistors M_1/M_2.

In the case of an IIR DFE, the current associated with the IIR tap must be a linear translation of the decaying exponential of the voltage signal. Therefore, IIR DFE taps are not fully switched as are FIR DFE taps. This means that noise from the differential pair transistors appears at the output of the summer and the differential pair must operate linearly.

The output of the summer is loaded by the drain-body capacitance of every differential pair as well as the input to the latch. This loading can be significant, requiring a smaller load resistance to allow sufficient bandwidth. Without sufficient bandwidth, the summer introduces ISI, negating some its equalization capability. An alternative approach is presented in Section 6.7.2.

6.7.2   Integrating Summer Implementation

Figure 6.12 shows a summer without load resistors. Instead, it integrates the currents from the differential pairs on the capacitance connected to nodes $V_{SUM+/-}$. So that this operation does not introduce ISI, the loads are reset each UI.

When a clock signal activates a sense-amplifier latch (Figure 3.14), currents are integrated on parasitic capacitance (at nodes $v_{O+/-}$), developing a differential voltage. As this voltage increases, regeneration causes the signal to grow exponentially. Rather than have a separate latch circuit to perform regeneration, regeneration can be built into the summer. Figure 6.12 [33] shows an example of a summer in an IIR DFE for well-behaved electrical link channels. Its differential input, $d_{i+/-}[n]$ is a sampled and held input, applied to the input for one UI at a time. The output of this differential pair is connected to three others:

- FIR tap h1: this tap, controlled by $d_{i-1+/i}$ is fully switched. Its weight is set by implementing I_{h1} as a digitally controlled current source, drawn as I_{h1}. All of

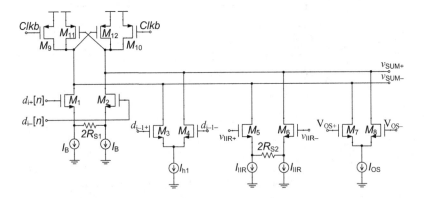

Figure 6.12 IIR DFE summer showing FIR and IIR taps as well as slicing built into the summer, based on [33].

this current is pulled from either V_{SUM} or $\overline{V_{SUM}}$ depending on the polarity of its input.

- IIR tap h2: this tap is driven by an IIR voltage ($v_{IIR+/-}$) and must linearly reproduce this voltage in a differential current applied to the summer's output. The transconductance of this differential pair is matched to that of the input pair.
- Offset compensation tap: this tap is controlled by a two-valued input ($V_{OS+/-}$) that sets the polarity of offset compensation. The amount of offset compensation is set by a digitally controlled current source (I_{OS}).

When the CLK signal is low, the output of the summer is reset. In this case, both outputs are pulled up to V_{DD} through PMOS transistors ($M_{9/10}$). When CLK is raised, the output of the summer will pull both output nodes down. However, one node will be pulled lower. The summer is loaded by a PMOS cross-coupled pair that will regenerate the differential output voltage.

A cross-coupled pair is linear for small input voltages and short regeneration times, although, since the voltage across the transistors saturates to V_{DD} and $V_{DD} - \Delta V$, the circuit is nonlinear as the regeneration time increases. It may seem counterintuitive to have a nonlinear circuit embedded in a circuit that we have argued must operate linearly to accurately cancel ISI. What is important is to have a differential current that is a linear representation of the input signal, from which the various taps can be added/subtracted. Also, any IIR tap must linearly convert the IIR voltage to a differential current for ISI cancellation in the summer. In a conventional summer the overall differential current is converted to a differential voltage and thresholded by a decision circuit. In Figure 6.12 the differential current is the signal that is thresholded by the cross-coupled PMOS transistors.

6.8   Inverter-Based Amplifiers

The use of inverters for electrical-link transmitters was presented in Section 5.6. Although output impedance is a significant consideration in that context, the general approach was to fully switch the transistors. Thus, a digital understanding of inverters explains the fundamentals of an SST driver. To appreciate the use of an inverter as an amplifier, the analog aspects of the circuit must be explored. Since inverter-based amplifiers are absent from many electronics courses, this section begins with the fundamentals.

Figure 6.13 (a) shows a CMOS inverter driven by a voltage source, v_{IN}, and loaded by a capacitor, C_L. Its voltage-transfer characteristic (VTC) is shown in Figure 6.13 (b), where the output voltage of the inverter is plotted as its input voltage is swept. The threshold of the inverter corresponds to the voltage where $V_{IN} = V_{OUT}$. The slope of the VTC reaches a maximum magnitude of A_V near its threshold. The small-signal gain of the circuit around an input dc bias of V_{IN} is the slope of its VTC at V_{IN}.

Figure 6.14 (a) shows the schematic of the inverter while Figure 6.14 (b) shows its ac small-signal model. The transconductances and output conductances of the NMOS

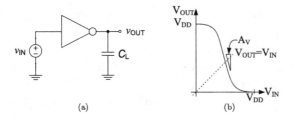

Figure 6.13 CMOS inverter (a) symbol and (b) static voltage-transfer characteristic.

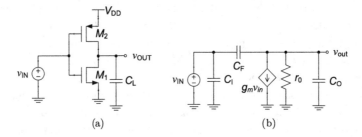

Figure 6.14 CMOS inverter: (a) schematic and (b) ac small-signal equivalent circuit.

and PMOS add to give an overall transconductance g_m and overall output conductance g_o, shown as a resistor r_0. The low-frequency gain of the circuit is given by:

$$A = -\frac{g_m}{g_o} = -g_m r_o, \tag{6.37}$$

where r_o is the parallel combination of the output resistances. The maximum achievable gain is limited by the MOSFETs' intrinsic gain. Ignoring the effect of C_F, the bandwidth of this circuit is:

$$f_{bw} = \frac{1}{r_o C_O}, \tag{6.38}$$

where C_O is the circuit's total load capacitance including self-loading (C_{DB} of the two MOSFETs) and the capacitance of the next stage. Since g_m and r_o are bias-dependent small-signal parameters, the inverter's gain and bandwidth, presented in (6.37) and (6.38) are also bias-dependent. The inverter's large-signal operation is presented in the next section.

6.8.1 Large-Signal Operation of a CMOS Inverter

To explore the use of inverters in a wider variety of contexts, we will derive the voltage-to-current conversion of an inverter when connected to a low impedance, as shown in Figure 6.15 (a). We assume that the NMOS and PMOS are matched, having $\beta = \mu_n C_{OX} \frac{W_N}{L_N} = \mu_p C_{OX} \frac{W_p}{L_p}$ and that there is no channel-length modulation ($g_o = 0$). Both PMOS and NMOS transistors are in saturation. The current sourced by the inverter is given by:

$$i_O = i_P - i_N, \tag{6.39}$$

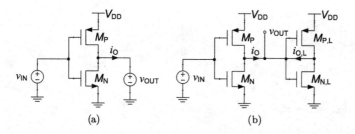

Figure 6.15 Schematics for CMOS inverter analysis: (a) I–V characteristic of one inverter and (b) one inverter loaded by a diode-connected inverter.

$$= \frac{\beta_p}{2} \left(V_{DD} - v_{IN} - V_{tp}\right)^2 - \frac{\beta_n}{2} \left(v_{IN} - V_{tn}\right)^2, \tag{6.40}$$

$$= \frac{\beta}{2} \left(V_{DD} - V_{tp} - V_{tn}\right)\left(V_{DD} - 2v_{IN} - V_{tp} + V_{tn}\right). \tag{6.41}$$

Although the transistors are quadratic, the output current is linear in v_{IN}. To make the expressions simpler, we will assume $V_{tn} = V_{tp} = V_t$, giving:

$$i_O = \frac{\beta}{2} \left(V_{DD} - 2V_t\right)\left(V_{DD} - 2v_{IN}\right). \tag{6.42}$$

This output current is 0 when $v_{IN} = V_{DD}/2$. This is the main term that will change if the threshold voltages are different. Now suppose, as shown in Figure 6.15 (b), that this current is applied to a diode-connected inverter acting as a load with the same matching assumptions, but with $\beta = \beta_L$. The output current of the load inverter $i_{O,L}$, is equal to the negated output current of the M_P/M_N inverter. That is, $i_{O,L} = -i_O$. We denote its voltage as v_{OUT}. This gives I–V expressions for both inverters as follows:

$$i_O = \frac{\beta}{2} \left(V_{DD} - 2V_t\right)\left(V_{DD} - 2v_{IN}\right), \tag{6.43}$$

$$i_{O,L} = \frac{\beta_L}{2} \left(V_{DD} - 2V_t\right)\left(V_{DD} - 2v_{OUT}\right). \tag{6.44}$$

which can be equated to give:

$$\frac{\beta_L}{2} \left(V_{DD} - 2V_t\right)\left(V_{DD} - 2v_{OUT}\right) = -\frac{\beta}{2} \left(V_{DD} - 2V_t\right)\left(V_{DD} - 2v_{IN}\right), \tag{6.45}$$

$$\left(V_{DD} - 2v_{OUT}\right) = -\frac{\beta}{\beta_L} \left(V_{DD} - 2v_{IN}\right). \tag{6.46}$$

Rewriting input and output voltages as signal voltages around a dc bias of $V_{DD}/2$ gives $v_{OUT} = v_{out} + V_{DD}/2$ and $v_{in} = v_{in} + V_{DD}/2$ resulting in:

$$-2v_{out} = -\frac{\beta}{\beta_L} \left(-2v_{in}\right), \tag{6.47}$$

$$\frac{v_{out}}{v_{in}} = -\frac{\beta}{\beta_L}, \tag{6.48}$$

which, because each inverter has linear I–V behaviour, is also linear. Although (6.48) is written using "signal" notation, frequently associated with small-signal analysis, the

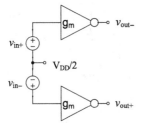

Figure 6.16 Two CMOS inverters used pseudo-differentially.

gain is linear so long as both transistors are in saturation. This criterion is explored in Section 6.9.

Note that if inverters exhibit nonlinear I–V behaviour, some of the nonlinearity is cancelled. The gain in (6.48) can be rewritten in terms of small-signal parameters as:

$$\frac{v_{out}}{v_{in}} = -\frac{g_m}{g_{m,L}},\tag{6.49}$$

where each g_m is the sum of the respective inverters NMOS and PMOS transconductances at the $v_{IN} = v_{OUT}$ bias point.

6.8.2   Techniques for Adding Common-Mode Rejection

Two instances of this single-ended circuit can be combined into a pseudo-differential circuit shown in Figure 6.16. Since each inverter is biased with the same input dc voltage of $V_{DD}/2$, each inverter has the same gain (A). The overall differential and common-mode gains of the pair of inverters is summarized as follows:

$$A_d = \frac{v_{out+} - v_{out-}}{v_{in+} - v_{in-}} = A,\tag{6.50}$$

$$A_{cm \to d} = \frac{v_{out+} - v_{out-}}{0.5(v_{in+} + v_{in-})} = A_1 - A_2 = 0,\tag{6.51}$$

$$A_{cm \to cm} = \frac{0.5(v_{out+} + v_{out-})}{0.5(v_{in+} + v_{in-})} = -A,\tag{6.52}$$

where $A_{cm \to d}$ is the gain from an input common-mode signal to the differential output and $A_{cm \to cm}$ is the gain from an input common-mode signal to the output common-mode signal. Using two inverters pseudo-differentially can eliminate $A_{cm \to d}$ provided the two inverters have the same gain, which requires there to be no mismatch between the two inverters, in addition to them having the same dc input voltage. Reasonably low conversion from common-mode inputs to differential outputs is possible. However, we see that the circuit has no inherent rejection of common-mode inputs.

The pseudo-differential arrangement in Figure 6.16 lacks common-mode rejection because its common-mode and differential half circuits are the same. To address the issue of common-mode rejection extra inverters can be added giving rise to the circuit in Figure 6.17 (a). The use of diode-connected and cross-coupled inverters

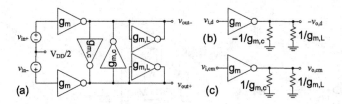

Figure 6.17 Analysis of Nauta's circuit: (a) complete circuit; (b) differential half circuit; (c) common-mode half circuit.

goes back to 1976 [11], but was analyzed in greater detail by Nauta, particularly as transconductors in [13] [14]. The circuit is frequently referred to as Nauta's circuit.

Figure 6.17 (a) adds diode-connected "load" inverters, denoted by $g_{m,L}$, and cross-coupled inverters, denoted by $g_{m,c}$, to the outputs of the pseudo-differential inverters. The differential and common-mode half circuits are shown in parts (b) and (c), respectively. Since each load inverters' input and output are connected to the same node, they present a resistance of $1/g_{m,L}$ for both common-mode and differential half circuits. The cross-coupled inverters, however, behave differently for differential and common-mode signals. They act like load inverters for common-mode voltages, since we assume that due to symmetry both outputs move together, giving the appearance that both input and output of each inverter are connected to the same terminal. Thus they load the output like a resistor of $1/g_{m,c}$. However, for differential signals, the signal at the input of the cross-coupled inverter moves opposite to the signal at the output, giving the appearance of a negative resistance equal to $-1/g_{m,c}$. The gains are derived as follows, ignoring the output conductance of each inverter:

$$A_d = \frac{v_{o,d}}{v_{i,d}} = \frac{g_m}{g_{m,L} - g_{m,c}}, \tag{6.53}$$

$$A_{cm \to cm} = \frac{v_{o,cm}}{v_{i,cm}} = \frac{g_m}{g_{m,L} + g_{m,c}}, \tag{6.54}$$

$$CMRR = \frac{A_d}{A_{cm \to cm}} = \frac{g_{m,L} + g_{m,c}}{g_{m,L} - g_{m,c}}. \tag{6.55}$$

This topology has the following features:

1. The common-mode gain is much smaller than the differential gain.
2. If $g_{m,L} = g_{m,c}$ large differential gain is possible, with the gain being limited by the output conductance of the inverters.
3. The effect of the inverters' output conductance can be removed by increasing $g_{m,c}$ to be slightly larger than $g_{m,L}$ coming at the risk of implementing a latch at the output.
4. Gains less than $\frac{g_m}{g_o}$ can be implemented predictably, since target ratios of g_m, determined by transistor width ratios, hold up well across process, voltage and temperature variations.
5. When used as a differential transconductor, the differential output current is linear even with inverters having mismatched β.

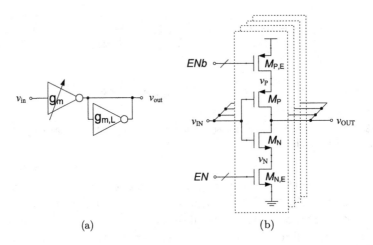

(a)   (b)

Figure 6.18 Varying the gain of inverter-based circuits: (a) complete circuit; (b) variable inverter.

6. Adding cross-coupled inverters adds extra capacitance to the output which will lower the circuit's bandwidth.

6.8.3 Adding Tunability to Inverter-Based Circuits

Before discussing the details of inverter-based amplifiers and CTLEs, the tools for varying the "strength" of an inverter must be discussed. In general, the adjustment of inverters is done by changing its transconductance.

Figure 6.18 (a) shows a single-ended portion of an inverter-based amplifier. As discussed above, its gain is $A = \frac{g_m}{g_{m,L}}$. However, g_m is an adjustable quantity. Figure 6.18 (b) shows how this is done. The overall inverter implementing g_m is subdivided into N slices, each enabled or disabled through $M_{P,E}$ and $M_{N,E}$, whose gates are controlled digitally. The NMOS network of each enabled slice has an effective transconductance ($g_{m,Ni}$) given by:

$$g_{m,Ni} = \frac{g_{m,N}}{1 + g_{m,N} r_{ds_{N,E}}},\tag{6.56}$$

$$\text{where } r_{ds_{N,E}} = \left(\frac{\partial i_d}{\partial v_{DS}}\right)^{-1} = \frac{1}{\mu_n C_{OX}\frac{W_{N,E}}{L_{N,E}}V_{OVE,N}}\tag{6.57}$$

$$\text{and } g_{m,N} = \mu_n C_{OX}\frac{W_N}{L_N}V_{OVN},\tag{6.58}$$

$$\text{simplifying to } g_{m,Ni} = \frac{\mu_n C_{OX}\frac{W_N}{L_N}V_{OVN}}{1 + \frac{W_N}{W_{N,E}}\frac{V_{OVN}}{V_{OVE,N}}}.\tag{6.59}$$

It is assumed minimum length transistors are used. Since the inverter's input is biased near $V_{DD}/2$, the overdrive voltage of the main NMOS transistor $V_{OVN} = V_{DD}/2 - V_N - V_t$ whereas $V_{OVN,E} = V_{DD} - V_t$. Therefore, $V_{OVN,E} \gg V_{OVN}$. Even for equal-sized

M_N and $M_{N,E}$, the effect of source degeneration on the inverter is relatively small. However, the overdrive voltage of M_N is reduced from what it was without an enable transistor because its source is no longer at ground, but sits at a voltage V_N.

The slice-based transconductor is designed with β matching meaning that:

$$\mu_n C_{OX} \frac{W_N}{L_N} = \mu_p C_{OX} \frac{W_P}{L_P} \tag{6.60}$$

$$\text{and } \mu_n C_{OX} \frac{W_{N,E}}{L_{N,E}} = \mu_p C_{OX} \frac{W_{P,E}}{L_{P,E}}. \tag{6.61}$$

Therefore, the overall transconductance of the slice is twice that given in (6.59).

The discussion above shows that a digitally controlled transconductor can be designed. Regardless of the state of the enable bits, the input sees the gate capacitance of all slices. The gate capacitance is larger when the channel is inverted, which occurs when the slice is active. Thus an OFF slice presents less input capacitance, but nevertheless loads the previous stage.

6.9   Inverter-Based Front-End Amplifiers

Two copies of the structure in Figure 6.18 (a) can be used as a differential programmable gain amplifier or PGA. Cross-coupled inverters can be added to the output to give common-mode rejection. Although it has been argued that the circuit is linear, the required assumptions are considered here. All discussion of inverter-based amplifiers required that the signal path transistors were in saturation. Assume the β-matched scenario where input and output are biased at $V_{DD}/2$ and the circuit has a gain of A. Transistor M_N is in saturation so long as:

$$V_{DS} > V_{GS} - V_t, \tag{6.62}$$

$$\text{or } V_D > V_G - V_t, \tag{6.63}$$

$$\text{where } v_{OUT} > v_{IN} - V_t, \tag{6.64}$$

$$V_{DD}/2 + v_{out} > V_{DD}/2 + v_{in} - V_t, \tag{6.65}$$

$$Av_{in} > v_{in} - V_t, \tag{6.66}$$

$$v_{in} < \frac{V_t}{1 - A}. \tag{6.67}$$

Therefore, a unity gain amplifier ($A = -1$) can only have an input amplitude of less than $V_t/2$. Some process technologies feature transistors with a standard V_t as well as high-V_t and low-V_t transistors. In some contexts, low-V_t transistors can offer higher speed because their maximum overdrive voltage is larger. However, for inverter-based designs low-V_t transistors have lower input range.

6.9.1   Input Impedance of the Diode-Connected Load Inverter

Figure 6.19 (a) shows the schematic of a diode-connected inverter. To understand the frequency response of inverter-based amplifiers we need to derive the input impedance

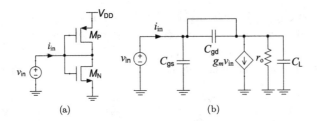

Figure 6.19 Input impedance of a diode-connect CMOS inverter: (a) schematic; (b) ac small-signal schematic.

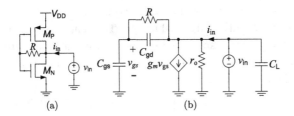

Figure 6.20 Input impedance of a CMOS inverter used as an active inductor: (a) schematic; (b) ac small-signal schematic

of the load inverter. In Section 6.9.2, an inductive-peaking technique that does not require passive inductors is presented. Here we explore a technique suitable for inverter-based amplifiers. Figure 6.19 (b) shows the ac small-signal schematic of the diode-connected inverter. Note that C_{gd} is shorted out and the circuit has one non-reference node, v_{in}. The load capacitance C_L includes C_{db} of the diode-connected inverter and whatever other capacitance is present. The input current is computed as:

$$i_{in} = v_{in} \left(s \left(C_{gs} + C_L \right) + g_o + g_m \right), \tag{6.68}$$

giving an input impedance of:

$$Z_{in} = \frac{v_{in}}{i_{in}} = \frac{1}{s \left(C_{gs} + C_L \right) + g_o + g_m}, \tag{6.69}$$

which gives a bandwidth of approximately $\frac{g_m}{C_{gs}+C_L}$. Assuming $C_L \approx C_{gs}$ simplifies the bandwidth to $\omega_T/2$ where ω_T is the unity-gain frequency of the combined NMOS and PMOS structure, biased at $V_{DD}/2$.

6.9.2   Active-Inductor-Based Bandwidth Extension in Inverter-Based Amplifiers

Figure 6.20 (a) shows the schematic of a bandwidth-extension technique. We will show that because the circuit's input impedance increases with frequency, exhibiting inductive behaviour, we can use it as an *active* inductor. To derive the input impedance, now looking into the drains, we will ignore r_o, C_{gd} and C_L. Now,

$$i_{in} = g_m v_{gs} + i_{rc}, \tag{6.70}$$

$$\text{where } i_{rc} = \frac{v_{in}}{R + 1/sC_{gs}} = v_{in}\frac{sC_{gs}}{sRC_{gs} + 1}, \tag{6.71}$$

$$\text{and } v_{gs} = i_{rc}\frac{1}{sC_{gs}} = v_{in}\frac{1}{sRC_{gs} + 1}, \tag{6.72}$$

$$\text{therefore: } i_{in} = v_{in}\left(\frac{g_m}{sRC_{gs} + 1} + \frac{sC_{gs}}{sRC_{gs} + 1}\right), \tag{6.73}$$

$$Z_{in} = \frac{v_{in}}{i_{in}} = \frac{1 + sRC_{gs}}{g_m + sC_{gs}}. \tag{6.74}$$

With these assumptions, the following conclusions can be drawn regarding the magnitude and the pole and zero of the impedance:

$$Z_{in}(0) = \frac{1}{g_m}, \tag{6.75}$$

$$Z_{in}(s \to \infty) = R, \tag{6.76}$$

$$\omega_z = \frac{1}{RC_{gs}}, \tag{6.77}$$

$$\omega_p = \frac{g_m}{C_{gs}}. \tag{6.78}$$

For $\omega < \omega_p$ we can approximate Z_{in} as:

$$Z_{in} \approx \frac{1 + sRC_{gs}}{g_m} = \frac{1}{g_m} + \frac{sRC_{gs}}{g_m}, \tag{6.79}$$

$$\approx \frac{1}{g_m} + sL_{eq}, \tag{6.80}$$

$$\text{where } L_{eq} = \frac{RC_{gs}}{g_m}. \tag{6.81}$$

Hence, based on (6.80) and (6.81) the structure behaves like the series combination of a resistance $(1/g_m)$ and an inductance $(\frac{RC_{gs}}{g_m})$.

To find the frequency response of the structure in Figure 6.18 (a) when $g_{m,L}$ is replaced by an active inductor, the effect of the parallel load capacitance (C_L) is added to (6.74), giving a total impedance of:

$$Z_{tot} = \frac{1 + sRC_{gs}}{g_m + sC_{gs}} \parallel \frac{1}{sC_L}, \tag{6.82}$$

$$= \left(\frac{g_m + sC_{gs}}{1 + sRC_{gs}} + sC_L\right)^{-1}, \tag{6.83}$$

$$= \left(\frac{g_m + sC_{gs}}{1 + sRC_{gs}} + \frac{(1 + sRC_{gs})sC_L}{1 + sRC_{gs}}\right)^{-1}, \tag{6.84}$$

$$= \left(\frac{g_m + s(C_L + C_{gs}) + s^2RC_{gs}C_L}{1 + sRC_{gs}}\right)^{-1}, \tag{6.85}$$

$$= \frac{1 + sRC_{gs}}{g_m + s(C_L + C_{gs}) + s^2RC_{gs}C_L}. \tag{6.86}$$

Rewriting it in standard form:

$$Z_{tot} = \frac{\frac{1}{RC_{gs}C_L}(1 + sRC_{gs})}{s^2 + s\frac{1}{RC_{eq}} + \frac{g_m}{RC_{gs}C_L}} = \frac{\frac{1}{RC_{gs}C_L}(1 + sRC_{gs})}{s^2 + 2\zeta\omega_n s + \omega_n^2}, \tag{6.87}$$

$$\text{where } C_{eq} = \frac{C_{gs}C_L}{C_{gs} + C_L}, \tag{6.88}$$

$$\omega_n = \sqrt{\frac{g_m}{RC_{gs}C_L}}, \tag{6.89}$$

$$\text{and } \zeta = \frac{C_{gs} + C_L}{2\sqrt{C_{gs}C_L g_m R}}. \tag{6.90}$$

Example 6.1   Consider a unity-gain buffer consisting of a transconductance inverter and a load inverter of equal g_m. Assume $C_{gs} = C_L$ and $\omega_T = \frac{g_m}{C_{gs}} = 2\pi \times 20$ GHz. Before employing the active-inductor technique for the load inverter, find the frequency response and pulse response for 50 Gb/s data. Determine the best value for R, relative to $1/g_m$ such that the VEO is maximized. What is the best-case normalized VEO?

Solution
For the case without the use of the active inductor, (6.69) showed that when $C_{gs} = C_L = C$ the load has a 3dB bandwidth of half the unity-gain frequency, ω_T. In this case, the bandwidth is only 10 GHz, which gives significant eye closure for 50 Gb/s data. Multiplying the impedance in (6.86) by g_m to get a voltage gain and rewriting it in terms of $\omega_T = \frac{g_m}{C}$ and $g_m R$, we get:

$$A_v(s) = \frac{\frac{\omega_T^2}{g_m R}(1 + s g_m R/\omega_T)}{s^2 + s\frac{2\omega_T}{g_m R} + \frac{\omega_T^2}{g_m R}} = \frac{\omega_n^2(1 + s/\omega_z)}{s^2 + 2\zeta\omega_n s + \omega_n^2} \tag{6.91}$$

$$\text{where } \omega_z = \frac{\omega_T}{g_m R}, \tag{6.92}$$

$$\omega_n = \frac{\omega_T}{\sqrt{g_m R}}, \tag{6.93}$$

$$\text{and } \zeta = \frac{1}{\sqrt{g_m R}}. \tag{6.94}$$

The particular case of $g_m R = 1$ is highlighted in [42]. This leads to $\omega_n = \omega_T$ and $\zeta = 1$. Thus the poles are real and repeated. However, one pole is cancelled by the zero, which is also equal to ω_T. The active inductor mitigates the C_{gs} loading of the diode-connected inverter. However, some additional bandwidth is possible, by slightly increasing $g_m R$.

Figure 6.21 (a) shows the frequency response of the unity-gain buffer with various values of $g_m R$. The thick dashed line shows the case for $R = 0$. Increasing R leads to more peaking and wider bandwidth. The pulse response is shown in Figure 6.21 (b). Since the denominator is second-order, the pulse response can have ringing for low ζ.

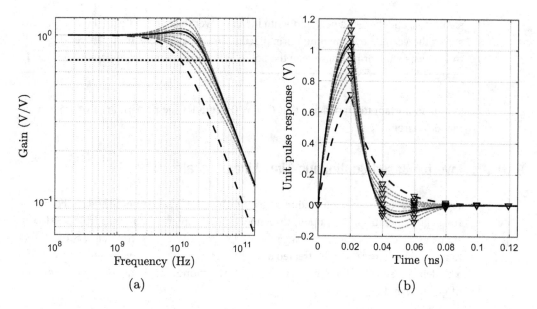

Figure 6.21 Behaviour of unity-gain inverter-based buffer using an active-inductor load. Thick dashed line for $R = 0$. Thin dashed lines for increasing R and the thick solid line is for the best response based on the pulse response: (a) frequency response; (b) pulse response for 50 Gb/s data.

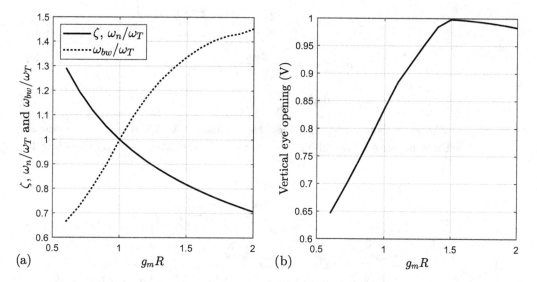

Figure 6.22 Behaviour of unity-gain inverter-based buffer using an active-inductor load: (a) ζ, ω_n/ω_T and ω_{bw}/ω_T as a function of $g_m R$; (b) vertical eye opening as a function of $g_m R$.

However, due to the zero, the dynamics at $t = 0^+$ resemble a first-order response with no pre-cursor ISI.

Figure 6.22 (a) shows how the second-order parameters vary with $g_m R$. Both ζ and the ratio of ω_n/ω_T follow the same curve. The effect of smaller ζ outweighs the smaller ω_n and there is a net bandwidth increase over this range of g_mR. Figure 6.22 (b) shows the vertical eye opening normalized to the peak-to-peak input. Increasing g_mR to 1.5 gives a better response than the pole-zero cancellation design point of $g_mR = 1$.

It is important to emphasize that $C_{gd} \neq 0$ degrades the performance of the active-inductor circuit.

6.10   Inverter-Based Continuous-Time Linear Equalizers

Figure 6.23 (a) shows an additive inverter-based CTLE from [42][43]. For low-frequency operation C_z is open, disconnecting the upper signal path from the lower signal path. Thus the circuit has a gain of $-g_{m1}/g_{m,L}$. When the frequency is increased to the point where C_z can be treated as a short circuit, the gain increases to $\frac{-g_{m1}+g_{m2}}{2g_{m,L}}$. The detailed transfer function can be found by writing nodal equations at the output of g_{m2} (v_2) and v_{out}:

$$\begin{bmatrix} sC_z + g_{mL} & -sC_z \\ -sC_z & sC_z + g_{mL} \end{bmatrix} \begin{bmatrix} v_2 \\ v_{out} \end{bmatrix} = \begin{bmatrix} -g_{m2}v_{in} \\ -g_{m1}v_{in} \end{bmatrix}. \tag{6.95}$$

Using Cramer's rule to solve for v_{out} leads to:

$$\frac{v_{out}}{v_{in}} = -\frac{g_{m1}}{g_{m,L}} \frac{sC_z(1 + \frac{g_{m2}}{g_{m1}})/g_{m,L} + 1}{s\frac{2C_z}{g_{m,L}} + 1}, \tag{6.96}$$

$$\text{giving } \omega_z = \frac{g_{m,L}}{C_z(1 + \frac{g_{m2}}{g_{m1}})} \tag{6.97}$$

$$\text{and } \omega_p = \frac{g_{m,L}}{2C_z}. \tag{6.98}$$

Zheng, the author of [42] argues that the additive CTLE is both power-efficient and suitable for NRZ signalling, but is not sufficiently linear for PAM4 signalling. He

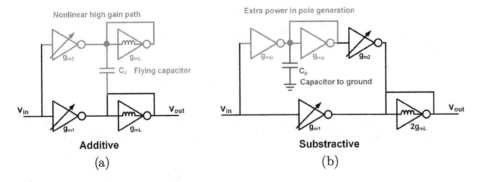

Figure 6.23 Inverter-based CTLEs that (a) add a high-gain signal path at high frequencies and that (b) subtract a low-frequency path from a wideband path. ©2018 IEEE. Reprinted, with permission, from [43].

proposes the use of the subtractive equalizer shown in Figure 6.23 (b). In this design, the upper path is a low-pass filter with a pole at $\omega_p = \frac{g_{mp}}{C_p}$. The upper and lower paths combine to give a total current into the load transconductor, g_{mL} given by:

$$i_{tot} = -g_{m1}v_{in} + g_{m2}\frac{1}{s/\omega_p + 1}v_{in}. \qquad (6.99)$$

The output voltage can be found by dividing i_{tot} by $2g_{m,L}$ to provide a gain of:

$$\frac{v_{out}}{v_{in}} = -\frac{g_{m1}}{2g_{m,L}}\frac{s/\omega_p + 1 - \frac{g_{m2}}{g_{m1}}}{s/\omega_p + 1}, \qquad (6.100)$$

$$\text{giving } \omega_p = \frac{g_{mp}}{C_p} \qquad (6.101)$$

$$\text{and } \omega_z = \omega_p\left(1 - \frac{g_{m2}}{g_{m1}}\right). \qquad (6.102)$$

Somewhat counterintuitive is that the additive equalizer requires a higher maximum stage gain than the subtractive equalizer. Consider the case of a low-frequency gain of unity (0 dB) and a high-frequency gain of A ($20\log_{10}A$ in dB). From (6.96) we see that for unity dc gain, $g_{m1} = g_{m,L}$ and as $s \to \infty$:

$$\frac{v_{out}}{v_{in}} = -\frac{g_{m1}}{2g_{m,L}}\left(1 + \frac{g_{m2}}{g_{m1}}\right). \qquad (6.103)$$

Setting $g_{m1} = g_{m,L}$ and $\left|\frac{v_{out}}{v_{in}}\right| = A$, we can write:

$$A = \frac{1}{2}\left(1 + \frac{g_{m2}}{g_{m1}}\right), \qquad (6.104)$$

$$g_{m2} = g_{m1}(2A - 1). \qquad (6.105)$$

This means that the upper signal path has a gain of $2A - 1$. On the other hand, the subtractive equalizer's lower path consisting of $g_{m1}/2g_{mL}$ implements a gain of A. Even for a small gain, such as $A = 2$, the gain of the additive equalizer's upper path is larger than that in the subtractive equalizer. Thus, the larger gain in the additive equalizer can lead to nonlinearity.

6.11 Mismatch and Offset Detection and Correction

The correction of offsets downstream in receivers was discussed in the context of decision circuits in Chapter 3. If all gain stages from the receiver's input to the latch were linear, offset correction could be performed in only one location in the receiver, such as the input of the latch. However, input-referred offsets from the first stage in the receiver can be comparable to the input signal. If left uncorrected, once amplified by the receiver, this offset will push the operating point of downstream circuits away from their optimum values or even saturate the receiver. Therefore, offset correction should be considered earlier in the receiver's signal path.

6.12   Summary

This chapter presented the transistor-level details of electrical-link receiver circuits, starting with input termination. We noted that establishing input common-mode voltages with resistive dividers can undo much of the power dissipation advantages of SST transmitters. The use of T-coils to give a wide-band resistive match was analyzed.

CML variable-gain amplifiers are often based on a degenerated differential pair, meriting their generalized analysis. The specific cases of VGAs and CTLEs were discussed, along with digitally controlled VGAs and hybrid implementations. CML implementation of the circuits in DFEs (summer and feedback taps) was also presented.

The next sections dug into the details of inverter-based amplifiers considering large-signal operation as linear transconductors and linear loads. The pseudo-differential implementation followed, allowing for common-mode rejection. Digitally controlled gain was implemented by enabling slices of a transconductor. To successfully use inverter-based amplifiers in a receiver front-end, their high-frequency behaviour was considered, along with an active inductor scheme. Inverter-based amplifiers were wrapped up with the presentation of two approaches to implement an inverter-based CTLE.

Problems

6.1   Design a T-coil input matching circuit for an ESD capacitance of 800 fF. Show the pulse response for a 40 ps UI and compute the eye closure from ISI.

6.2   Consider splitting the ESD capacitance used in Problem 6.1 into two capacitors, each of 400 fF. Apply T-coils to each as shown in Figure 6.24. Since Z_1 was shown to be R across all frequencies, the right-hand T-coil loads the left-hand one as a resistor R. Compute values for the elements and explore the pulse response to v_i, $v_{o,3}$ and $v_{o,4}$. If ESD considerations allow the receiver input to be taken from any of these nodes, where should we connect the rest of the receiver?

6.3   Use a voltage divider to generate $v_{RX,CM}$ in Figure 6.2 as shown in Figure 6.25. In this problem, the sizing of the capacitor C is considered. It was argued in Section 6.2 that the receiver's input impedance in Figure 6.1 was infinite for common-mode (CM)

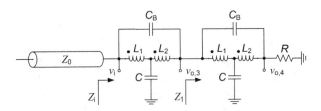

Figure 6.24 Use of T-coils for input matching where the ESD capacitance is split into two equal capacitors of C.

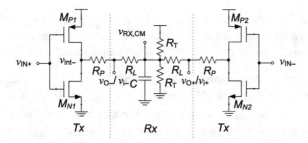

Figure 6.25 SST driver connected to a receiver input termination with input resistors of R_L biased with a voltage divider and a by-pass capacitor C.

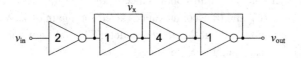

Figure 6.26 Inverter-based amplifier chain. Numbers indicate the relative width of the transistors in each inverter.

inputs. At high-frequency, where the capacitor can be treated as a short circuit, the CM input impedance of Figure 6.25 is R_L but increases to $R_L + R_T$ at dc. Note that the parallel combination of the two R_T resistors is split in two when we consider the common-mode half circuit. If $R_L = 50\ \Omega$ and $R_T = 500\ \Omega$ find the value of C required if $|Z_i| \leq \sqrt{2} \times 50\ \Omega$ at 10 MHz.

6.4 Replace the voltage dividers in Figure 6.3 with diode-connected inverters, sized with matched P/N and so that $R_L = \frac{1}{g_m} = 50\ \Omega$ at the inverters' threshold. If square-law operation is assumed with $V_{DD} = 4V_t$ by how much does the termination's small-signal impedance change as the receiver's input voltage varies from $V_{DD}/4$ to $3V_{DD}/4$?

6.5 Design a differential-pair-based CTLE (Figure 6.7) for unity dc gain and a zero at 500 MHz with at least 10 dB of peaking. Aim for an octave (a factor of 2 in frequency) between the poles. Assume the circuit is loaded by a differential pair of equal transistor sizes.

(a) Give element values and determine the frequency of the poles.

(b) What is the maximum amplitude of input sinusoid that gives 5% THD at frequencies below the zero?

(c) What is the maximum amplitude of input sinusoid that gives 5% THD at the geometric mean of the poles?

6.6 The circuit shown in Figure 6.26 is used as an amplifier. The inverters labelled with "1" have transistors with $W_P = 2W_N = 2\ \mu$m. The inverter labelled with "2" has transistors that are twice as wide. All transistors are minimum length. The "1" inverter has a total transconductance of 1 mA/V. It has an input (C_{gs}) and output capacitance (C_{db}) of 3 fF. For this problem you can assume $r_o \rightarrow \infty$.

(a) Find the small-signal low-frequency gain of the circuit $A_v = \frac{v_{out}}{v_{in}}$.

(b)   Determine the frequency-dependent gain of $A_{v1} = \frac{v_x}{v_{in}}$. What is the bandwidth of
      the circuit from input to v_x if you ignore gate-to-drain capacitance?

(c)   What is the bandwidth of $A_{v1} = \frac{v_x}{v_{in}}$ if C_{gs} is unchanged but $C_{gd} = \frac{C_{gs}}{3}$?

6.7   In Example 6.1 the output capacitance of the first inverter (C_L) was equal to the
gate capacitance of the load inverter (C_{gs}) giving rise to a particular conclusion for the
optimum $g_m R$. Note that this g_m is the transconductance of the load inverter. Revisit
this example but with $C_{gs} = 0.5C_L$ and transconductances sized for a gain of 2. What
is the resulting bandwidth with and without the active inductor?

6.8   In Example 6.1 the output capacitance of the first inverter (C_L) was equal to the
gate capacitance of the load inverter (C_{gs}) giving rise to a particular conclusion for the
optimum $g_m R$. Note that this is the transconductance of the load inverter. Revisit this
example but with $C_{gs} = 2C_L$ and transconductances sized for a gain of 1. What is the
resulting bandwidth with and without the active inductor? This design point is what
you might expect when C_{db} is small relative to C_{gs}.

6.9   Repeat Problem 6.5 but use an inverter-based design of your choice.

6.10   Compare the two inverter-based CTLEs in Figure 6.23 in terms of linearity and
bandwidth for matched low frequency gain and zero frequency.

6.11   This problem explores the effect on bandwidth of adding cross-coupled inverters in inverter-based amplifiers. Consider two sizing scenarios for the circuit in
Figure 6.17 (a). In design 1, $g_m = 2g_{m.L}$ and $g_{m,c} = 0$ whereas in design 2, $g_m = g_{m.L}$
and $g_{m,c} = 0.5g_{m.L}$. Assume for all inverters that output capacitance (C_{db}) is half of
its input capacitance (C_{gs}) and that $\frac{g_m}{2\pi C_{gs}} = 100$ GHz.

(a)   Show that the two designs have the same differential small-signal gain.

(b)   Compute the 3dB bandwidth for each design.

7   Optical Channels and Components

Every designer of integrated circuits for optical transceivers needs to be familiar with the fundamentals of optical channels and the devices that convert electrical signals to optical signals and vice versa. This chapter provides a concise overview starting with optical fibre. Single-mode and multi-mode fibre are described as well as the characteristics of on-chip optical channels. Optical-to-electrical conversion through photodiodes is discussed along with simple electrical models. Considerations for implementing photodiodes entirely in silicon are included in a separate section. On the transmitter side, both direct modulation and indirect modulation are presented. This chapter summarizes the operation of laser diodes and Mach–Zehnder interferometer-based modulators. It closes with an overview of silicon photonics in which the behaviour of microring resonators is introduced.

7.1   Optical Channels

Optical fibre has become the main transmission medium for long-distance terrestrial communication. Cross sections of optical fibre are shown in Figure 7.1 (a) and (b). For both types of fibre, total internal reflection confines light in the core. Total internal reflection occurs since the index of refraction of the core is higher than that of the cladding. A detailed presentation of fibre types and manufacturing can be found in [44].

Silica (SiO_2) is the primary material in fibres for long-distance communication, though plastic can be used for some very-short-reach applications. The index contrast in silica is achieved by doping the core, the cladding or both so as to give a slightly higher index in the core. When the core of the fibre has a diameter much larger than the wavelength of light, as shown in Figure 7.1 (a), multiple propagation modes can be supported; hence, this type of fibre is referred to as multi-mode fibre (MMF). The advantage of a larger core is that alignment tolerances at the ends of the fibre are wider, reducing assembly cost. However, different modes experience different effective indices of refraction and hence different velocity, a phenomenon known as *modal dispersion*. This means that an NRZ pulse launched into several modes will spread out as it propagates down the fibre, giving rise to a limit to the product of distance x data rate that a link can support. Typically MMF is limited to distances of less than

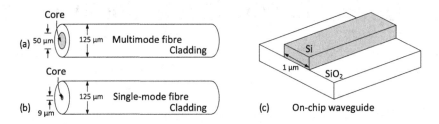

Figure 7.1 Various optical channels: (a) multimode fibre; (b) single-mode fibre; (c) on-chip waveguide used in silicon photonics.

300 m, making it suitable to replace copper cables in a data centre, satellite, airplane or spacecraft.

Figure 7.1 (b) shows a core diameter that is in the order of the wavelength of light. Such fibre will only support one optical mode and is thus referred to as single-mode fibre (SMF). While single-mode fibre is immune to modal dispersion, it still suffers from a variation in velocity as a function of wavelength, known as chromatic dispersion. The wavelength of light generated by a laser varies slightly with its bias current, causing a pulse corresponding to a 1 to travel at a different velocity from a 0. This can limit transmission to 10s of km at wavelengths near 1 310 nm. At this wavelength, SMF exhibits its minimum dispersion. The wavelength shifts induced by modulating lasers motivate using a laser with constant bias and an external modulator. A laser operating under constant bias current produces a near impulse in the frequency (or wavelength) domain. However, it still has non-zero linewidth, meaning that pulses will still spread out over long distances. Single-mode fibre can have a loss as low as 0.2 dB/km.

Wave-guiding structures can also be built on integrated circuits, such as those used in silicon photonics, introduced briefly in Section 7.5. An example of such a waveguide is in Figure 7.1 (c). The narrow ridge of high-index Si confines a propagation mode of light. Various cross-sectional shapes are used.

7.2   Fundamentals of Photodiodes

Optical receivers use a photodiode to convert an incoming optical signal to an electrical current. The physical structure of a commonly used configuration is shown in Figure 7.2, known as a p-i-n diode. It consists of p-type and n-type regions separated by an intrinsic layer. Incident photons, if absorbed, produce electron-hole pairs. The diode is reverse biased; therefore pairs generated in the intrinsic layer can be swept away by the electric field with electrons entering the n-region and holes entering the p-region.

The received optical power, $p(t)$, produces a current given by:

$$i_{PD}(t) = Rp(t), \tag{7.1}$$

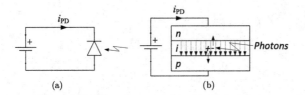

Figure 7.2 Photodiode: (a) symbol and biasing; (b) cross section.

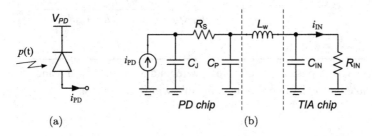

Figure 7.3 Electrical model of a photodiode (PD) and its interface to a transimpedance amplifier: (a) PD biasing; (b) equivalent circuit.

where R is the responsivity of the photodiode with units of A per W. Values between 0.5 and 1 A/W are common but depend on the diode's materials and the wavelength of light. An electrical model of a photodiode is shown in Figure 7.3 (b). The current source i_{PD} models the generated photocurrent as per (7.1); C_J represents the depletion capacitance of the photodiode; R_S models the series resistance between the intrinsic region and the pads of the photodiode. The pad of the photodiode is modelled by the capacitor C_P. The bondwire between the photodiode and the optical receiver's transimpedance amplifier (TIA) is modelled using L_W. The combined ESD, pad, and input capacitance of the TIA's transistors is modelled using C_{IN}. For preliminary analysis and simulation, R_S and L_w are often assumed to be small enough that the photodiode can be modelled as a current source in parallel with a single capacitor, $C = C_J + C_P + C_{IN}$.

If we assume the TIA has an input resistance of R_{IN}, without L_w and R_S the transfer function from I_{pd} to I_{in} is given by:

$$\frac{I_{in}(s)}{I_{pd}(s)} = \frac{1}{sR_{IN}C + 1},$$
(7.2)

which gives a pole at:

$$\omega_{in} = \frac{1}{R_{IN}C}.$$
(7.3)

This first-order system becomes third-order with the introduction of the inductor:

$$\frac{I_{in}(s)}{I_{pd}(s)} = \frac{Z_C}{Z_C + Z_{LRC}} \frac{Z_{C_{IN}}}{Z_{C_{IN}} + R_{IN}},$$
(7.4)

$$\text{where } Z_C = \frac{1}{s(C_J + C_P)} = \frac{1}{sC_1},$$
(7.5)

$$ Z_{LRC} = sL_w + \frac{R_{IN}}{sR_{IN}C_{IN} + 1}, \tag{7.6} $$

$$ Z_{C_{IN}} = \frac{1}{sC_{IN}}. \tag{7.7} $$

Therefore $\dfrac{I_{in}(s)}{I_{pd}(s)} = \dfrac{\frac{1}{sC_1}}{\frac{1}{sC_1} + sL_w + \frac{R_{IN}}{sR_{IN}C_{IN}+1}} \dfrac{1}{sR_{IN}C_{IN} + 1},$ \hfill (7.8)

$$ = \frac{1}{s^3 C_1 L_w R_{IN} C_{IN} + s^2 C_1 L_w + sR_{IN}C + 1}. \tag{7.9} $$

Example 7.1   Consider the effect of a wirebond on the bandwidth and pulse response of a photodiode/TIA interface with the following parameters:

- $R_{IN} = 100\ \Omega$
- $C_1 = 100\ \text{fF}$
- $C_{IN} = 100\ \text{fF}$

Compute the bandwidth when $L_w = 0$. Find the best value of L_w.

Solution

$C = 200$ fF. Without the inductor the first-order circuit has a bandwidth of

$$ f_{bw} = \frac{1}{2\pi R_{IN} C} = \frac{1}{100 \times 200 \times 10^{-15}} \approx 8\ \text{GHz}. \tag{7.10} $$

With $C_{IN} = C_1 = C/2$, (7.9) simplifies to:

$$ \frac{I_{in}(s)}{I_{pd}(s)} = \frac{1}{s^3 C_1 L_w R_{IN} C_{IN} + s^2 C_1 L_w + sR_{IN}C + 1}, \tag{7.11} $$

$$ = \frac{1}{s^3 C^2 L_w R_{IN}/4 + s^2 C L_w/2 + sR_{IN}C + 1}. \tag{7.12} $$

The behaviour of this interface is summarized in Figures 7.4 and 7.5. Figure 7.4 (a) shows the frequency response for L_w from 0 to 1 nH in steps of 0.1 nH. The response with small L_w clearly shows the effect of a real pole near 8 GHz, as well as complex poles with high Q (low ζ since $\zeta = \frac{1}{2Q}$). As L_w increases the location of the peak moves to lower frequency and the damping factor of the complex poles increases. The pulse response is shown in Figure 7.4 (b) for $T_b = 35$ ps. The pole locations are shown in Figures 7.5 (a). Larger L_w increases the magnitude of the real pole, decreases the magnitude of the complex poles and increases their damping factor. The complex poles for practical values of L_w have low damping factor. The dashed line shows pole locations with equal magnitude of real and imaginary parts, corresponding to $Q = \zeta = 1/\sqrt{2}$. In a second-order system, this gives the maximally flat (Butterworth) response. In all cases, the damping factor is much smaller. Figure 7.5 (b) shows the VEO taking into account ISI. It reaches a maximum for $L_w = 0.8$ nH, giving a bandwidth of approximately 27 GHz, much higher than the 8 GHz bandwidth achieved for $L_w = 0$.

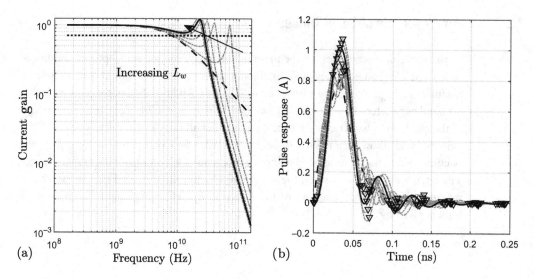

Figure 7.4 Frequency and pulse response of a wirebonded photodiode: (a) frequency response of current gain for various values of L_w; (b) pulse response for $T_b = 35$ ps.

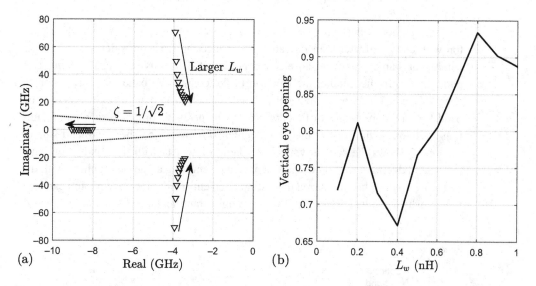

Figure 7.5 Details of wirebonded photodiode interface: (a) pole locations; (b) resulting *VEO* for $T_b = 35$ ps.

Example 7.1 showed that a model of the interface between a photodiode and an IC that includes a wirebond generated a third-order model. Including R_S increases the order of the system further, while adding frequency-dependent resistance of the wirebond will increase the damping factor of complex poles.

7.3   Fundamentals of Laser Diodes

Semiconductor laser diodes are the main light source used for fibre-optic communication. Depending on how they are manufactured, they can produce light over a range of useful wavelengths covering 850 nm up to 1.55 μm. All laser diodes are built from direct band gap materials, a sketch of which is shown in Figure 7.6. The minimum energy of the conduction band occurs for the same momentum as the maximum energy of the valence band, allowing a photon to be emitted when an electron moves from the conduction band to the valence band without a momentum transfer from the crystal lattice. The energy difference between the conduction-band minimum and the valence-band maximum defines the band gap of the material, E_g. The wavelength (λ) of emitted light is given by:

$$\lambda = \frac{hc}{E_g}, \tag{7.13}$$

where h is Planck's constant with a value of $4.135667696 \times 10^{-15}$ eV $\cdot$ s, $c = 2.99792458 \times 10^8$ m/s is the speed of light in a vacuum (often rounded to 3×10^8 m/s) and E_g is the band gap in eV. A wavelength of 850 nm is associated with a band gap of 1.46 eV.

Various photon/electron processes are depicted in Figure 7.7. On the left, a photon with energy greater than E_g can raise an electron from the valence band to the conduction band, producing an electron/hole pair. This was seen in the context of photodiodes. In the middle, the electron in the conduction band emits a photon of energy E_g when it drops to the valence band. The electron remains in the conduction band for a random period of time before "spontaneously" dropping down and emitting a photon, giving rise to the term spontaneous emission. The right-hand side of Figure 7.7 shows the case in which one photon interacts with the conduction band electron, inducing it to drop to the valence band, emitting a photon, in phase with the incident photon. This is referred to as stimulated emission. Since one photon can lead to two, which can lead to four, etc, we have light amplification. The acronym LASER

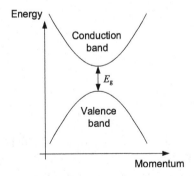

Figure 7.6 Conduction and valence bands in a direct band gap material.

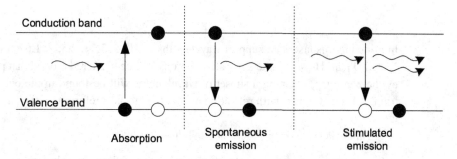

Conduction band

Valence band

Absorption

Spontaneous emission

Stimulated emission

Figure 7.7 Possible photon/electron interactions. Absorption occurs when a photon raises an electron to the conduction band. In spontaneous emission a photon is emitted when the electron returns to the valence band. In stimulated emission an energy-matched photon induces an electron to drop from the conduction to the valence band.

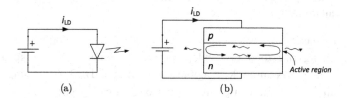

(a) (b)

Figure 7.8 Electrical model of a laser diode.

refers to the light amplification through the stimulated emission of radiation. Hence, stimulated emission is the basis of the lasing process.

Absorption and stimulated emission both involve incident photons. One might ask what determines whether a photon is absorbed or stimulates emission. The probability of each is determined by the population of electrons in the conduction band relative to that in the valence band. To have significant stimulated emission, we want "population inversion" to occur, a scenario in which conduction band electrons outnumber those in the valence band. This can occur in a region of lightly doped material sandwiched between highly doped p- and n-type materials, as depicted in Figure 7.8 (b).

Figure 7.8 (a) shows the schematic of a laser diode connected to a bias voltage, leading to a current i_{LD} to flow. This current provides a steady supply of conduction band electrons in the active region, which generate photons through initially spontaneous emission and then stimulated emission. Figure 7.8 (b) shows emitted photons moving left and right in the active region. Due to index contrast with the heavily doped p/n regions, photons are confined to the active region. At the facets of the active region (ends) some photons are emitted while others are reflected. The wavelengths that see round-trip path lengths that are integer multiples of λ are amplified and build up, becoming the dominant wavelengths that are emitted. For example, if the length of the active region is 5 μm, giving rise to a roundtrip path length of 10 μm, the following wavelengths can be supported:

$$\lambda_n = \frac{10 \ \mu m}{n}. \tag{7.14}$$

In principle this laser can support wavelengths of 10, 5, 3.33, 2.5, 2, 1.67, 1.43, 1.25, 1.11, 1.0 μm. However, the generated photons will be clustered around that predicted by the band gap, meaning a subset of wavelengths will be found in the output spectrum. A laser structure with reflection only at the facets is referred to as a Fabry–Perot laser.

A Fabry–Perot laser has the following disadvantages:

- Its wide spectrum prevents closely spaced wavelengths from being used in the same medium for transmission. That is, wavelength division multiplexing can only be done using widely spaced wavelengths.
- Since the signal spans a wide optical frequency range, it is more sensitive to chromatic dispersion. Pulses spread out, limiting transmission range.
- The output, composed of multiple wavelengths, cannot be used for high-capacity phase modulation.

To make the laser more wavelength-selective, reflections can be distributed along the cavity, giving rise to a distributed feedback (DFB) laser. Unlike a Fabry–Perot laser, which has multiple longitudinal modes, a DFB laser supports only one longitudinal mode and can produce a narrow optical spectrum around a single wavelength.

Although Figure 7.8 (b) is a stylized sketch of a laser diode, it accurately depicts that light is emitted out of the ends of the cavity. This can require the cleaving of a wafer to expose the facets. Additional structures are used to confine the optical signal laterally so that the emitted light does not spread out too much. Cleaving the laser diodes and coupling to the edge is a comparatively expensive process.

Another category of laser diode is a vertical-cavity surface-emitting laser (VCSEL). Built up in layers, VCSELs emit light vertically, without the need for precision cutting of the wafer. This allows testing to occur before additional cleaving is done. Small arrays of lasers can be cut from the wafer allowing coupling to fibres also manufactured in arrays.

The layered structure of VCSELs allows for both Fabry–Perot and DFB approaches. For 850 nm applications where larger core MMF is used, VCSELs are multi-mode whereas single-mode VCSELs are used in other applications.

The light output of a laser as a function of its bias current typically follows that shown in Figure 7.9. The minimum current required to get light output is the threshold current, I_{th}, beyond which optical output power grows approximately linearly with current. Self heating makes the slope decrease for high currents. This roll-off is more pronounced at higher ambient temperature.

There are two main approaches to generate an NRZ data stream in an optical signal. The most straightforward is direct modulation in which the current to the laser is modulated between I_0 and I_1 to generate optical powers P_0 and P_1. The second approach is to use the laser to generate a constant wave (CW) optical signal that is subsequently modulated using an external modulator.

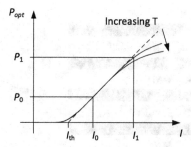

Figure 7.9 Laser diode output power, P_{opt}, vs laser current, I.

When directly modulated, I_0 is selected to be above the laser diode's threshold current since the onset of lasing can be slow. If the zero current is less than the threshold current, the output power will take more time to reach the one level and, due to the random nature of spontaneous emission, will take a varying time to reach the one level, introducing jitter.

Figure 7.9 shows the static operation of a laser diode. For high-speed links, the dynamic performance is also critical if the laser is directly modulated. Dynamic modelling is presented in Section 8.1.5 where laser-diode drivers are discussed. Another issue with direct modulation of laser diodes is that its wavelength depends on carrier concentration, meaning that the wavelength of a 1 may be slightly different from that of a 0. Therefore, signals originating from a directly modulated laser will be more sensitive to chromatic dispersion. Long-reach links use externally modulated lasers using a CW laser and a modulator such as a Mach–Zehnder interferometer-based modulator.

7.4   Fundamentals of Mach–Zehnder Interferometers

In a Mach–Zehnder interferometer (MZI), light is split into two paths and then recombined. Figure 7.10 shows an example. If the length and propagation velocity of the two branches are equal, the phase of the signals P_{1o} and P_{2o} are equal and the signals combine constructively. In this case $P_{out} = P_{in}$, ignoring losses in the material. If there is a difference in length and/or propagation velocity that gives rise to a phase difference of π the signals combine destructively. In this case, $P_{out} = 0$.

To create a modulator, we need a phenomenon that converts a particular voltage amplitude to a change in index n and ultimately a π-phase shift. The general case considers lengths L_1 and L_2 for the two branches and an index $n(V_{1/2})$ for each of the two branches. The input waves' electric-field intensities are written as:

$$E_{i1} = E_{i2} = |E_i| \cos(\omega_c t + \phi_0), \tag{7.15}$$

where ω_c is the angular frequency of the input optical signal ($\omega_c = 2\pi c/\lambda$) and ϕ_0 is its initial phase. By substituting $t - \frac{L_{1/2}}{v}$ we can write the output signals as:

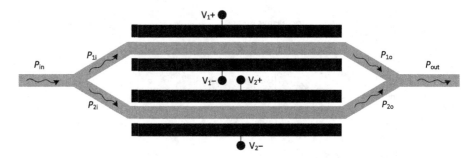

Figure 7.10 Mach–Zehnder interferometer.

$$E_{o1} = |E_i| \cos(\omega_c t - \omega_c \frac{L_1}{v_1} + \phi_0), \tag{7.16}$$

$$= |E_i| \cos(\omega_c t - 2\pi \frac{n(V_1)L_1}{\lambda_0} + \phi_0), \tag{7.17}$$

$$E_{o2} = |E_i| \cos(\omega_c t - 2\pi \frac{n(V_2)L_2}{\lambda_0} + \phi_0), \tag{7.18}$$

where λ_0 is the signal's free-space wavelength. When the intensities add at the output, E_{out} is given by:

$$E_{out} = \frac{1}{\sqrt{2}} (E_{o1} + E_{o2}) \tag{7.19}$$

$$= \frac{|E_i|}{\sqrt{2}} \left(\cos(\omega_c t - 2\pi \frac{n(V_1)L_1}{\lambda_0} + \phi_0) + \cos(\omega_c t - 2\pi \frac{n(V_2)L_2}{\lambda_0} + \phi_0) \right). \tag{7.20}$$

Using the identity $\cos A + \cos B = 2 \cos \frac{A-B}{2} \cos \frac{A+B}{2}$ we get:

$$E_{out} = \sqrt{2}|E_i| \left(\cos \left(\pi \frac{n(V_2)L_2 - n(V_1)L_1}{\lambda_0} \right) \cos(\omega_c t + \phi_{out}) \right), \tag{7.21}$$

where $\phi_{out} = \phi_0 - \pi \frac{n(V_2)L_2 + n(V_1)L_1}{\lambda_0}$. Here we further assume $n(V_{1/2}) = n + \Delta n V_{1/2}$. Assuming a lossless splitter at the input, $\sqrt{2}|E_i| = |E_{in}|$. In intensity-modulated systems we are primarily concerned with the output power, which is proportional to the squared magnitude of the E-field, giving:

$$P_{out} = P_{in} \cos^2 \left(\pi \frac{(n + \Delta n V_2) L_2 - (n + \Delta n V_1) L_1}{\lambda_0} \right) \tag{7.22}$$

$$= \frac{P_{in}}{2} \left(1 + \cos \left(2\pi \frac{(n + \Delta n V_2) L_2 - (n + \Delta n V_1) L_1}{\lambda_0} \right) \right). \tag{7.23}$$

To better understand (7.23) consider a structure in which $V_2 = 0$ and $L_1 = L_2 = L$. Under these circumstances, (7.23) becomes:

$$P_{out} = \frac{P_{in}}{2} \left(1 + \cos \left(2\pi \frac{\Delta n V_1 L}{\lambda_0} \right) \right). \tag{7.24}$$

For $V_1 = 0$, $P_{out} = P_{in}$, ignoring loss. Equation (7.24) can be solved to find V_π, the value of V_1 that causes a π phase shift and for this configuration gives $P_{out} = 0$.

$$\cos\left(2\pi\frac{\Delta nV_\pi L}{\lambda_0}\right) = -1, \tag{7.25}$$

$$V_\pi = \frac{\lambda_0}{2\Delta nL}. \tag{7.26}$$

In general, the electronics driving the modulator will be more efficient if V_π and L are smaller. A large V_π can exceed the breakdown voltage of scaled CMOS technology, meaning that modulator drivers either use thicker oxide CMOS devices at the output or use a higher voltage bipolar transistor technology. Equation (7.26) shows that V_π can be reduced by using a material that shows a larger index dependence on voltage or by using a longer L, the latter being available to the component designer. However, increasing L comes at the cost of increased physical size, more optical loss and also larger capacitance, which must be driven by the electronics.

The preceding discussion gave the classical derivation that applies to MZIs made of $LiNbO_3$. However, MZIs have been successfully ported to silicon photonic technologies. In this platform, the index is changed by changing the carrier concentration of the waveguides.

Other biasing arrangements are possible. For example if $L_1 = L_2 = L$ but the MZI is driven differentially with $V_2 = -V_1$, P_{out} becomes

$$P_{out} = \frac{P_{in}}{2}\left(1 + \cos\left(2\pi\frac{(n - \Delta nV_1)L - (n + \Delta nV_1)L}{\lambda_0}\right)\right), \tag{7.27}$$

leading to a differential V_π of:

$$V_{\pi,diff} = \frac{\lambda_0}{\Delta nL}. \tag{7.28}$$

Thus far, the input to the MZI has been presented as a voltage ranging from 0 to V_π. By modifying the length of one branch, we can shift the required modulation signal to toggle between $-\frac{V_\pi}{2}$ to $\frac{V_\pi}{2}$.

Example 7.2 Find the required length L that gives V_π calculated by (7.26) of 8 V for $\Delta n = 20$ ppmV^{-1} and $\lambda_0 = 1.55$ μm. Show how this voltage can be applied while keeping the maximum per-node voltage swing to only 2V.

Solution
Using (7.26), the length can found as:

$$L = \frac{\lambda_0}{2\Delta nV_\pi} \tag{7.29}$$

$$= \frac{1.55}{2 \times (20 \times 10^{-6}) \times 8} = 4.8 \text{ mm}. \tag{7.30}$$

The voltage swing of 8 V can be realized by keeping three of the four terminals in Figure 7.10 at ground and toggling the fourth between 0 and 8 V. On the other hand, by

Table 7.1 Voltages for MZI modulation: $V_1 = V_{1+} - V_{1-}$ and $V_2 = V_{2+} - V_{2-}$.

Logic level	V_{1+} (V)	V_{1-} (V)	V_1 (V)	V_{2+} (V)	V_{2-} (V)	V_2 (V)	$V_1 - V_2$ (V)
0	0	2	-2	2	0	2	-4
1	2	0	2	0	2	-2	4
Peak-to-peak signal (V)	2	2	4	2	2	4	8

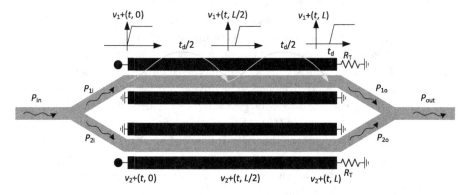

Figure 7.11 Mach–Zehnder interferometer: dynamic modulation of electrode voltage.

using the voltages presented in Table 7.1, the same peak-to-peak differential voltage of 8 V is achieved with each node only toggling between 0 and 2 V. This eases the voltage strain on the driver's output transistors.

7.4.1   Considerations for MZI Modulation

The ability of the modulating voltage to vary the output power based on constructive and destructive interference has been presented in the previous section. However, thus far, only quasi-static manipulation of the relative phase of the signals in each branch has been considered. Here we consider non-zero propagation delay of light in the arms of the structure and how that interacts with the modulation of phase.

To motivate this, consider an MZI built from silicon with a nominal index of 3.48. This corresponds to a phase velocity of 8.62×10^7 m/s or 86.2 μm/ps. A length of 3 mm corresponds to an optical time delay (t_d) of 34.8 ps, which is more than 1 UI for modern data rates. Therefore, at any given moment there is light in the MZI from more than 1 UI. If a spatially uniform but time-varying voltage is applied along the length of the electrodes, optical power near the data edges will not see the same index along the waveguide since the applied voltage will vary as the light travels down the modulator. Instead, spatially varying signals should be applied, as shown in Figure 7.11. In this scenario, we modulate light at $t = 0$ by applying a rising edge to v_1+ at the left side of the electrode, whose voltage is now specified with respect to time and distance in the x-direction. The objective is to ensure that incident light arriving for $t < 0$ sees the effect of low electrode voltage as it propagates down the upper arm and light arriving for $t > 0$ sees the high voltage. Thus, the rising edge on v_1+ must spatially follow the incident light as it propagates through the waveguide. This is shown in the figure where the voltage at the middle of the electrode is delayed by $t_d/2$ with respect to the voltage at $x = 0$. Finally, the voltage at $x = L$ is delayed by t_d compared to the voltage at $x = 0$.

As the notation for v_1+ suggests, one way to construct an electrode voltage that varies in both space and time is to use a terminated transmission line, giving rise to a "travelling wave" modulator drive. This works provided the transmission-line phase velocity (v) matches the velocity of light in the waveguide, given by c/n. In general, there is a mismatch. Techniques for correcting this are discussed in Section 8.2. A second issue with travelling wave modulators arises if the electrodes present loss over their length, decreasing the signal voltage at $x = L$ relative to $x = 0$. To overcome transmission-line attenuation, other schemes to appropriately delay the electrical signal are presented in Section 8.2.

7.5 Overview of Silicon Photonic Integration

Since the early 2000s silicon photonics has become a massive research and development focus for short-reach and long-haul transceiver designers. As discussed in Section 7.1, to confine light, optical channels rely on an index contrast. Higher contrast allows for tighter bends but also requires smaller cross-sectional area if single-mode operation is to be guaranteed. Silicon-on-insulator CMOS was identified as a potential photonic process. With a thin layer of silicon on top of a thicker buried-oxide layer (known as BOX), various structures can be built, including waveguides, grating couplers and directional couplers/interferometers. By leveraging silicon's carrier-dependent index (free carrier plasma dispersion effect), MZIs and ring resonators can be built.

By allowing several optical components to be built on a common substrate, photonic integration results in a miniaturized system. The optical components in coherent transceivers have been greatly reduced in size. The context in which several optical components are integrated together, but reside on a separate die from the driving electronics, is referred to as hybrid integration.

Since the optical process is a Si process, Si photonics have been presented as a means to integrate photonic components with electronics in what is known as monolithic integration. This has the potential to reduce interconnection parasitics between electronic and photonic components such as a photodiode and a transimpedance amplifier as well as increase the interconnection density between electronics and photonics. Finally, instead of cores on a large microprocessor connected by electrical buses, advocates for monolithic integration propose optical buses embedded in larger chips.

To fully leverage the capabilities of monolithic integration, process steps that enable photonic components need to be added to each successive CMOS process, adding cost and delaying its availability. Also, optical components such as MZIs are large (mm) relative to advanced CMOS, effectively wasting expensive area.

Today, commercial products lean toward hybrid integration for applications that require single-mode operation such as links beyond 300 m up to coherent transceivers for long-haul applications. Since silicon is not a direct band gap material, additional materials must be added to make on-chip lasers. This requires materials that have a direct band gap and are lattice matched to the substrate. Si-photonic friendly lasers comprise an active area of research. Generally, the laser is also hybrid-integrated alongside the photonics chip.

7.6   Microring Resonators and Modulators

An alternative to the comparatively large (mm scale) MZI is the microring resonator. It consists of a directional coupler and a closed ring. What makes a ring advantageous is that it can be in the order of 10 μm in radius compared to the mm-scale MZI. When used as a modulator, it has a mechanism for varying the effective index of the ring.

A directional coupler is shown in Figure 7.12. It has two input ports (left) and two output ports (right) with intensity relationships as given by:

$$E_{o1} = tE_{i1} - kjE_{i2}, \tag{7.31}$$

$$E_{o2} = tE_{i2} - kjE_{i1}, \tag{7.32}$$

where t and k are the scaler values of the transmission and coupling coefficients, respectively. Here, $j = \sqrt{-1}$, indicating a $\pi/2$ phase shift of the coupled signal. The inherent phase delay of the structure is omitted from (7.31) and (7.32). If the structure is lossless, then:

$$t^2 + k^2 = 1. \tag{7.33}$$

A practical directional coupler will introduce some loss. A good reference for how directional coupler parameters depend on the physical structure and material properties is [45].

The microring resonator is formed by looping port two back on itself as shown in Figure 7.13 (a). For certain wavelengths, the ring's length is an integer number of wavelengths, giving rise to resonance. For continuous input, the intensity in the ring

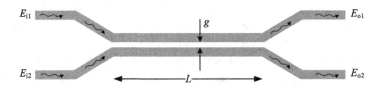

Figure 7.12 Directional coupler of length L and gap spacing g.

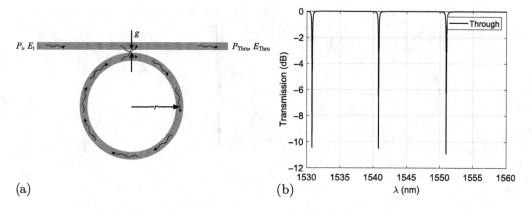

Figure 7.13 Microring resonator: (a) structure; (b) transmission.

increases until the inputted light balances the optical loss and the light coupled back out of the ring. As a result of the $\pi/2$ shifts for light entering the ring and coupling back out of the ring to the through port, the wavelengths for which the ring is in resonance correspond to destructive interference at the through port. The transmission spectrum is shown in Figure 7.13 (b). The vertical axis shows "normalized" transmission, normalized to the transmission loss away from resonance. The spacing between the nulls, known as the free-spectral range (FSR) is given by:

$$FSR = \frac{\lambda^2}{nL}. \tag{7.34}$$

where $L = 2\pi r$ is the circumference of the ring and n is the effective index.

The structure from Figure 7.13 (a) can be extended by adding another directional coupler as shown in Figure 7.14 (a). When the ring is in resonance, intensity in the ring reaches a maximum. Some of this light couples out to the drop point, giving a spectral peak corresponding in wavelength to the through port's null. However, due to loss in the ring, the drop port's peak will not reach as high as the through port's maximum transmission. This structure can be used as a filter for selecting a particular wavelength.

To use a microring resonator as a modulator, voltage control must be added. In Figure 7.15 (a) a portion of the ring is embedded in a p-n junction. As with a silicon photonics (SiPh) MZI, the free-carrier plasma dispersion effect is used to modulate the index of the ring. This shifts the spectrum to the right as the applied reverse-bias voltage, v_{Mod}, increases, as depicted in Figure 7.15 (b). The vertical line indicates a suitable laser wavelength. If modulation voltages of 0 and 4 V are used to encode binary 0s and 1s, the resulting extinction ratio (ER_{dB}) is the vertical distance between the logic 1 and logic 0 horizontal lines, since the vertical axis is on a logarithmic scale. At this wavelength, the 4 V modulation voltage gives an 8 dB extinction ratio. The insertion loss (IL_{dB}) is also indicated. This is the distance down to the logic 1 level, approximately 1.5 dB in this case. Note that this insertion loss is on top of the loss to which the measured transmission was normalized in the generation of the plot.

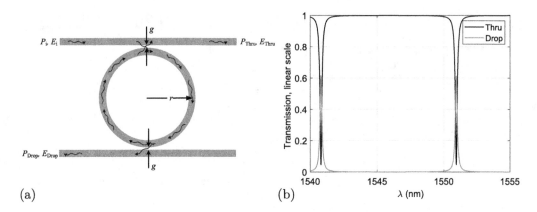

(a)   (b)

Figure 7.14 Microring resonator with a drop port: (a) structure; (b) transmission.

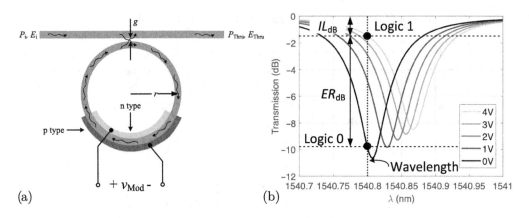

(a)   (b)

Figure 7.15 Microring resonator with a p-n junction, used as a phase-shifting element: (a) structure and (b) transmission under different bias voltages. Note that IL_{dB} and ER_{dB} are labelled assuming the ring is modulated between 0 and 4 V.

The index of the ring is a function not only of the applied voltage but also temperature. Given how wavelength-selective the ring is, small fluctuations in temperature cause significant shifts in the transmission spectrum as shown in Figure 7.16 (a). These curves are the same shape as those in Figure 7.15 (b) but have been shifted to the right, by a distance determined by the rise in temperature and the ring's thermal sensitivity. Thermal sensitivity is on the order of 10s of pm/°C. Comparing this to Figure 7.15 (b) at the same operation wavelength, we see that both the extinction ratio (ER) and insertion loss have reduced. Although lower insertion loss is preferred (now less than 1 dB), a reduced extinction ratio (now only about 4 dB) can degrade link performance. To combat this temperature sensitivity, rings are equipped with heaters to allow thermal tuning. If the ring's response is locked to the laser's wavelength by heating it above the ambient temperature, the power delivered to the heater can be lowered when the ambient temperature rises. Similarly, as the ambient temperature drops,

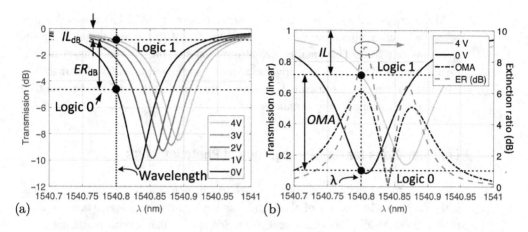

Figure 7.16 Transmission response of the ring from Figure 7.15: (a) shift due to a temperature rise of 1 °C; (b) 0 and 4 V response along with OMA and ER at original temperature.

more power can be delivered to the heater. This comes at the cost of 10s of mW of power per ring.

The optimum transmitted signal and hence ring operation point must take into account the trade-off between extinction ratio and insertion loss. The amplitude of the generated photocurrent in the receiver is proportional to the received optical modulation amplitude (OMA). Of course, lower extinction ratio for a given OMA increases the dc bias of the photocurrent, which in turn increases noise generated by the photodiode. This noise, which has a power-spectral density proportional to the current level, known as shot noise, is usually small for intensity-modulation direct detection links.

To find the operation point for the ring that maximizes OMA, transmission is plotted on a linear vertical scale in Figure 7.16 (b). For small modulation voltages, we can look for the point where the slope of the transmission response is steepest. For example, if a driver could only vary the applied voltage by a small excursion around 0 V, we would look at the 0 V transmission plot and identify the point with the steepest slope, near 1 540.77 nm in this case. For small modulation this wavelength would have the largest OMA. For operation at another wavelength, we would want the thermal tuning to shift the ring response such that this region of steep slope was aligned with the desired wavelength of operation.

When a larger amplitude of modulation is used, such as 0 to 4 V, we look for the largest vertical separation of the transmission characteristic for the two voltages. This is the operation point around which the ring will give the largest OMA for a given modulation signal. The OMA and ER_{dB} are plotted in Figure 7.16 (b), which shows that OMA reaches a maximum for approximately 1 540.81 nm. If operated with a wavelength on the right-hand side of the transmission peak 4 V would be used to encode a logic 0. The wavelengths giving the largest OMA and ER_{dB} are close to one another but are not exactly aligned.

Microring modulators offer two main advantages over MZIs. The first is that they occupy a much smaller area. With a radius in the 10s of μm, rings are tiny compared to a mm-scale MZI. This smaller size also means they present much lower loading capacitance to their respective driver electronics compared to MZIs. The main disadvantage of rings is the thermal sensitivity, necessitating thermal tuning, and their sensitivity to manufacturing variations.

7.7   Considerations for All-Si Integration of Photodetectors

Silicon photonics enables integration, which has the potential to improve performance and reduce cost. Integration of photodiodes and optical receivers is also being pursued in standard CMOS processes. Recall from Section 7.5 that silicon is transparent to light near $1\,550\ \mu$m. However, it absorbs light at shorter wavelengths such as 850 nm, making it possible to build photodiodes in CMOS processes, without requiring the addition of other materials such as germanium (Ge). By integrating the PD on the same die as the rest of the optical receiver, parasitic capacitance of bond pads on both the PD and optical receiver chips can be eliminated. Also, ESD protection is no longer needed on the input of the transimpedance amplifier further reducing capacitance. As presented in Chapter 9, reduced interface capacitance allows lower noise, lower power dissipation and higher gain for the same bandwidth. Integration can also reduce cost.

Figure 7.17 [46] shows a CMOS-friendly implementation of a photodiode. An IC designer can easily recognize the N-well in which PMOS transistors are built along with the N-well's N+ contact. The substrate's P+ contact is also shown. When reverse biased, the depletion region is within both the N-well and substrate. This structure is illuminated vertically, with horizontal dimensions chosen based on the anticipated beam width. Figure 7.17 also shows how an incident power P_0 is absorbed as a

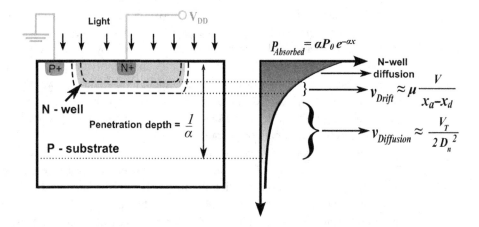

Figure 7.17 Vertical illumination of an N-well in a typical CMOS process showing the structure on the left and the absorption profile of photons on the right. ©The Optical Society. Reprinted with permission from [46].

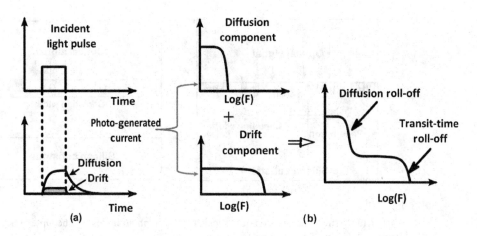

Figure 7.18 Diffusion and drift currents in a Si PD: (a) time domain plot showing how a 1 UI incident light pulse leads to a small amplitude but fast drift current pulse along with a larger amplitude, slower diffusion current pulse and (b) frequency domain representation showing the frequency content of drift and diffusion components as well as the frequency content of the complete response, formed by summing the responses of the drift and diffusion currents. ©The Optical Society. Reprinted with permission from [46].

function of depth, x. Carriers generated in the depletion region lead to a fast drift current. However, this can be a small percentage of the total generated carriers. Light absorbed outside the depletion region generates carriers that diffuse more slowly.

Figure 7.18 (a) [46] shows the time-domain response of these two currents. The incident light pulse generates a fast, but smaller drift component and a larger but slower diffusion component. Figure 7.18 (b) summarizes these phenomena in the frequency domain. The intrinsic bandwidth can be relatively low (100s of MHz). Several strategies have been developed to increase the bandwidth of all-Si receivers. These include:

1. Remove the diffusion current in the photodiode structure using a silicon on insulator (SOI) process.
2. Remove the diffusion current using two PD structures in a technique called spatially modulated light (SML).
3. Increase the depletion-region absorption by widening the region using a higher voltage.
4. Equalize the overall pulse response in the optical receiver.
5. Use a grating coupler to direct light laterally, keeping it in the depletion region.

Each of these approaches is discussed in the following sections.

7.7.1   Diffusion Current Elimination using an SOI Process

The first approach is shown in Figure 7.19 (b). In this case the p-i-n structure is lateral. By having a thin structure on top of an oxide, diffusion currents generated in the

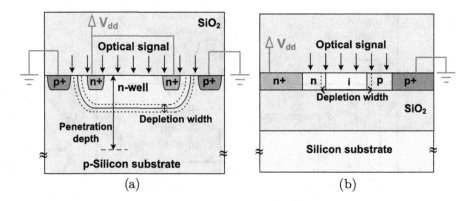

Figure 7.19 Vertical illumination of vertical and lateral photodiodes. ©The Optical Society. Reprinted with permission from [47].

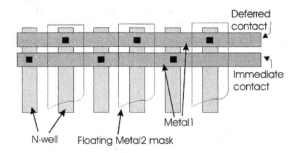

Figure 7.20 Photodiode layout using spatially modulated light in a 1-D array. Every second photodiode is covered by metal. ©1998 IEEE. Reprinted, with permission, from [48].

substrate do not flow to the terminals of the PD. However, the relatively thin silicon layer means that a small percentage of light is actually absorbed by the PD, giving rise to a high-bandwidth, but low-responsivity PD.

7.7.2   Diffusion Current Elimination Using Spatially Modulated Light

Figure 7.20 (a) shows a 1-D array of photodiodes. Every second PD is connected to the "Immediate contact." These PDs receive the optical signal. The "Deferred contact" is connected to the alternate PDs, but these PDs are covered with metal, which blocks incident light. Thus, these PDs do not directly receive the optical signal. Figure 7.21 (a) shows the density of photon-generated minority carriers below the PDs' depletion regions as a result of a short optical pulse applied at $t = t_0$. At $t = t_0^+$, the carriers have a near-rectangular distribution, following the profile of absorbed photons. However, this distribution tends toward a uniform distribution with respect to x as t increases. The near-uniform distribution of carriers diffuses toward immediate and defered contacts nearly equally. By connecting these contacts to a differential TIA, slow diffusion currents, which become a common-mode signal, are cancelled.

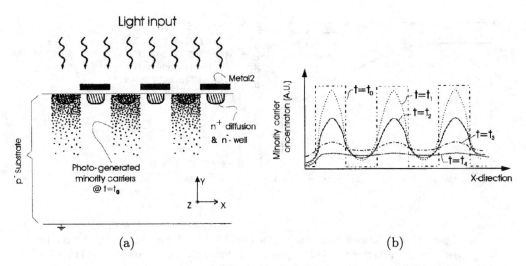

Figure 7.21 Carrier concentration due to an optical impulse at $t = t_0$: (a) illustrated for $t = t_0$; (b) shown along a cross section below the depletion areas for various $t \geq t_0$. ©1998 IEEE. Reprinted, with permission, from [48].

The TIA then responds only to the differential, high-bandwidth drift current, but with a lower overall responsivity.

    This approach has been extended to a 2-D array in [49]. The net responsivity is low.

7.7.3   Equalization of the Photodiode's Pulse Response in the Optical Receiver

An example of equalizing the PD response is presented in [50]. The frequency response of the PD is shown in Figure 7.22 (a). The PD in this work is designed for plastic optical fibre, having a larger core than glass MMF. Here, 4 Gb/s was achieved using a CTLE having a typical boost at $f_{bit}/2$. The overall channel, consisting of the VCSEL, PD, TIA and VGA is inverted and plotted in Figure 7.22 (b), for operation at 670 nm. Similar to the case of a lossy transmission line, the channel loss is less than 20 dB/dec. Four zero/pole pairs are used to implement the equalizer. The overall receiver architecture is shown in Figure 7.23.

7.7.4   Grating-Assisted Si PDs

The problems arising from light's long penetration depth in Si are mitigated if the light is absorbed in a depletion region. Grating couplers are commonly used structures in Si photonics to direct near normal incident light into on-chip waveguides at wavelengths where Si is transparent (i.e., 1 550 nm). For wavelengths near 850 nm, a grating coupler has been used to direct light laterally into depletion regions in [51] and [47]. Figure 7.24 (a) shows a 3-D view of the structure while (b) and (c) show top and side views, respectively. The approach was extended to a grating coupler that focuses

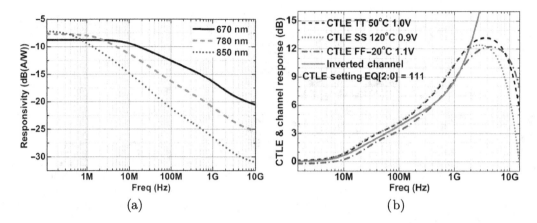

Figure 7.22 Equalization-based receiver for all-Si PD: (a) SiPD frequency response at different wavelengths; (b) required CTLE response to invert the channel and simulated CTLE response. Channel includes VCSEL, PD, TIA and VGA. ©2013 IEEE. Reprinted, with permission, from [50].

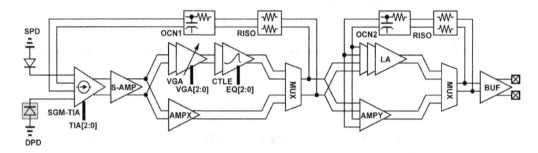

Figure 7.23 CTLE-based receiver to compensate for roll-off of Si PD. ©2013 IEEE. Reprinted, with permission, from [50].

light shown in Figure 7.25. The structures [47] were built in an SOI photonics process while similar structures have been built in a CMOS SOI process [51].

7.8   Summary

This chapter qualitatively described single- and multi-mode fibre, noting their respective advantages and disadvantages. A simple electrical model of a photodiode was introduced and the impact of series inductance between the photodiode and optical receiver IC was considered. Laser diodes were presented next, based on three photon/electron interactions, namely, absorption, spontaneous emission and stimulated emission, the last of which gives rise to lasing. Typical current to output power behaviour was shown along with the effect of temperature. The dependence of wavelength on bias current was discussed, giving motivation for external modulation, leading to an overview of Mach–Zehnder interferometers.

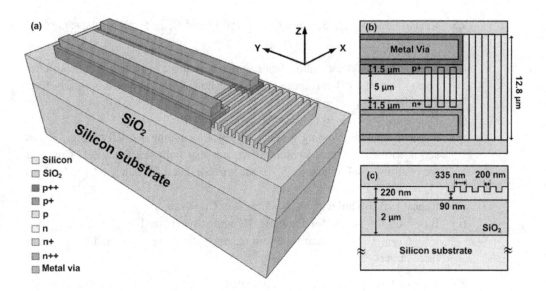

Figure 7.24 Drawing of a grating-assisted all-Si PD on an SOI process: (a) 3-D view; (b) top view; (c) side view. ©The Optical Society. Reprinted with permission from [47].

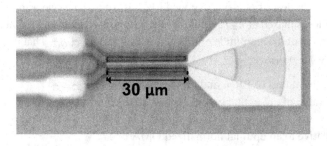

Figure 7.25 All-Si photodiode with focusing grating coupler. ©The Optical Society. Reprinted with permission from [47].

Silicon photonic integration was introduced very briefly before presenting the operation of microring resonators. Their use as modulators and their sensitivity to temperature were explained. Continuing with integration, the implementation of photodiodes in CMOS processes was explored next. Various techniques were presented to address the large penetration depth of light in Si, including spatially modulated light, equalization and grating-assisted schemes.

Problems

7.1 What characteristic of a silica fibre determines if it supports one mode (single-mode fibre) many modes?

7.2 What are the advantages and disadvantages of single-mode operation compared to multi-mode operation?

7.3 Revisit Example 7.1 but with $C_{IN} = 150$ fF and $C_1 = 50$ fF for the same total capacitance.

(a) What is the 3 dB bandwidth when $L_w = 0$? Remember, the relevant transfer function is from the PD current to the current flowing into the resistive part of the TIA, denoted by i_{IN} in Figure 7.3.

(b) Determine the optimal L_w considering a 35 ps UI.

(c) Experiment with damping the high-Q poles by adding a small resistor, R_w, in series with L_w. Is there a value of resistor that improves the VEO? Compute the resulting quality factor, Q_w, of the wirebond at a frequency of $f_{bit}/2$, where $Q_w = \frac{\omega L_w}{R_w}$ and $\omega = \pi f_{bit}$.

(d) Repeat part (b), but with $C_{IN} = 50$ fF and $C_1 = 150$.

(e) Based on the results for Example 7.1 and parts (b) and (d), for a given total $C = C_{IN} + C_1$, is it better to have it equally split between C_{IN} and C_1 or one of them larger?

7.4 When a laser diode is directly modulated, why is the current used to encode a 0 such that $P_0 > 0$?

7.5 Explain the current-dependent characteristics of a laser diode that motivate the use of an external modulator.

7.6 Equation 7.26 describes a single-ended MZI. If we assume one arm is biased at 0 V and the other is biased at $V_\pi/2$, plot the extinction ratio in dB and the normalized OMA as a function of the amplitude of the modulating signal, v_{MOD}. That is, $V_2 = 0$ V and $V_1 = \frac{V_\pi}{2} + / - v_{MOD}$.

(a) What are the values of ER_{dB} and OMA if $v_{MOD} = \frac{V_\pi}{4}$?

(b) Find the value of v_{MOD} that gives $ER_{dB} = 10$ dB.

(c) Find the value of v_{MOD} that give a normalized OMA of 0.8. How many dB down is this relative to an optimal normalized OMA of 1?

7.7 For the ring modulator characteristic plotted in Figure 7.15 (b), what is the maximum extinction ratio if only a 2 V swing can be generated? Consider modulation between 0 and 2, 1 and 3, and 2 and 4 V. What is the maximum normalized OMA?

8   Optical-Link Transmitter Circuits

This chapter builds on the basic transmitter building blocks presented in Chapter 5 and applies them to optical links. The section on direct modulation discusses CML and SST circuits for driving laser diodes, along with how equalization can be added, incorporating concepts from Chapter 4. One aspect of laser diodes that is presented is their nonlinear behaviour that favours having different equalization for rising and falling edges, something not typical of electrical links where channels are linear. Circuit design for Mach–Zehnder modulator drivers is discussed in Section 8.2. Finally, microring modulator drivers are presented in Section 8.3. Since the basic IC building blocks have already been introduced in Chapter 5, this chapter emphasizes aspects particular to driving optical devices, such as the electrical modelling of the electronics/photonic packaging and the challenges in producing voltage swings beyond the breakdown voltage of the underlying CMOS technology.

8.1   Laser-Diode Drivers

The traditional approach to driving a laser diode involves a current-mode logic circuit such as the ones described in Section 5.3, with an additional bias current for the laser.

Two configurations are shown in Figure 8.1. In both cases, it is assumed that the laser diode is a VCSEL within an array, in which VCSELs are connected together through either their anodes (a) or cathodes (b). For Figure 8.1 (a), the current to the VCSEL (i_{LD}) ranges from I_0 to $I_0 + I_{Mod}$ whereas in Figure 8.1 (b) i_{LD} ranges from $I_1 - I_{Mod}$ to I_1. The voltage across the VCSEL can be in the range of 1.5 V and up. Thus, $V_{DD,LD}$ and $V_{B,LD}$ must be chosen so that v_{O+} is held within a CMOS-friendly voltage range. Until recently, VCSEL drivers have been implemented using SiGe technology, operating from a 3.3 V supply. In this case, the common-cathode configuration in Figure 8.1 (b) can be used where $V_{B,LD}$ is grounded, leading to a simple biasing arrangement. This has favoured the use of common-cathode VCSEL arrays.

8.1.1   General Considerations for Laser-Diode Drivers

Although the core of a current-mode laser-diode driver is a familiar differential pair, there are several additional design considerations, namely:

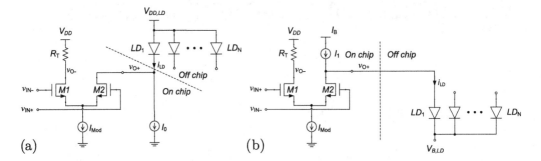

Figure 8.1 Current-mode logic VCSEL drivers: (a) common-anode connection; (b) common-cathode connection.

1. The input signal to the laser diode is a current.
2. The I–V characteristic of the laser is nonlinear.
3. Laser diodes can exhibit second-order carrier/photon dynamics whose parameters depend on the bias current.
4. The single-ended nature of array-based laser diodes prevents differential connection from driver to diode.
5. Large dc voltage bias of VCSELs requires the use of bias voltages beyond the breakdown voltage of scaled CMOS.
6. The effect of supply inductance, in particular inductance shared among several VCSELs, must be taken into account.
7. Laser diodes are usually connected to their driver without a transmission line, freeing the design from impedance-matching constraints, allowing the driver's output impedance to be selected based on the overall dynamics of the driver and laser diode.

Electrical-link transmitters drive linear channels, ideally matched to a real resistance over a reasonable frequency range. Thus, current and voltage mode signalling are equivalent, related through Z_0. In the case of a laser diode, the relevant input signal is current (issue 1 in the preceding list), which is related to the voltage across the laser diode by its nonlinear I–V characteristic (issue 2). For small signals this can linearized by a bias-dependent impedance.

As introduced in issue 3, above, and discussed in detail in Section 8.1.5 a VCSEL model consists of electrical aspects and optical aspects that account for second-order carrier/photon dynamics. These dynamics can be underdamped with bias-dependent natural frequency and damping factor. This has two implications. The first is that equalization must account for both an electrical and an optical model. The second is that equalization needs to asymmetrically emphasize rising and falling edges.

Issue 4 implies that, unlike a differential channel, the current delivered to the VCSEL has a non-zero average. As a result of issue 5 the VCSEL may need a separate supply, meaning that the current drawn from the on-chip supply is not constant, as it is for an electrical-link CML transmitter. Recall that constant supply current is one of the purported advantages of differential circuits. With time-varying current drawn from the supply or sunk into the local ground, supply inductance must be carefully considered (issue 6).

In the list above, issue 7 is the only aspect that potentially makes driving a laser diode more straightforward than an electrical link. Without a transmission line between the driver and the laser diode, the output impedance of the driver can be chosen without worrying about reflections. However, the combined electrical/optical dynamics of the driver and laser diode must be carefully considered. One might assume that without an impedance-matching constraint, open-drain CML drivers would be preferred due to their improved energy efficiency. However, a high output resistance combined with wirebond inductance can have excessive ringing in the pulse response. Without a rigid standard of a 50 Ω transmission line, however, VCSELs themselves may vary widely in their impedance, complicating the design of a driver if it is to serve many different VCSELs.

Each of these issues is discussed further in the sections that follow.

8.1.2   Differential vs Single-Ended Connection to VCSELs

When both terminals of the VCSEL are accessible, greater power efficiency and lower supply noise are possible. This is only an option if the devices are neither common-anode nor common-cathode VCSELs, meaning that both terminals are accessible without connections to other VCSELs. When driven differentially, as shown in Figure 8.2, ac-coupling capacitors are required to separate the dc operating point of the VCSEL from the driver circuit in order to forward bias the laser. These capacitors invariably introduce a high-pass response to the circuit which can lead to current droop during long runs of consecutive identical digits (CIDs). There are two approaches to address this, namely:

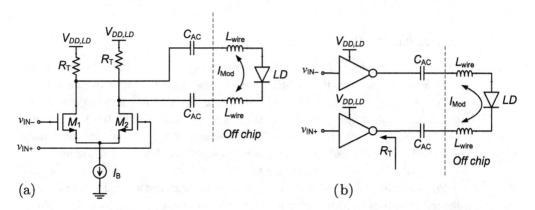

Figure 8.2 Differential connection of VCSELs adapted from [52]: (a) current-mode driver; (b) voltage-mode driver.

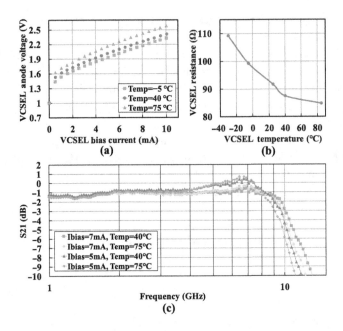

Figure 8.3 Low-cost 10 Gb/s VCSEL characteristics across temperature: (a) measured I–V characteristic; (b) measured low-frequency incremental resistance; (c) measured ac response. ©2019 IEEE. Reprinted, with permission, from [54].

- Choose a large enough capacitor such that $R_{EQ}C \gg NT_b$ where R_{EQ} is the equivalent resistance seen by the capacitor and N is the maximum number of CIDs expected.
- Add a low-pass signal path using thick-oxide transistors as in [53]. The thick-oxide transistors can tolerate the VCSEL's required dc bias voltage, but cannot support the required bandwidth. In this approach the ac coupled thin-oxide devices provide a high-pass path to the VCSEL.

8.1.3   DC Bias Voltage of VCSELs

The dc I–V behaviour of a VCSEL is shown in Figure 8.3 (a) [54]. As an example, consider a case in which the VCSEL's current is modulated between and 2 and 8 mA. This gives rise to a voltage ranging from 1.6 to 2.2 V at low temperature and 1.8 to 2.4 V at high temperature. The required signal swing of about 0.6 V is manageable in low-supply technologies (<1 V) but the absolute dc bias voltage can damage advanced CMOS technologies. In the common-anode configuration in Figure 8.1 (a), a tolerable voltage is presented to the CMOS chip if $V_{DD,LD}$ is raised above the technology's nominal supply. To maintain symmetry in the circuit and reduce supply noise the left branch of the differential pair can be connected to $V_{DD,LD}$ as well, as shown in Figure 8.9. In the common-cathode configuration the cathodes can be biased below 0 V to prevent the CMOS chip's output from seeing a voltage of > 1.5 V.

Table 8.1 Example of VCSEL model electrical parameter values [6].

Parameter	Symbol	Value
Junction capacitance	C_J	110–117 fF
Junction resistance	R_J	150–180 Ω
DBR resistance	R_S	50 Ω
Pad capacitance	C_P	10 fF

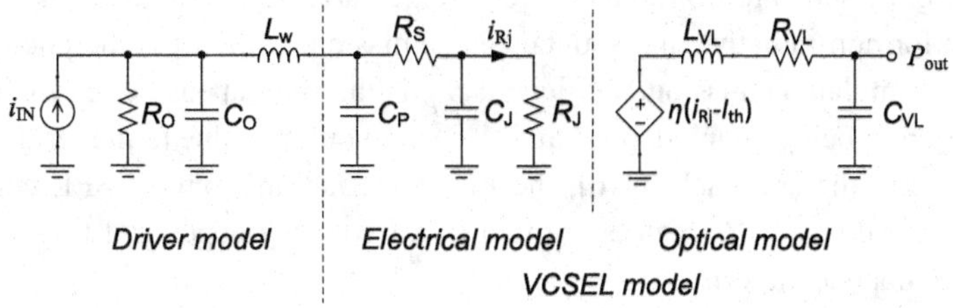

Figure 8.4 VCSEL model for circuit simulation, based on the model in [6].

8.1.4   Supply Inductance

Differential drive reduces the effect of some inductors. However, array-based VCSELs often allow independent access to only one terminal, requiring single-ended driving. It is critically important to include all inductances of bondwires in simulations in order to predict their impact.

8.1.5   VCSEL Modelling

Before discussing equalization, a simplified modelling strategy is introduced in this subsection. A VCSEL model involves the modelling of both the electrical behaviour (i.e., its I–V characteristics and dynamic response) and its optical behaviour. Such a model is required to predict the VCSEL's optical output power for a given input current supplied by its driver.

Figure 8.4 shows a complete model of a VCSEL and its driver [6]. The driver is modelled as a current source in parallel with an output resistance (R_O) and capacitance (C_O). The VCSEL is connected to the driver through a bondwire (L_w). The middle portion models the VCSEL's electrical dynamics where R_J and C_j are the diode junction resistance and capacitance. The current through R_J denoted as (i_{Rj}) is the current that ultimately generates photons. Although R_J models a nonlinear diode's I–V characteristic, a linear model is sufficiently accurate. Here, C_P models the bondpad's capacitance while R_S is the series resistance of the VCSEL between the pad and the active region. Example values are summarized in Table 8.1 [6]. The dc bias voltage of the VCSEL is omitted from this model, but can be added as a voltage source in series with R_J.

The photon-electron dynamics are governed by the following differential equations [55]:

$$\frac{dN}{dt} = \frac{I_{R_j}}{qV} - \frac{N}{\tau_{sp}} - GNN_p, \tag{8.1}$$

$$\frac{dN_p}{dt} = GNN_p + \beta_{sp}\frac{N}{\tau_{sp}} - \frac{N_p}{\tau_p}, \tag{8.2}$$

where N and N_p are the electron and photon densities in the laser's cavity, having a volume V, and I_{R_j} is the VCSEL current, defined as the current flowing through the active region, separate from the charge-storage current flowing through C_J; τ_{sp} is the lifetime for non-radiative and spontaneous emission, while τ_p is the photon lifetime; G is the stimulated emission coefficient; β_{sp} is the spontaneous emission coefficient. In addition to being coupled, (8.1) and (8.2) are nonlinear due to the term GNN_p.

Following the approach in [6], the transfer function from VCSEL current I_{R_j} to output optical power P_{op} can be written as a frequency-dependent transfer function, with bias-dependent terms:

$$\frac{P_{op}}{I_{R_j}}(f) = \frac{f_r^2}{f_r^2 - f^2 + j\left(\frac{f}{2\pi}\right)\gamma}, \tag{8.3}$$

$$f_r = \frac{1}{2\pi}\sqrt{\frac{GN_p}{\tau_p}}, \tag{8.4}$$

$$\gamma = Kf_r^2 + \gamma_0, \tag{8.5}$$

where f_r is the relaxation oscillation frequency and γ is the damping factor; K maps relaxation frequency to damping factor while γ_0 is an offset in damping factor. Equation (8.4) can be written in terms of VCSEL current as:

$$f_r = D\sqrt{I_{R_j} - I_{th}}, \tag{8.6}$$

where D shows how quickly f_r increases for current above the threshold current, I_{th}. The transfer function in (8.3) and the dependency on current in (8.5) and (8.6) are captured by the series RLC circuit in the right-hand side of Figure 8.4. This model consists of a current-controlled voltage source that excites the second-order circuit with a voltage of $\eta(I_{R_j} - I_{th})$. The voltage across the capacitor is a proxy for the output optical power of the VCSEL. Since we do not care about the impedance level in the right-hand portion of the circuit (we want a voltage to voltage transfer function), we can follow the approach in [6] where the capacitor value is fixed at 100 fF. The remaining elements have values that depend on the value of the current ($I_{R_j} - I_{th}$) as:

$$L_{VL} = \frac{1}{4\pi^2 C_{VL}D^2(I_{R_j} - I_{th})}, \tag{8.7}$$

$$R_{VL} = (Kf_r^2 + \gamma_0)L_{VL} = \frac{Kf_r^2 + \gamma_0}{4\pi^2 C_{VL}D^2(I_{R_j} - I_{th})},$$

$$= \frac{1}{4\pi^2 C_{VL}}\left(K + \frac{\gamma_0}{D^2(I_{R_j} - I_{th})}\right). \tag{8.8}$$

Table 8.2 Example of VCSEL model optical parameter values [6].

Parameter	Symbol	Value
Threshold current	I_{th}	0.6 mA
Slope efficiency	η	0.78 mW/mA
D-factor	D	7.6 GHz/$\sqrt{mA}$
K-factor	K	0.25 ns
Damping factor offset	γ_0	37 ns^{-1}

Table 8.3 VCSEL model optical model circuit element values [6].

I_{R_j} (mA)	L_{LV} (nH)	R_{LV} (Ω)	f_r (GHz)	ζ
2	3.13	179	8.99	0.506
5	0.997	100	15.9	0.502
11.5	0.402	78.2	25.1	0.617

Examples of optical parameters are shown in Table 8.2. The critical parameters of the transfer function (circuit element values) change significantly as the bias current increases from 2 to 11.5 mA, as shown in Figure 8.5. The corresponding RLC element values are summarized in Table 8.3. As the current to the active region increases, the natural frequency does as well. The damping factor also increases for higher current. Figure 8.5 shows significant flattening of the frequency response for higher current. This is due to the slight increase in damping factor seen in Table 8.3 but also due to the interaction between the peaking in the optical model and the low-pass roll-off of the electrical model. If the driver model in Figure 8.4 is replaced by an ideal current source, and if C_P is ignored, the electrical model behaves like a first-order circuit with bandwidth given by:

$$f_{3dB} = \frac{1}{2\pi R_J C_J} \approx 9.6 \text{ GHz}, \tag{8.9}$$

where $R_J = 150\ \Omega$ and $C_J = 110$ fF. Since the input is fed by a current, the equivalent resistance is R_j and not $R_j \parallel R_s$ as it would be if the equivalent circuit were driven by a voltage source. A voltage input, or at least a low-impedance driver, would increase the bandwidth of the electrical portion of the VCSEL model. Also, since the driver is connected to the VCSEL by an inductor, a lower output impedance favourably increases the damping factor in the electrical model. It must be reiterated that the element values in Table 8.3 are instantaneous functions of I_{R_j}. This is shown in Figure 8.6 where part (a) shows the output optical power varying due to a modulated input current. Part (b) shows the nonlinear, signal-dependent variation of R_{VL} and L_{VL}.

Because of the nonlinearity of the VCSEL we cannot compute its output due to a particular stream of input data using the pulse response approach in (1.9). To illustrate

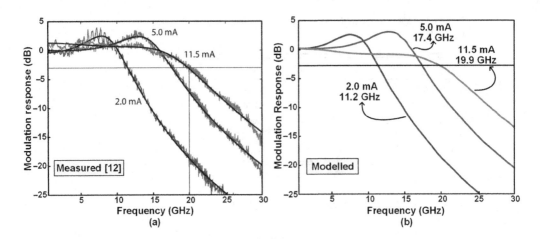

Figure 8.5 Measured and simulated VCSEL frequency response for bias currents of 2, 5 and 11.5 mA. Measured data in (a) is from [56]. © 2016 IEEE. Reprinted, with permission, from [6].

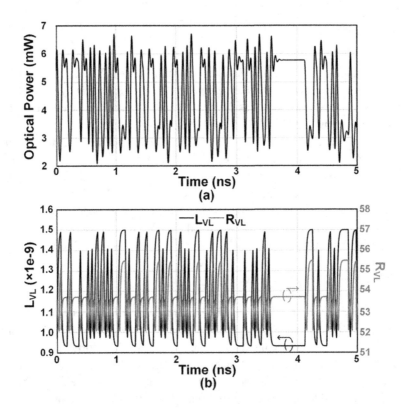

Figure 8.6 Transient simulation using the VCSEL model in Figure 8.4: (a) VCSEL power and (b) variation in L_{VL} and R_{VL}. ©2016 IEEE. Reprinted, with permission, from [6].

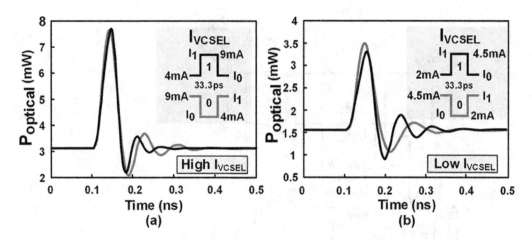

Figure 8.7 Pulse responses for isolated 0 and 1. Response for 0 has been inverted to align with pulse for isolated 1. ©2016 IEEE. Reprinted, with permission, from [6].

the variation in VCSEL dynamics, consider the pulse responses in Figure 8.7. The difference in behaviour for an isolated 1 compared to an isolated 0 motivates a different amount of equalization for rising edges compared to falling edges as presented in Section 8.1.6.

8.1.6 Equalization

Due to the nonlinear behaviour of the VCSEL, its behaviour for an input rising edge is different from its falling edge behaviour. To address this asymmetry, circuits that independently detect rising and falling edges generate short pulses allowing unequal rising/falling edge equalization. The concept is depicted in Figure 8.8. The input signal $Data_{in}$ is delayed by t_{eq}. $Data_{in}$ and its delayed version are applied to rising edge and falling edge detectors that generate pulses of duration t_{eq}. Although both edges are equalized for the same time duration t_{eq}, the magnitude and polarity of equalization is independently controllable through the current in the blocks I_r and I_f. The circuit implementation of the VCSEL driver is shown in Figure 8.9. Its operation is as follows:

- The current source I_{bias} sets the minimum current through the VCSEL. This is the current applied when a 0 is transmitted.
- Since the VCSEL will have a voltage drop across it in the range of 1.5 V, it is connected to a higher voltage supply, $V_{DD,LD}$.
- Power-supply noise is reduced if both sides of the differential pair are connected to the same supply voltage.
- Diode-connected transistors M_{D1} and M_{D2} are added to reduce the voltage stress on the left-hand side of the differential pair.

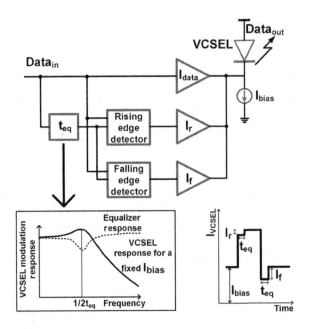

Figure 8.8 Unequal rising/falling edge equalization. ©2016 IEEE. Reprinted, with permission, from [6].

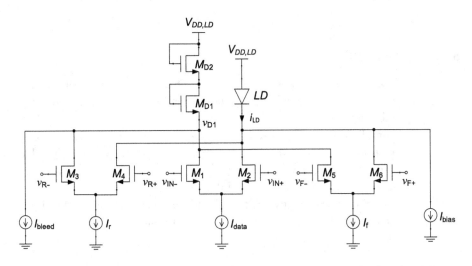

Figure 8.9 Circuit implementation of unequal rising/falling edge equalization based on [6].

- A current source I_{bleed} is added to forward bias the diode-connected transistors so that the voltage across them does not collapse when the three differential pairs are switched to send current to the VCSEL.
- The main differential pair controlled by $v_{IN+/-}$ adds a modulation current of I_{data} to the VCSEL.

- Differential pairs controlled by $v_{R+/-}$ and $v_{F+/-}$ can add a current of I_r/I_f to the output during an interval of t_{eq}.

In Figure 8.8 the current to the VCSEL is *reduced* by I_r/I_f during rising and falling edges. This means that outside of interval t_{eq} around edges, I_r/I_f is added to the VCSEL current. Therefore the VCSEL current for runs of 0s is $I_{bias} + I_r + I_f$ and for runs of 1s is $I_{bias} + I_{data} + I_r + I_f$. More details on the efficacy of equalization can be found in [6].

The delay t_{eq} is a continuous-time delay, implemented using tuneable buffers. Since the amount of equalization depends on the VCSEL's dynamics and not the data rate, the value of t_{eq} required is independent of the data rate.

8.1.7   Voltage-Mode VCSEL Drivers

Following the approach taken in SST drivers for electrical links, presented in Section 5.6, VCSEL drivers can be implemented using a voltage-mode approach, either in a push-pull (differential) connection (Figure 8.2) or single-ended approach. The core of a voltage-mode driver is an inverter as shown in Figures 5.18, 5.19 or 5.20. An appropriately biased VCSEL could be connected directly to v_{OUT-}. Alternatively, AC coupling can be used to separate the dc biasing of the VCSEL from the driver. For example, in [53], a common-cathode VCSEL array with 0 V-biased cathodes is driven in an ac-coupled fashion. Also, the series resistor may be eliminated so as to reduce the output impedance of the driver. If a VCSEL is driven differentially with two voltage-mode drivers, the VCSEL is connected across the V_{OUT} terminals using ac coupling capacitors.

The main design parameters in any voltage-mode driver are the width of the MOSFETs, the value of any series resistors and the supply voltage. Invariably, tuning of output impedance is needed, either through enabling more output-connected inverters or by varying the supply voltage. Enabling fewer or more inverters is a "digital-friendly" approach. In Figure 5.20 the logic gates preceding the output transistors allow the inverter to be turned off, thereby presenting an open circuit to the output V_O.

When a driver is ac-coupled, a high-pass response is introduced with a cut-off frequency given by:

$$f_{-3dB} = \frac{1}{2\pi R_{EQ} C},$$ (8.10)

where R_{EQ} is the series combination of the VCSEL's resistances and the output resistance of the driver. Making the driver output resistance lower was beneficial for increasing the damping factor of resonances introduced by wirebonds but it increases the cut-off frequency of the ac-coupling circuit. Instead of increasing the value of the capacitance, an auxiliary high-voltage-tolerant dc-coupled signal path can be introduced, using thick-oxide transistors. Invariably, these devices are slower, giving rise to a low-frequency driver. The combination of high-frequency and low-frequency drivers has been used in [53] and [57], the latter implementation shown in Figure 8.10.

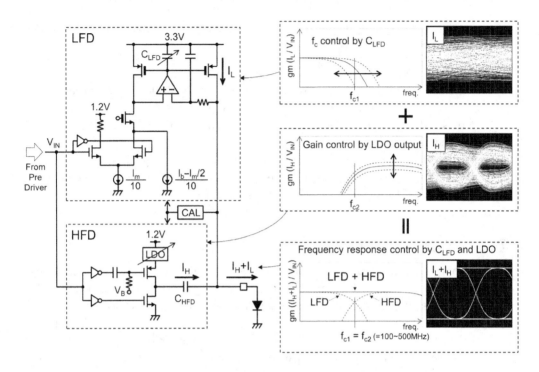

Figure 8.10 VCSEL driver with ac-coupled low-voltage transistors and dc-coupled low-frequency signal path. ©2014 IEEE. Reprinted, with permission, from [57].

8.2   Mach–Zehnder Modulator Drivers

The approach to driving a Mach–Zehnder interferometer-based modulator differs from that for a VCSEL for two main reasons, namely:

1. **Physical size:** an MZI can be comparatively large, with a length in the range of a few mms. The propagation time of signals along its electrodes can be sufficiently long that each electrode must be treated as a distributed circuit. As introduced in Section 7.4.1, edges of the voltage signal applied to the electrodes should propagate down the electrode in phase with the optical signal in the waveguide.
2. **Voltage swing:** the required drive voltage, related to the MZI's V_π, can be larger than the breakdown voltage of an advanced CMOS process. Whereas a laser diode required a dc bias larger than the breakdown of advanced CMOS, an MZI requires a voltage swing that is large.

8.2.1   Overview of MZI Driver Architectures

Figure 8.11 shows three approaches to delay matching. In part (a), a travelling wave structure is used. Since the optical signal is confined in Si with $\epsilon_r \approx 11.47$, whereas the electrical signal propagates in SiO$_2$ with $\epsilon_r \approx 3.5$, the electrical signal travels faster, requiring either lumped capacitance to slow it, or meanders in the transmission

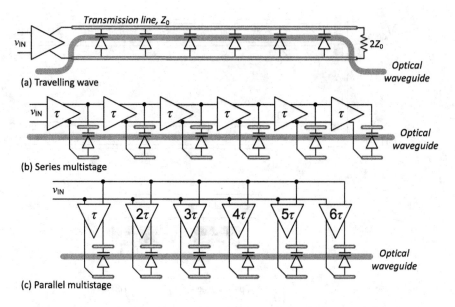

Figure 8.11 Various approaches for matching the phase of electrical and optical signals, adapted from [58]. Diode-like symbols represent p-n junctions used in reverse-bias as high-speed phase modulators (HSPMs): (a) travelling wave approach; (b) series multistage approach; and (c) parallel multistage approach.

line between MZI segments. If the electrodes each have a characteristic impedance of Z_0, the structure is terminated differentially with a resistance of $2Z_0$. Longer electrode length allows for lower V_π but introduces more attenuation for the electrical signal in the electrode and more optical loss.

Figure 8.12 (a) shows a dual-drive travelling-wave modulator. Such a device is designed to have a $\pi/2$ shift when the driver output is zero. A positive V_{Drive} pushes the modulator toward a π phase shift and a negative drive voltage toward 0 phase shift. The differential, dual-drive approach allows for lower modulation voltage, as discussed in Example 7.2. The lumped capacitance from the segments of high-speed phase modulators slow down the electrical signals to match the propagation of the optical signals in each branch.

Transmission-line loss can be eliminated in a multistage approach. Figure 8.11 (b) shows a series multistage design in which the MZI is separated into electrically distinct sub-electrodes. If the optical signal sees a delay of τ between MZI sections, the electrical signal is driven by a cascade of buffers each introducing a delay of τ. Since each sub-electrode is shorter than the overall electrode in the distributed case, it is treated as a lumped capacitance requiring no termination resistance. The challenge in this approach is implementing a small enough τ. However, all electrodes see the same voltage swing, unlike in the travelling-wave design where the portion of the electrode far from the driver sees a reduced voltage swing.

In a parallel multistage approach, shown in Figure 8.11 (c), the MZI electrodes are driven in parallel by buffers having delays increasing in steps of τ. Although depicted

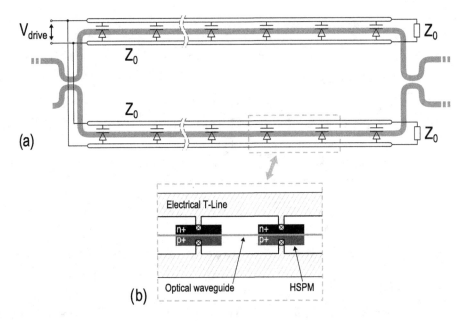

Figure 8.12 Dual-drive travelling-wave modulator shown with periodic spacing of high-speed phase modulators (HSPMs): (a) dual-drive architecture and (b) expanded view of waveguide, transmission line and HSPMs. ©2016 IEEE. Reprinted, with permission, from [59].

with a minimum delay of τ this approach works equally well if the smallest delay is larger than τ, provided each parallel stage has a delay increased by τ. The delay increment must be matched across PVT variation. Thus, a robust way to implement a delay of 6τ is to cascade six copies of the circuit that implements the delay of τ. This is power-inefficient and still requires the use of a circuit with delay of only τ. Regardless of how the delays are implemented, the parallel structure loads the preceding stage with a larger capacitance. A discussion and comparison of travelling-wave and segmented drivers can be found in [59].

If active circuits with only a delay as short as 2τ are feasible, a parallel approach shown in Figure 8.13 uses a parallel scheme of active circuits to generate delays of $0, 2\tau, 4\tau$ while a passive circuit generates delays of τ. The 2τ circuits, shown in the inset in the lower left of Figure 8.13, use current-starved inverters. The first inverter is current-starved while the second has no starving to ensure fast rise and fall times.

8.2.2   Circuits for Driving Lumped Electrodes above Typical CMOS Voltage Limits

The MZI presented in [58] uses a silicon photonic structure based on the free carrier plasma dispersion effect. The optical waveguide has diode structures across it. The effective index is modulated by operating the diode in reverse bias with voltages ranging from 0 to 2.5 V. This is approximately 2x the safe operating range for 65 nm

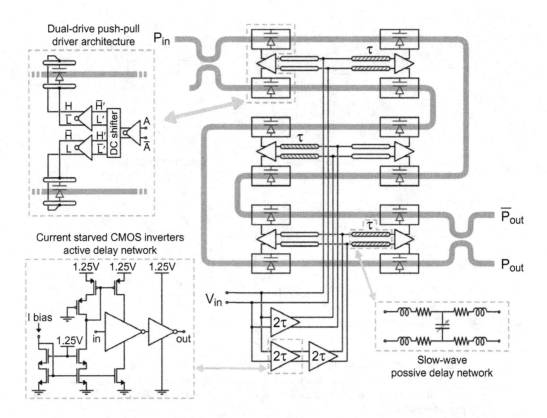

Figure 8.13 Parallel MZI driver approach, using active and passive delays. ©2015 IEEE. Reprinted, with permission, from [58].

CMOS. To safely generate this high voltage swing, the circuits in Figure 8.14 are used, the core of which is a stack of two inverters. When L' is high, $\overline{L}$ is pulled to ground. Likewise, $\overline{H}' = 1.25$ V turns on the uppermost PMOS, raising H to 2.5 V. The MZI section has 2.5 V across it. Since the uppermost PMOS's source is connected to 2.5 V, its gate is only pulled down to 1.25 V, rather than ground. To discharge the voltage across the MZI section to 0, $\overline{H}'$ is raised to 2.5 V turning on the upper NMOS. L' is lowered to ground, turning on the lower PMOS. This creates a low resistance path across the MZI section, discharging the voltage across it to 0 V. To keep the MOSFETs operating safely, we further require that $H = \overline{L} = 1.25$ V. This is achieved if the parasitic capacitances at those nodes are equal so that the charge redistribution between those nodes leads to their voltages meeting half-way between 0 and 2.5 V.

The drive voltages for the upper inverter in the stack are generated using the level-shifter scheme in Figure 8.14 (bottom). At its core is a CMOS latch formed by back-to-back inverters operating from the high-supply domain (2.5/1.25 V). The state of the latch is toggled by inverters operating in the low-supply domain (1.25/0 V) through coupling capacitors $C1$ and $C2$.

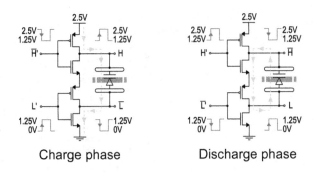

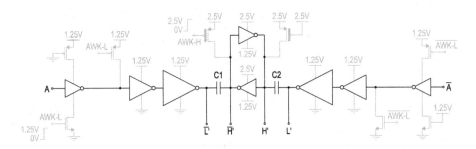

Figure 8.14 CMOS circuits for generating a 2.5 V swing. The top section shows stacked inverters. The bottom section shows level shifters. Transistors for pre-charging nodes are shown in grey. Thick-oxide devices are drawn with thick gates. ©2015 IEEE. Reprinted, with permission, from [58].

8.3   Microring Modulator Drivers

With the fundamentals of mircoring resonators when used as modulators described in Section 7.6, this section presents considerations for designing the driver electronics. To predict the overall electro-optical behaviour of a driver and modulator, we need a circuit simulator model of the MRM that includes electrical parasitics, diode parameters and the carrier/photon interactions in the ring resonator. Section 8.3.1 presents suitable models following a framework similar to that used for VCSELs.

A second challenge in designing MRM drivers is the need to have voltages beyond the safe operating voltages of scaled CMOS, presented in Section 8.3.2.

8.3.1   Microring Modulator Models

Figure 8.15 [60], [61] shows an equivalent circuit that models the complete electrical to optical conversion of a microring modulator. V_{Signal} is the signal voltage generated by the driver. The "Parasitics" block includes a lumped model of the pad capacitance and any wirebond inductance in the signal path. In the case of a driver wirebonded to a ring, there are pad capacitances on both sides of the wirebond. Here, R_S and C_j model the series resistance and junction capacitance of the reverse-biased p-n-junction in the

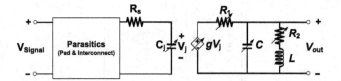

Figure 8.15 Equivalent circuit of a microring modulator. Reprinted with permission from [60].

ring. Its capacitance is voltage-dependent, captured by the arrow across the symbol. The voltage V_j determines the width of the depletion region through which the optical signal circulates. The remaining elements model the dynamics of how optical power is delivered to the ring's through port. These element values depend on the voltage applied to the p-n junction and on the wavelength relative to the ring's transmission minimum.

The modelling approach is explained in more detail below. The normalized electrical to optical transfer function of the ring has been modelled as:

$$\frac{P_{out}}{V_j} = K\frac{s + 2/\tau_l}{s^2 + (2/\tau)s + D^2 + 1/\tau^2},$$ (8.11)

where $D = |\omega_{in} - \omega_r|$ quantifies how off-resonance the ring's optical input (ω_{in}) is from its resonance (ω_r); τ_l is a decay time of the ring due to round-trip optical loss; t_e is another decay time related to coupling to the ring's bus. The overall decay time τ is found as $\tau^{-1} = \tau_l^{-1} + \tau_e^{-1}$. Additionally, K is a gain term which normalizes the transfer function. These parameters can be found from ring measurements that sweep the optical transmission spectrum of the ring.

The circuit in Figure 8.15 has a transfer function from V_j to V_{out}:

$$\frac{V_{out}}{V_j} = \frac{g}{R_1 C}\frac{s + \frac{R_2}{L}}{s^2 + \left(\frac{1}{R_1 C} + \frac{R_2}{L}\right)s + \frac{1}{LC}\left(1 + \frac{R_2}{R_1}\right)}.$$ (8.12)

Knowing the parameters of (8.11) from measurements, circuit element values in (8.12) can be found by matching terms. An example from [61] is shown in Table 8.4.

8.3.2 Microring Modulator Driver Circuits

As presented in Section 7.6, microring modulators need voltages with an amplitude of a few volts, well beyond the safe operating range of modern CMOS transistors. Figure 8.16 shows three approaches to overcome the voltage limits of a given process technology. In Figure 8.16 (a), a cascode structure is employed where the cascode transistors connected to V_{B1} and V_{B2} are thick-oxide transistors available in most CMOS technologies. Since these transistors usually have longer minimum channel length than the thin-oxide transistors, this approach is limited by the larger capacitance of these transistors. High-speed operation can be challenging. The resistively loaded common-source amplifier also dissipates static power. In part (b), static power is eliminated by using an inverter. Voltage tolerance requires cascoding transistors and the use of

Table 8.4 MRM model element values. ©Chinese Laser Press. Reprinted with permission from [61].

V_{Bias} (V)	R_s (Ω) $D_\lambda = 40$ and 70 pm	C_j (fF) $D_\lambda = 40$ and 70 pm	R_1 (kΩ) $D_\lambda = 40$ pm	R_1 (kΩ) $D_\lambda = 70$ pm	C (fF) $D_\lambda = 40$ pm	C (fF) $D_\lambda = 70$ pm	R_2 (kΩ) $D_\lambda = 40$ and 70 pm	L (nH) $D_\lambda = 40$ and 70 pm
0		14.26	0.35	2.07	42.15	7.14	10.00	
−1		10.95	0.75	3.15	19.65	4.70	9.96	
−2	249	9.47	1.77	5.15	8.37	2.87	9.71	114.41
−3		8.55	3.01	7.19	4.92	2.06	9.71	
−4		7.90	4.20	8.99	3.52	1.65	9.71	

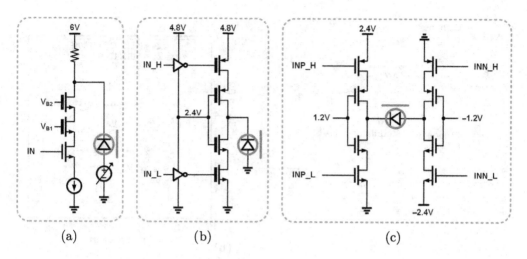

Figure 8.16 Microring modulator driver candidates: (a) double cascode using thick-oxide transistors; (b) cascoded inverter using thick-oxide transistors; (c) pseudo differential thin-oxide inverters. ©2015 IEEE. Reprinted, with permission, from [62].

thick-oxide devices. The output can be driven between 0 and 4.8 V, while worst-case V_{DS}/V_{GS} is limited to 2.4 V. Figure 8.16 (c) shows how two inverters can differentially drive a MRM. The left-hand inverter produces output voltages of 0 and 2.4 V while the right-hand inverter, biased from ground and −2.4 V, generates output voltages of 0 and −2.4 V. Thus, the ring is driven between a neutral bias (0 V) and a reverse bias of 4.8 V. Through the use of level shifters, gate voltages INP_H, INP_L, INN_H and INN_L toggle such that no transistor's $|V_{GS}|$ exceeds 1.2 V. Therefore, high-speed thin-oxide transistors can be used throughout.

Biasing part of the circuit with a negative supply complicates the design. Instead, as shown in Figure 8.17 (a) [62], ac-coupling capacitors C_C can be used between the driver and the MRM. Each driver generates a voltage swing of $2V_{DD}$. To reproduce the differential voltages generated in Figure 8.16 (c), the anode bias is set at V_{DD} whereas the cathode is biased at $3V_{DD}$. The frequency response plots in Figure 8.17 (b) show the importance of selecting the correct driver output resistance. These plots show the electrical to electrical behaviour of a linear model of the driver and the electrical parasitics of the interconnection and the ring. A full electrical to optical simulation including a model from Section 8.3.1 is advisable.

8.4 Summary

Building on the basic transmitter building blocks introduced in Chapter 5, this chapter presented laser diode drivers, Mach–Zehnder modulator drivers and microring modulator drivers. Several differences between electrical link drivers and laser-diode drivers were discussed, among which was the need for nonlinear equalization. A model for

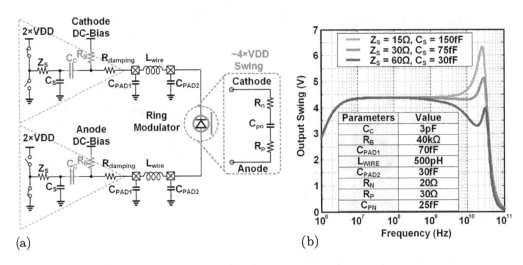

(a)   (b)

Figure 8.17 Microring modulator driver: (a) connected through ac coupling capacitors to ring; (b) simulated frequency response. ©2015 IEEE. Reprinted, with permission, from [62].

the electrical interface to the laser diode and its electron-photon dynamics was discussed. Circuits based on edge detectors that perform asymmetrical equalization were introduced. Both CML and SST laser-diode drivers were considered.

The main architectures for driving Mach–Zehnder modulators were compared: travelling wave, parallel multistage, series multistage and hybrid schemes. Techniques for generating voltage signals beyond a CMOS technology's standard V_{DD} were explained. The chapter closed with an overview of microring modulator models and driver circuits.

Problems

8.1 Explain the impact of bondwire inductance on the electrical performance of laser-diode drivers. Under what circumstances is the inductance helpful?

8.2 The asymmetry of driving an off-chip VCSEL makes differential CML VCSEL drivers sensitive to supply inductance. Simulate the impact of a 1 nH supply inductance on the driver in Figure 8.1. Add one between R_T and V_{DD} and one in the $V_{DD,LD}$ supply. Explain how the electrical eye (current through VCSEL) is degraded. Does the eye improve using the topology in Figure 8.9 where the currents through the LD and M_{D2} share the same wirebond? For this question, use a resistor to model the IV behaviour of the VCSEL.

8.3 Explore, through simulation, the MZI driver presented in Figure 8.13. Clearly, correct operation assumes that the circuit implements the correct value of τ (6.4 ps in the case of [58]) and that the τ of the passive delay networks matches the τ of the current-starved CMOS inverters. These delays must also match the optical delay of the line. If your CMOS technology is too slow to achieve a small enough delay, for the rest of this problem use what you can achieve. Explore the mismatch in τ as process parameters, supply voltage and temperature are adjusted. Can you readjust I_{BIAS} to tune the delay for the extremes of PVT variation?

8.4 This question investigates the stacked ring driver in Figure 8.16 (c). Design the circuit for a supply voltage of twice the technology's V_{DD}. Verify through transient simulation that no v_{GS} or v_{DS} exceeds the technology's standard V_{DD}.

9 Optical Receivers

This chapter begins with a recapitulation of an optical link and what the general requirements of an optical receiver are. The discussion of optical receivers starts with a brief analysis of a passive current-voltage converter (i.e., a resistor) in terms of its gain, bandwidth and input-referred noise. This section proposes reasonable bandwidth requirements for optical receivers that do not use equalizers, so-called low-ISI systems. Open-loop and feedback amplifiers are considered. Additional amplification through main amplifier design is explained, starting with the effect on bandwidth of cascading multiple first-order stages. Behaviour of second- and third-order systems are also presented. Examples of Cherry-Hooper, second-order active feedback and third-order active feedback as well as interleaving feedback are presented. In addition, CMOS inverter-based designs are discussed.

If the gain of an optical receiver is increased, its bandwidth and input-referred noise will decrease, introducing ISI that must be removed. Chapter 10 presents this approach.

9.1   Metrics for Optical Receivers

Figure 9.1 shows an optical link. Recall that optical power incident on the photodiode (PD) produces a proportional current. The optical receiver converts this current to a voltage. A circuit that performs this conversion is referred to as a transimpedance amplifier (TIA), with units of amplification of Ω. Usually, further amplification in the main amplifier (MA) is required before applying the received signal to a decision circuit in the clock and data recovery unit. An optical receiver front-end should have:

- **High sensitivity:** the sensitivity of an optical receiver is defined as the minimum incident optical power required for the receiver to provide a BER less than a specified value. When no additional error control coding is used, a BER of 10^{-12} is usually targeted. Sensitivity is quantified either in terms of an OMA or an average power. When average power is used, the assumed extinction ratio must also be stated. In this context, "high sensitivity" is used to suggest a highly sensitive receiver capable of receiving a low optical power. Sensitivity is a function of both noise and gain. Thus, each of these is discussed below.

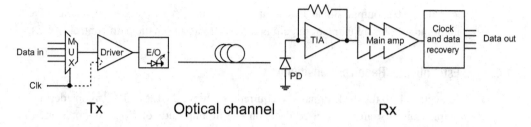

Tx   Optical channel   Rx

Figure 9.1 Block diagram of a generic optical link.

- **Low noise:** the noise added to the signal by the receiver should be minimized, particularly in the first stage of the receiver. The input-referred noise dictates a minimum input signal level required to meet the link's target BER.
- **High gain:** so that small photocurrents give rise to latch-friendly voltages, a gain in the $k\Omega$ range is common. This gain is traditionally achieved by a transimpedance amplifier (gain in Ω) followed by a main amplifier (gain in V/V). For a given topology, increasing TIA or main amplifier (MA) gain will reduce its bandwidth.
- **Bandwidth:** optical receivers fall into two broad categories, namely those that are designed to introduce a small amount of ISI and those that favour a smaller number of high-gain stages, thereby introducing ISI. This second category uses an equalizer to remove the ISI. Both types are discussed in this book. In the case of low-ISI systems, a bandwidth of approximately 50 to 70% of the data rate is required.

The list above applies to all optical receivers. Depending on the particular application, other aspects of a front-end are important such as linearity, automatic-gain control and recovery from overload conditions.

The linearity requirements of an optical receiver front-end vary considerably depending on its application. In an NRZ link the signal is resolved into 1s and 0s by directly applying the amplified input signal to a decision circuit. In this case, when designed for low-ISI, the gain of the front-end can be so large that the later stages of the main amplifier will saturate. In this scenario, the amplitude of the output will not change if the input signal increases. The main amplifier is said to be a "limiting amplifier." On the other hand, if the optical receiver is a low-bandwidth implementation where ISI is introduced, the circuit must be linear until the ISI is sufficiently removed. When PAM4 modulation, or more complex modulation, is used, detection of the transmitted 1s and 0s is usually done in the digital domain. The output of the main amplifier is quantized by a high-speed ADC. This requires a linear receiver to avoid bit errors.

The design of high-resolution ADCs in the 10s of GSamples/s range is challenging. Current designs at these speeds are limited to approximately 7 bits of resolution. To reduce the impact of quantization error, the received signal must occupy the entire ADC input range, without exceeding it. Therefore, in addition to high linearity, the front-end must automatically adjust its gain so that whatever the input signal amplitude, the output signal is a relatively constant, high percentage of the ADC's input range. This adjustment in gain is referred to as automatic gain control (AGC).

9.1.1   Estimation of Receiver Sensitivity

In Section 1.10 the peak signal V_P required to achieve a BER of 10^{-12} considering an output noise σ_N and a required decision circuit amplitude of V_{MIN} was presented in (1.31), rewritten here in terms of signal to noise ratio (SNR):

$$V_P = \sigma_N SNR + V_{MIN}. \tag{9.1}$$

This can be modified to use peak-to-peak quantities:

$$V_{PP} = \sigma_N SNR_{PP} + V_{MIN_{PP}}, \tag{9.2}$$

where V_{PP} is the peak-to-peak output swing of the analog part of the receiver, then applied to the input of the decision circuit, SNR_{PP} is the required SNR for the target BER, computed based on the peak-to-peak output signal (14 in the case of 10^{-12}) and $V_{MIN_{PP}}$ is the peak-to-peak signal required for correct decision circuit operation. This expression can be referred to receiver input quantities by dividing by the gain. We use the midband gain, assuming we introduce negligible ISI:

$$I_{PP} = \frac{\sigma_N SNR_{PP} + V_{MIN_{PP}}}{Z_{FE}}, \tag{9.3}$$

where I_{PP} is the required peak-to-peak input current signal and Z_{FE} is the overall gain of the front-end (TIA/MA combined). We can rewrite (9.3) in terms of input-referred noise as:

$$I_{PP} = \sigma_{i,N} SNR_{PP} + \frac{V_{MIN_{PP}}}{Z_{FE}}. \tag{9.4}$$

To convert the required input current to an incident optical power, we divide by the photodiode's responsivity, R_{PD}, to compute OMA as:

$$OMA_{MIN} = \frac{I_{PP}}{R_{PD}} = \frac{1}{R_{PD}} \left(\sigma_{i,N} SNR_{PP} + \frac{V_{MIN_{PP}}}{Z_{FE}} \right). \tag{9.5}$$

From (9.5) we see that to reduce OMA_{MIN} we need to reduce input-referred noise $(\sigma_{N,i})$ and either reduce the required voltage to the latch $(V_{MIN_{PP}})$ or increase the gain (Z_{FE}). Thus, both low noise and high gain are needed.

The effect of finite gain on input sensitivity is sometimes expressed as a power penalty:

$$OMA_{MIN} = \frac{1}{R_{PD}} \sigma_{i,N} SNR_{PP} PP, \tag{9.6}$$

where PP is a power penalty given by:

$$PP = 1 + \frac{V_{MIN_{PP}}}{\sigma_{i,N} SNR_{PP} Z_{FE}}. \tag{9.7}$$

Power penalty computes the ratio of the OMA needed in the case of a practical decision circuit to that for an ideal decision circuit ($V_{MIN,PP} = 0$).

Example 9.1   Find the input sensitivity of an optical receiver having the following parameters:

- Input-referred noise current of 1.5 μA (rms)
- Front-end gain $Z_{FE} = 5\,000\ \Omega$
- Decision circuit voltage $V_{MIN_{PP}} = 35$ mV
- Photodiode responsivity of 0.7 A/W

Solution

Using (9.5) we compute:

$$OMA_{MIN} = \frac{1}{R_{PD}} \left(\sigma_{i,N} SNR_{PP} + \frac{V_{MIN_{PP}}}{Z_{FE}} \right) \tag{9.8}$$

$$= \frac{1}{0.7 \text{ A/W}} \left((1.5\ \mu\text{A})(14) + \frac{35 \text{ mV}}{5000\ \Omega} \right) \tag{9.9}$$

$$= \frac{1}{0.7 \text{ A/W}} (21\ \mu\text{A} + 7\ \mu\text{A}) \tag{9.10}$$

$$= 40\ \mu\text{W} \tag{9.11}$$

$$OMA_{MIN_{dBm}} = -14 \text{ dBm.} \tag{9.12}$$

The power penalty can be computed using (9.13):

$$PP = 1 + \frac{35 \text{ mV}}{(1.5\ \mu\text{A})(14)(5000\ \Omega)} = 1 + \frac{35}{105} = 1.33. \tag{9.13}$$

The computed PP means that with an ideal decision circuit, the input power could be reduced by a factor of 1.33. In dB, this means the input power can be reduced by $10\log_{10} 1.33 = 1.25$ dB to -15.25 dBm.

The example above showed how to estimate input sensitivity. The power penalty was relatively small for this combination of decision circuit sensitivity and gain. If a better decision circuit were available, the same overall sensitivity could be achieved with a lower gain front-end. This is beneficial since, as we will see later in this chapter, higher gain requires more power dissipation.

9.2   Transimpedance Amplifiers for Low-ISI Systems

In this section we consider TIAs for use in front-ends that do not include equalization to remove ISI. Therefore, we must design the TIA to have sufficient bandwidth, usually taken as being between 50 and 70% of the bit rate. For a 56 Gb/s NRZ system, the TIA should have a bandwidth between 28 and 39 GHz.

We start by considering a resistor as a TIA. After discussing its trade-offs, active TIAs are presented.

9.2.1   Resistors as TIAs

Consider the schematic shown in Figure 9.2. The photocurrent is converted to a voltage using a load resistor, denoted by R_L. The capacitance C is the sum of the photodiode's junction and pad capacitance along with the parasitic capacitance of the receiver chip. Whatever downstream circuitry processes the signal v_{OUT} will load this circuit requiring its capacitance to be added to C. This circuit provides a transfer function, bandwidth, low-frequency gain and mean-squared output noise given by:

$$Z_{TIA}(s) = \frac{V_{OUT}(s)}{I_{PD}(s)} = \frac{R_L}{sR_LC + 1},\qquad(9.14)$$

$$f_{-3dB} = \frac{1}{2\pi R_L C},\qquad(9.15)$$

$$A_{dc} = R_L,\qquad(9.16)$$

$$\overline{v_{n_{out}}^2} = \frac{kT}{C},\qquad(9.17)$$

where k is the Boltzmann constant and T is the absolute temperature in Kelvin. See Appendix Section B.3 for more details on the calculation of mean-squared noise. From the equations above, we can see that increasing R_L gives higher gain, but lower bandwidth. For this topology, we can estimate its input current sensitivity for a given BER, assuming an ideal latch. In using A_{dc} to refer the output noise to the input, we implicitly assume that the input signal is at a low enough data rate such that negligible ISI is introduced. For a first-order system, we will assume that the bandwidth of the circuit must be at least 50% of the data rate (f_b). From this we calculate the input-referred noise current $(\overline{i_{n_{in}}^2})$ as:

$$\overline{i_{n_{in}}^2} = \frac{\overline{v_{n,out}^2}}{|Z_{TIA}(0)|^2} = \frac{kT}{CR_L^2},\qquad(9.18)$$

$$f_b = 2f_{-3dB} = \frac{1}{\pi R_L C}.\qquad(9.19)$$

If the load resistor is the only source of noise in this system, we need a peak-to-peak input current 14x the input referred noise for a BER of 10^{-12}. Using (9.18) and (9.19) the relationship between capacitance, data rate, input-referred noise ($i_{n_{in}}$) and

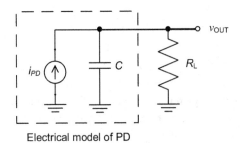

Electrical model of PD

Figure 9.2 Photodiode and resistive TIA.

sensitivity ($i_{pp_{min}}$) are calculated as:

$$\overline{i_{n_{in}}^2} = \pi^2 kTCf_b^2 = \frac{kT\pi f_b}{R_L}, \tag{9.20}$$

$$i_{n_{in}} = \pi f_b \sqrt{kTC}, \tag{9.21}$$

$$i_{pp_{min}} = 14\pi f_b \sqrt{kTC}. \tag{9.22}$$

Equation (9.22) shows the important relationship between data rate, capacitance and noise. Smaller photodiodes with reduced parasitic capacitance will improve sensitivity. Although this is illustrated for a passive TIA, the trend extends to amplifier-based TIAs as well. Similarly, increasing the data rate degrades sensitivity.

Example 9.2   At a data rate (f_b) of 10 Gb/s with a capacitance (C) of 200 fF, find the input-referred rms noise current, dc gain, output swing for a noise-limited input.

Solution
Substituting into (9.21) the input-referred rms noise current is 0.9 μA. Multiplying by 14 we get an input sensitivity of approximately 13 μA. By rearranging (9.19) we can solve for $R_L = 318\ \Omega$. Using this gain, the noise-limited peak-to-peak input current of 13 μA leads to a peak-to-peak output voltage of about 4 mV. This voltage is too small to be reliably latched in a 100 ps UI, even with sub-rate clocking. Additional amplification is needed. However, with a relatively low TIA gain, the noise of subsequent stages will degrade sensitivity.

Notice that for the passive TIA, its "input resistance" and its gain are both R_L. Active circuits can be designed such that their input resistance, which will form a time constant with C, will be smaller than their TIA gain. Ultimately, active circuits can be used to achieve higher gain for a given input-referred noise.

9.2.2   Open-Loop TIAs Focusing on the Common-Gate TIA

Regardless of the topology of the TIA, its input resistance and the combined capacitance of the photodiode, pad and TIA input form an RC time constant. Therefore, any candidate for a wide-band TIA must have low input resistance. Of the three single-transistor amplifiers, when used in an open-loop configuration, the common-gate amplifier has the lowest input resistance, approximately equal to the $\frac{1}{g_m}$ of the gain transistor. A common-gate TIA is shown in Figure 9.3.

Transistor M_B biases the common-gate transistor M_{CG}, while V_B is set by a current mirror. Provided V_G is high enough, M_B is in saturation and acts like a current source. The photodiode's current, modelled by input source i_{in}, flows through M_{CG} and develops a signal voltage across R, leading to a low-frequency transimpedance gain of R. The high-frequency gain is limited by capacitance at the input and output. Here, C_I is the sum of the photodiode capacitance (C_J), pad capacitance (C_P), C_{db} and C_{gd}

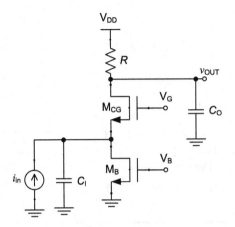

Figure 9.3 Common-gate transimpedance amplifier.

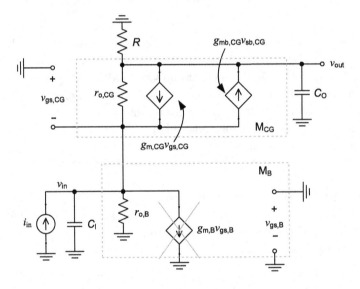

Figure 9.4 An ac small-signal equivalent circuit of the common-gate TIA.

of M_B and C_{gs} and C_{sb} of M_{CG}; C_O is the sum of whatever load capacitance is present (C_L) and C_{gd} and C_{db} of M_{CG}. The ac small-signal equivalent circuit is shown in Figure 9.4.

Nodal analysis of this circuit using v_{in} and v_{out} as the unknown node voltages leads to the following nodal equations, initially including transistor terminal voltages:

$$sC_I v_{in} + g_{o,B} v_{in} + g_{o,CG}(v_{in} - v_{out}) -$$
$$g_{m,CG} v_{gs,CG} + g_{mb,CG} v_{sb,CG} = i_{in}, \qquad (9.23)$$
$$sC_O v_{out} + g_{o,CG}(v_{out} - v_{in}) + g_{m,CG} v_{gs,CG} -$$
$$g_{mb,CG} v_{sb,CG} + G v_{out} = 0, \qquad (9.24)$$

where $G = 1/R$, $g_{o,CG} = 1/r_{o,CG}$ and $g_{o,B} = 1/r_{o,B}$. Since the gate of M_{CG} is grounded in an ac sense, we can substitute $v_{gs,CG} = -v_{in}$. Also, $v_{sb,CG} = v_{in}$. With these substitutions, the equations can be written in matrix form as:

$$\begin{bmatrix} sC_I + g_{o,B} + g_{o,CG} + g_{m,CG} + g_{mb,CG} & -g_{o,CG} \\ -g_{o,CG} - g_{m,CG} - g_{mb,CG} & sC_O + g_{o,CG} + G \end{bmatrix} \begin{bmatrix} v_{in} \\ v_{out} \end{bmatrix} = \begin{bmatrix} i_{in} \\ 0 \end{bmatrix}.$$

(9.25)

The nodal equations are now in the form $Av = J$ where A is the admittance matrix, v is the vector of unknown node voltages and J is the source vector. One convenient method to solve such equations in the s-domain is Cramer's rule. A summary of this approach is presented in Appendix Section A.1.

$$v_{out} = \frac{|A_2|}{|A|},$$

(9.26)

where $|\ |$ denotes the determinant operator and A_2 is a modified admittance matrix in which the second column in A is replaced by the source vector J. For the system described in (9.25) $|A_2|$ is given by:

$$|A_2| = \begin{vmatrix} sC_I + g_{o,B} + g_{o,CG} + g_{m,CG} + g_{mb,CG} & i_{in} \\ -g_{o,CG} - g_{m,CG} - g_{mb,CG} & 0 \end{vmatrix} = i_{in}(g_{o,CG} + G_m), \quad (9.27)$$

where $G_m = g_{m,CG} + g_{mb,CG}$. $|A|$ is found as:

$$|A| = (sC_I + g_{o,B} + g_{o,CG} + G_m)(sC_{OU} + g_{o,CG} + G)$$
$$-(-g_{o,CG})(-g_{o,CG} - G_m)$$
$$= s^2 C_I C_O + s\left(g_{o,CG}(C_I + C_O) + C_O(g_{o,B} + G_m,) + sC_I G\right)$$
$$+G\left(G_{m,CG} + g_{o,B}\right) + g_{o,CG} G_{m,CG}.$$

(9.28)

Therefore, $v_{out} = \frac{|A_2|}{|A|}$ and the transimpedance gain is given by $\frac{v_{out}}{i_{in}} = \frac{|A_2|}{i_{in}|A|}$:

$$\frac{v_{out}}{i_{in}}$$

$$= \frac{(g_{o,CG}+G_m)}{s^2 C_I C_O + s\left(g_{o,CG}(C_I+C_O) + C_O(g_{o,B}+G_m) + sC_I G\right) + G\left(G_m+g_{o,B}\right) + g_{o,CG} G_m}.$$

(9.29)

The expression above does not provide much insight. One might prefer having a unity coefficient for the s^2 term in the denominator or a constant term in the denominator of unity. Notice that if $g_{o,CG} = g_{o,B} = 0$ (9.29) is greatly simplified to:

$$Z_{CG}(s) = \frac{v_{out}}{i_{in}} = \frac{G_m}{s^2 C_I C_O + s(G_M C_O + sC_I G) + GG_m}$$

$$Z_{CG}(s) = \frac{G_m}{(sC_I + G_m)(sC_O + G)}.$$

(9.30)

Since the denominator of (9.30) can be easily written in factored form, the poles of the common-gate TIA have simple expressions. The relevant parameters of the transfer function can be written as:

$$Z_{CG}(0) = \frac{G_m}{G_m G} = R, \tag{9.31}$$

$$f_{p_i} = -\frac{G_m}{2\pi C_I}, \tag{9.32}$$

$$f_{p_o} = -\frac{1}{2\pi R C_O}, \tag{9.33}$$

where $Z_{CG}(0)$ is the dc gain of the TIA and f_{p_i} and f_{p_o} are the input and output poles, respectively. Note that the two poles are always real. In fact, even if we include non-zero g_o for the transistors, the poles are real. The comparison between the gain and poles of the common-gate TIA when non-zero g_o is taken into account is left as an exercise to the reader. The bandwidth limitation of the circuit is usually set by the input pole. This is because C_I can be much larger than C_O since the former is due to elements such as ESD protection, bondpads and a potentially large junction capacitance of the photodiode. The common-gate transistor is usually biased to achieve as large a G_m as possible given the supply limitations and given that increasing its width will further increase C_I. With the input pole set, a designer will usually select R to be as large as possible without bringing f_{p_o} below f_{p_i}. If $f_{p_o} \gg f_{p_i}$ then the bandwidth of the circuit can be approximated as $f_{-3dB} = f_{p_i}$. As R is increased, reducing f_{p_o}, the bandwidth can be approximated a couple of ways, presented in Appendix Section A.2. The first method uses the coefficients of the denominator directly, without requiring expressions or values for the poles, useful if we want to consider the bandwidth of (9.29). The second approach uses the actual values of the poles.

For the -3 dB bandwidth of the common-gate TIA to be 90% of the dominant pole we need the non-dominant pole to be 2.77 times larger. Although the bandwidth will approach 100% of the dominant pole's magnitude if we push the non-dominant pole higher, it is not advantageous, as this will come at the cost of lowering the gain. Also, as we will see in the following subsection, the noise performance degrades as the non-dominant increases.

9.2.2.1   Noise Analysis of the Common-Gate TIA

The noise analysis follows the approach outlined in Appendix B. The ac small-signal circuit with noise sources added is shown in Figure 9.5. For the circuit analysis given below, output conductances of the MOSFETs are assumed to be zero. To allow all three noise sources to be treated in a similar fashion computationally, the current-source model of the resistor's noise has been used. For each noise source the transfer function from the noise current source to the output voltage, v_{out}, must be calculated. All three noise transfer functions (NTFs) are found by solving the equation below for different source vectors, J_i:

$$\begin{bmatrix} sC_I + G_m & 0 \\ -G_m & sC_O + G \end{bmatrix} \begin{bmatrix} v_{in} \\ v_{out} \end{bmatrix} = J_i. \tag{9.34}$$

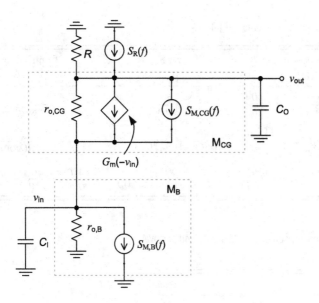

Figure 9.5 An ac small-signal equivalent circuit of the common-gate TIA (simplified for noise analysis).

The source vectors and other relevant aspects of the noise analysis are shown in Table 9.1. Each NTF is found by solving for v_{out}/i_t where i_t is the respective noise current source. Since the resistor's noise flows into node v_{out}, i_t appears in the second row of its source vector. Its contribution to the output is low-pass, seeing only the output pole. The noise from M_B is applied at the input. Therefore, it sees the signal transfer function. The noise from the common-gate transistor flows from the node v_{out} to v_{in}. Hence it appears in both rows of its source vector, but with opposite polarity, ultimately having a band-pass response. Notice that at dc, its NTF is 0. This means that the noise current does not flow through the load resistor R. Rather, it flows through transistor M_{CG}. As the frequency increases, more noise current flows through C_I and less into the source of M_{CG}. The same magnitude of current that flows into C_I also flows through R, developing a noise voltage. Only above the output pole frequency does this noise voltage drop due to the noise current flowing through the capacitor C_O.

The overall output noise is found by summing the individual contributions and integrating:

$$\overline{v_{n,o}^2} = \int_0^\infty 4kT \left(\frac{R}{(2\pi fRC_O)^2 + 1} + \right.$$

$$\left. \frac{\gamma \left(g_{m,B} R^2 + g_{m,CG} \left(2\pi fR \frac{C_I}{G_m} \right)^2 \right)}{\left((2\pi fRC_O)^2 + 1 \right) \left((2\pi f \frac{C_I}{G_m})^2 + 1 \right)} \right) df. \qquad (9.35)$$

Table 9.1 Noise analysis of a common-gate TIA.

Noise source	PSD $[\text{A}^2/\text{Hz}]$	J_i	NTF $[\Omega]$	Contribution $[\text{V}^2/\text{Hz}]$
R	$\frac{4kT}{R}$	$\begin{bmatrix} 0 \\ i_t \end{bmatrix}$	$\frac{R}{sRC_O+1}$	$\frac{4kTR}{(2\pi f\,RC_O)^2+1}$
M_B	$4kT\gamma g_{m,B}$	$\begin{bmatrix} -i_t \\ 0 \end{bmatrix}$	$\frac{R}{(sRC_O+1)(sC_I/G_m+1)}$	$\frac{4kT\gamma g_{m,B}R^2}{\left((2\pi f\,RC_O)^2+1\right)\left((2\pi fC_I/G_m)^2+1\right)}$
M_{CG}	$4kT\gamma g_{m,CG}$	$\begin{bmatrix} i_t \\ -i_t \end{bmatrix}$	$\frac{sC_IR/G_m}{(sRC_O+1)(sC_I/G_m+1)}$	$\frac{4kT\gamma g_{m,CG}(2\pi f\,RC_I/G_m)^2}{\left((2\pi f\,RC_O)^2+1\right)\left((2\pi fC_I/G_m)^2+1\right)}$

Table 9.2 Noise comparison of common-gate and passive TIAs.

	CG	Resistive
Gain	R	R_L
Input noise	$2kT\pi\gamma g_{m,CG}f_b$	$\frac{kT\pi f_b}{R_L}$

According to [63], the integral in (9.35) can be evaluated and referred to the input giving:

$$\overline{i_{n,i}^2} \approx 4kT\gamma\left(\frac{1}{4}g_{m,CG}\omega_{p_o} + \frac{1}{2}g_{m,B}\omega_{p_i}\right), \qquad (9.36)$$

assuming R is large enough that its contribution can be neglected. Equation (9.36) gives us design guidelines. In particular, the bias transistor's transconductance should be minimized by decreasing its $\left(\frac{W}{L}\right)$ ratio. This will also slightly reduce C_I but comes at a cost of spending more of the supply voltage across the V_{DS} of M_B. Equation (9.36) must not be interpreted as suggesting $g_{m,CG}$ should also be reduced, since doing so will reduce the bandwidth (ω_{p_i}). On the other hand, we can see that having $f_{p_o} \gg f_{p_i}$ contributes to noise, without adding significant bandwidth since the bandwidth is usually set by f_{p_i}.

The discussion of a passive TIA ended with the claim that an active TIA would generally be superior. This claim is investigated by comparing (9.36) and (9.18). For the common-gate TIA, we consider the best-case scenario of small $g_{m,B}$ leaving only the first term of (9.36). The passive TIA is inherently first-order. To have similar bandwidth, we will assume $f_{p_o} = 2f_{p_i}$ for the common-gate TIA. Also, we assume that the data rate can be twice the input pole frequency. Putting this together, the input-referred noise of the common-gate TIA and the passive TIA are summarized in Table 9.2.

Assuming only a small amount of the capacitance at the input node comes from the common-gate transistor, we require both TIAs to have the same input resistance so that they have the same input pole. Therefore $g_{m,CG} = 1/R_L$. Ultimately, the common-gate

TIA has a higher input-referred noise than the passive TIA by a factor of 2γ. However, whereas the passive TIA's gain is limited to its input resistance (R_L) the common-gate TIA will have much higher gain. Without adequate gain, the noise of the main amplifier and the minimum input signal requirements of the latch will degrade the sensitivity of the receiver. The larger the input capacitance, the larger the gain improvement of the common-gate TIA. It is this decoupling of gain from input resistance that favours active TIAs for all but the smallest photodiode capacitances.

9.2.3   Shunt-Feedback TIAs

9.2.3.1   Analysis of the Shunt-Feedback TIA Assuming an Ideal Voltage Amplifier

Figure 9.6 shows a schematic of a TIA that uses shunt-shunt feedback. The feedback resistor R_F senses the output voltage, v_{out}, converts it to a current through R_F and feeds it back to the input where it is summed with the input current i_{in}. The voltage amplifier has a gain of A such that $v_{out} = -Av_{in}$. The circuit is analyzed as follows:

$$\begin{bmatrix} sC + G_F & -G_F \\ A & 1 \end{bmatrix} \begin{bmatrix} v_{in} \\ v_{out} \end{bmatrix} = \begin{bmatrix} i_{in} \\ 0 \end{bmatrix}, \tag{9.37}$$

where $G_F = 1/R_F$. The first row in (9.37) is a nodal equation at v_{in}. The second row describes the gain of the amplifier. These can be solved to give the gain and input impedance of the amplifier:

$$Z_{in}(s) = \frac{v_{in}}{i_{in}} = \frac{R_F}{sR_FC + 1 + A}, \tag{9.38}$$

$$Z_{TIA}(s) = \frac{v_{out}}{i_{in}} = -\frac{AR_F}{sR_FC + 1 + A}. \tag{9.39}$$

From (9.39) we see that $Z_{TIA}(0) \to -R_F$ for large A. However, rather than the pole being set by C in parallel with R_F, feedback reduces the resistance seen by the capacitor to $R_F/(1 + A)$, as shown by evaluating (9.38) at $s = 0$. This moves the input pole to $\frac{1+A}{R_FC}$. For any active TIA, we expect that the bandwidth will be improved from $\frac{1}{RC}$ where C is the input capacitance and R is the element converting input current to output voltage. For the common-gate TIA, the improvement was seen by separating R from C using the common-gate transistor as a current buffer. The gain element was

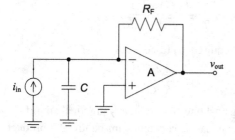

Figure 9.6 Shunt-feedback TIA.

still in parallel with the load capacitance, C_O. In the case of the shunt-feedback TIA, the use of feedback means that neither the input nor the output capacitance sees a parallel resistance of R_F.

9.2.3.2   Analysis of the Shunt-Feedback TIA Assuming a First-Order Voltage Amplifier

In the analysis presented thus far, the amplifier's gain A has been presented as being frequency-independent. A more realistic model is to replace A with:

$$A(s) = \frac{A_0}{\frac{s}{\omega_0} + 1}. \tag{9.40}$$

Substituting (9.40) into (9.39) and rearranging gives:

$$Z_{TIA}(s) = -\frac{A_0 R_F}{\frac{R_F C}{\omega_0}s^2 + \left(R_F C + \frac{1}{\omega_0}\right)s + 1 + A_0}$$

$$Z_{TIA}(s) = -\frac{\frac{A_0 \omega_0}{C}}{s^2 + \left(\omega_0 + \frac{1}{R_F C}\right)s + \frac{(1+A_0)\omega_0}{R_F C}}. \tag{9.41}$$

Applying (A.10) we see that the dominant pole is still $\frac{1+A_0}{R_F C}$ provided $\omega_0 \gg (1 + A_0)/(R_F C)$. The non-dominant pole can be estimated by dividing the coefficient of s by the coefficient of s^2 yielding $\omega_0 + \frac{1}{R_F C}$. Designing the forward amplifier to have such a large ω_0 may not be possible. Also, a second-order system with complex poles can have faster settling than widely spaced real poles. Hence the shunt-feedback TIA is often designed to have a damping factor $\zeta = 1/\sqrt{2}$. Recall that a second-order system can be written in the form:

$$H(s) = \frac{H_0 \omega_n^2}{s^2 + 2\zeta \omega_n s + \omega_n^2}. \tag{9.42}$$

Matching coefficients between (9.41) and (9.42) we find:

$$H_0 = -\frac{R_F}{1 + \frac{1}{A_0}}, \tag{9.43}$$

$$\omega_n = \sqrt{\frac{(1 + A_0)\omega_0}{R_F C}}, \tag{9.44}$$

$$\zeta = \frac{R_F C \omega_0 + 1}{2\sqrt{(1 + A_0)\omega_0 R_F C}}. \tag{9.45}$$

Setting $\zeta = 1/\sqrt{2}$, we find the required ω_0 to be:

$$\omega_0 \approx \frac{2A_0}{R_F C}. \tag{9.46}$$

Note that ω_0 and A_0 trade-off against one another in any practical amplifier design. For a simple voltage amplifier, we may have a constant gain-bandwidth product, meaning:

$$\omega_{GBW} = A_0 \omega_0, \tag{9.47}$$

where ω_{GBW} is the forward amplifier's gain-bandwidth product. In the case of a first-order amplifier ω_{GBW} is also the unity-gain frequency for a particular A_0. In this analysis, we must not lose sight of what parameters we are stuck with and what parameters we are choosing. For example, suppose C is predetermined and we require a certain bandwidth, ω_{-3dB}. If we fix $\zeta = 1/\sqrt{2}$, the bandwidth, ω_{-3dB} is equal to the natural frequency, ω_n. By approximating $A_0 \approx A_0 + 1$, R_F can be found substituting (9.47) into (9.44) to give:

$$R_F = \frac{\omega_{GBW}}{C\omega_{-3dB}^2}.$$

(9.48)

The relationship in (9.48), known as the transimpedance limit [64] shows an important trade-off in shunt-feedback TIA design. Whereas designers are used to seeing gain and bandwidth be inversely proportionate with one another, (9.48) shows that the gain decreases with the square of the bandwidth.

9.2.3.3 Inverter-Based Transistor-Level Implementation of the Shunt-Feedback TIA

There are several transistor-level implementations of the shunt-feedback TIA, although the most common implementation in CMOS uses one, or in some cases three inverters, to implement the voltage amplifier. A schematic of an inverter-based TIA implemented as a single inverter is shown in Figure 9.7 (a). Here, M_1 and M_2 form a static-CMOS inverter. When $i_{in} = 0$, $V_I = V_0$. This voltage is referred to as the inverter's threshold, not to be confused with the threshold voltage of the transistors. The ac small-signal circuit is shown in Figure 9.7 (b). Here, the following simplifications were made:

$$g_m = g_{m1} + g_{m2},$$

(9.49)

$$R_0 = r_{o1}\|r_{o2},$$

(9.50)

$$C_I = C_D + C_{GS1} + C_{GS2},$$

(9.51)

$$C_F = C_{GD1} + C_{GD2},$$

(9.52)

$$C_O = C_{DB1} + C_{DB2} + C_L,$$

(9.53)

where C_D is the combined capacitance of the photodetector and bond pads. The nodal equations can be written as:

$$\begin{bmatrix} G_F + s(C_I + C_F) & -G_F - sC_F \\ g_m - G_F - sC_F & G_F + G_0 + s(C_O + C_F) \end{bmatrix} \begin{bmatrix} v_i \\ v_o \end{bmatrix} = \begin{bmatrix} i_{in} \\ 0 \end{bmatrix}.$$

(9.54)

These can be solved to find $Z_T(s) = \frac{v_o}{i_{in}}$ as:

$$Z_T(s) = \frac{(1 - g_m R_F + sC_F R_F)R_0}{s^2 R_F R_0 \left(\dfrac{C_I C_O +}{C_I C_F + C_F C_O} \right) + s \left(\dfrac{R_0 C_O + (R_0 + R_F)C_I}{+R_F(1 + g_m R_0)C_F} \right) + 1 + g_m R_0}.$$

(9.55)

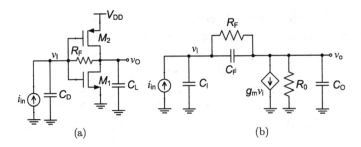

Figure 9.7 CMOS inverter-based shunt-feedback TIA: (a) schematic; (b) ac small-signal schematic.

The transfer function's dc gain, damping factor and natural frequency are given by:

$$Z_T(0) = \frac{(1 - g_m R_F)R_0}{1 + g_m R_0} \approx -R_F, \tag{9.56}$$

$$\omega_n^2 = \frac{1 + g_m R_0}{R_F R_0 (C_I C_O + C_I C_F + C_F C_O)}, \tag{9.57}$$

$$\zeta = \frac{R_0 C_O + (R_0 + R_F)C_I + R_F(1 + g_m R_0)C_F}{2\sqrt{R_F R_0 (1 + g_m R_0)(C_I C_O + C_I C_F + C_F C_O)}}. \tag{9.58}$$

This section began with a shunt-feedback TIA built around a voltage amplifier, without a feedback capacitor. How well does that model describe the behaviour of the actual inverter-based circuit, given that the latter behaves more like a voltage-controlled current source rather than the voltage-controlled voltage source in the original model? If we set $C_F = 0$, replace $g_m R_0$ with A_0 and $1/R_0 C_O$ with ω_0, (9.57) and (9.58) become:

$$\omega_n^2 = \frac{1 + g_m R_0}{R_F R_0 C_I C_O} = \frac{(1 + A_0)\omega_0}{R_F C_I}, \tag{9.59}$$

$$\zeta = \frac{R_0 C_O + (R_0 + R_F)C_I}{2\sqrt{R_F R_0 C_I C_O (1 + g_m R_0)}} = \frac{\omega_0^{-1} + (R_0 + R_F)C_I}{2\sqrt{R_F C_I \omega_0^{-1}(1 + A_0)}},$$

$$= \frac{1 + \omega_0(R_0 + R_F)C_I}{2\sqrt{(1 + A_0)\omega_0 R_F C_I}}. \tag{9.60}$$

Equation (9.59) is the same as (9.44). However, the numerator of (9.60) differs from (9.45). The relatively good agreement between the voltage amplifier and transconductance amplifier-based analysis is due to the effect of feedback on the output impedance of the transconductance amplifier. Solving (9.54) for v_i gives the input impedance of:

$$Z_{IN}(s) = \frac{R_0 + R_F + s R_F R_0 (C_O + C_F)}{s^2 R_F R_0 \left(\frac{C_I C_O +}{C_I C_F + C_F C_O}\right) + s \left(\frac{R_0 C_O + (R_0 + R_F)C_I}{+ R_F(1 + g_m R_0)C_F}\right) + 1 + g_m R_0}, \tag{9.61}$$

which for modest values of R_F gives a low-frequency input resistance of:

$$Z_{IN}(0) = \frac{R_0 + R_F}{1 + g_m R_0} \approx \frac{1}{g_m}. \tag{9.62}$$

Notice, that although the poles of Z_{IN} are the same as those of Z_T, Z_{IN} has a zero given by:

$$\omega_z = \frac{1}{R_F||R_0(C_O + C_F)}, \tag{9.63}$$

The output impedance of the circuit is found by solving (9.54) for v_o but with a source vector $J = \begin{bmatrix} 0 \\ i_t \end{bmatrix}$, giving:

$$Z_O(s) = \frac{(1 + sR_F(C_F + C_I))R_0}{s^2 R_F R_0 \begin{pmatrix} C_I C_O + \\ C_I C_F + C_F C_O \end{pmatrix} + s \begin{pmatrix} R_0 C_O + (R_0 + R_F)C_I \\ + R_F(1 + g_m R_0)C_F \end{pmatrix} + 1 + g_m R_0}. \tag{9.64}$$

This is also approximately equal to $1/g_m$ at low frequencies but increases starting at $\omega_z = \frac{1}{R_F(C_F + C_I)}$. This increase in output impedance is relevant to the noise performance since the thermal noise current of the MOSFETs is injected at the output of the circuit.

9.2.3.4 Design Considerations for Inverter-Based Shunt-Feedback TIAs

Tuning the damping factor of the circuit requires independent control of A_0, which is not a free parameter in inverter-based designs. If we can tolerate the resulting, non-ideal ζ, the design of an inverter-based TIA can be reduced to the following design choices:

- Determine the ratio between the width of the PMOS transistor and the width of the NMOS transistor. It is assumed that both transistors are minimum length.
- Determine the combination of R_F and transistor widths that gives the required bandwidth and noise performance.

At this point in the text it is assumed that bandwidth is a non-negotiable but achievable specification. At the target bandwidth, the designer will maximize the gain and minimize noise. However, noise-optimal and gain-optimal sizing will not be the same. Finally, the power dissipation of the circuit is proportional to the choice of transistor width.

In considering the size ratio between PMOS and NMOS, digital designers will recall various sizing criteria for an inverter. Notably, setting $\frac{W_P}{W_N} = \frac{\mu_N}{\mu_P}$ matches rising and falling delay, a criterion also useful for preserving duty cycle in clock distribution networks. Alternatively, setting $\frac{W_P}{W_N} = \sqrt{\frac{\mu_N}{\mu_P}}$ minimizes the sum of rising and falling delays. In TIA design, the sizing objective is to maximize the total transconductance given a fixed total input capacitance. The input capacitance is fixed by assuming a fixed total transistor, width since all transistor capacitances are proportional to width. Therefore, keeping total width ($W_1 + W_2$) constant keeps the capacitances constant.

For this analysis, we assume a square-law model for the MOSFET, therefore:

$$g_{m1} = \mu_n C_{OX} \frac{W_1}{L_1} (V_0 - V_{tn}),$$
(9.65)

$$g_{m2} = \mu_p C_{OX} \frac{W_2}{L_2} (V_{DD} - V_0 - |V_{tp}|),$$
(9.66)

where V_0 is the voltage at the input and output of the inverter, made equal to one another by the feedback resistor. If W_1 is increased while W_2 is decreased, V_0 decreases. To maximize g_m we need to include the dependence of V_0 on the sizing choice. This is found by equating the current through M_1 and M_2 and noting that $V_{GS1} = V_0$ and $V_{SG2} = V_{DD} - V_0$, following analysis similar to that in [65]:

$$I_{D1} = \frac{1}{2} \mu_n C_{OX} \frac{W_1}{L_1} (V_0 - V_{tn})^2,$$
(9.67)

$$I_{D2} = \frac{1}{2} \mu_p C_{OX} \frac{W_2}{L_2} (V_{DD} - V_0 - |V_{tp}|)^2,$$
(9.68)

$$\frac{1}{2} \mu_n C_{OX} \frac{W_1}{L_1} (V_0 - V_{tn})^2 = \frac{1}{2} \mu_p C_{OX} \frac{W_2}{L_2} (V_{DD} - V_0 - |V_{tp}|)^2,$$
(9.69)

$$V_0 = \frac{r(V_{DD} - |V_{tp}|) + V_{tn}}{1 + r}, \text{ where}$$
(9.70)

$$r = \sqrt{\frac{\mu_p C_{OX} \frac{W_2}{L_2}}{\mu_n C_{OX} \frac{W_1}{L_1}}}.$$
(9.71)

Writing g_{m2} in terms of r and the parameters of M_1, the total transconductance can be written as:

$$g_m = g_{m1} + g_{m2},$$
(9.72)

$$= \mu_n C_{OX} \frac{W_1}{L_1} (V_0 - V_{tn}) + \mu_p C_{OX} \frac{W_2}{L_2} (V_{DD} - V_0 - |V_{tp}|),$$
(9.73)

$$= \mu_n C_{OX} \frac{W_1}{L_1} (V_0 - V_{tn}) + r^2 \mu_n C_{OX} \frac{W_1}{L_1} (V_{DD} - V_0 - |V_{tp}|),$$
(9.74)

$$= \mu_n C_{OX} \frac{W_1}{L_1} \left(V_0 - V_{tn} + r^2 (V_{DD} - V_0 - |V_{tp}|) \right).$$
(9.75)

If we assume $V_{tn} = |V_{tp}|$ and we substitute (9.70) in (9.75), after simplification we can write:

$$g_m = \frac{\mu_n C_{OX}}{L_1} (V_{DD} - 2V_t) W_1 r.$$
(9.76)

To maximize g_m as a function of W_1 we differentiate with respect to W_1, bearing in mind that r also depends on W_1.

$$\frac{dg_m}{dW_1} = \frac{\mu_n C_{OX}}{L_1} (V_{DD} - 2V_t) \left(r + W_1 \frac{dr}{dW_1} \right) \text{ and}$$
(9.77)

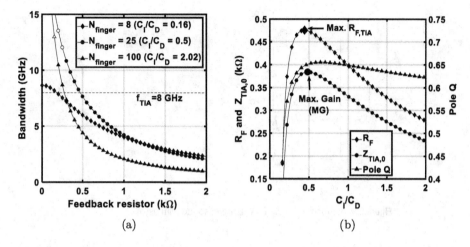

Figure 9.8 Transistor width and resistor sweep of an inverter-based shunt-feedback TIA, taken from [66]: (a) inverter-based TIA bandwidth as a function of R_F; (b) required R_F and resulting gain and pole Q as a function of transistor width for an 8 GHz bandwidth.

$$\frac{dr}{dW_1} = -\sqrt{\frac{\mu_p}{W_2 W_1 \mu_n}}. \tag{9.78}$$

Setting $r + W_1 \dfrac{dr}{dW_1} = 0$, we get $\hspace{2cm}$ (9.79)

$$r + W_1 \frac{dr}{dW_1} = \sqrt{\frac{W_2 \mu_p}{W_1 \mu_n}} - W_1 \sqrt{\frac{\mu_p}{W_2 W_1 \mu_n}} = 0, \tag{9.80}$$

$$W_1 = W_2. \tag{9.81}$$

The result that total g_m is maximized by setting the width of the PMOS equal to the width of the NMOS is somewhat counterintuitive. Although the derivation began with a square-law model for the MOSFETs, simulations done in sub-micron technologies that are heavily velocity saturated (not square-law) agree with this finding.

Once the ratio between W_2 and W_1 has been selected, designers can sweep the remaining parameters of width and resistor value. An example is shown in Figure 9.8 where the combined photodiode and pad capacitance is 200 fF. In Figure 9.8 (a) transistor widths of 8, 25 and 100 μm are considered. Increasing R_F reduces the bandwidth. However, for a target bandwidth of 8 GHz, 25 μm transistors support larger R_F than the other choices, which will give rise to larger gain. In Figure 9.8 (b) the value of R_F that gives an 8 GHz bandwidth is found as transistor width is swept. Thus, R_F, $Z_{TIA}(0)$ and the resulting pole Q are plotted. We can see that for this scenario a maximum gain is achieved when the TIA's input capacitance, C_I, is about half of the combined detector/pad capacitance corresponding to transistors of about 25 μm each in width. Larger transistors can give slightly better noise performance at the cost of linearly increasing power dissipation and slightly reduced gain.

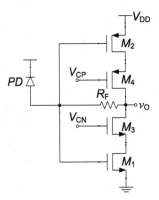

Figure 9.9 Shunt-feedback TIA using cascode transistors.

In the initial derivation of the transfer function of an idealized shunt-feedback TIA, we assumed a constant gain-bandwidth product (GBW) for the voltage-amplifier. Trading of its gain for its bandwidth subject to a GBW constraint allows us to adjust the damping factor of the second-order poles. However, a static CMOS inverter has a particular gain, set by the ratio of the MOSFETs' output resistance and transconductance:

$$A_0 = (g_{m1} + g_{m2})(r_{o1}||r_{o2}). \tag{9.82}$$

As the total width increases, the transconductances increase while the output resistances decrease. For a given process technology a designer is stuck with a particular value in the range of 5–10 for modern processes. Despite CMOS scaling this value will not decrease further since A_0 is the slope of an inverter's dc transfer characteristic. Should it decrease further, static-noise margins, which are critical to the reliability of logic circuits, become unacceptable. Process developers fight to ensure adequate noise margins. Nevertheless a value of 5–6 may not be optimal for TIA design. In Figure 9.8 (b), the resulting pole Q was slightly less than the ideal Butterworth value of 0.707. Analog designers will note that if we increase W and L, maintaining the same aspect ratio, g_m stays relatively constant but $r_o \propto L$. However, the resultant increase in capacitance will offset any gains in performance due to larger A_0.

There are four variants of the inverter-based TIA that increase A_0. The first, shown in Figure 9.9, involves the use of cascode transistors. Transistors M_4 and M_3 are biased to keep them as well as M_2 and M_1 in saturation. The open-loop gain of the voltage amplifier increases from $g_m r_o$ to $(g_m r_o)^2$ where these terms are the combined transconductance and output resistance of the NMOS and PMOS transistors in the inverter. Although this scheme improves A_0, cascode transistors add noise.

The second variant involves putting multiple inverters (such as three) inside the feedback loop as shown in Figure 9.10. This must be done carefully to ensure stability. This can be achieved by putting resistive feedback around some of the inverters as done in [67].

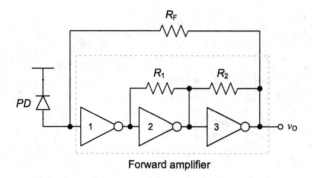

Figure 9.10 Shunt-feedback TIA with three inverters within the feedback loop.

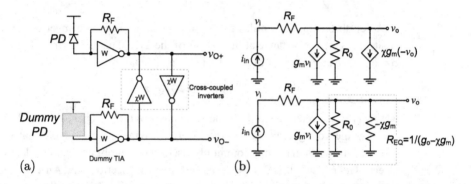

Figure 9.11 Pseudo-differential inverter-based shunt-feedback TIA with cross-coupled inverters at the output: (a) schematic and (b) (top) AC small-signal equivalent circuit for low-frequency operation and (bottom) circuit shown with output resistance calculation.

The third approach uses two cross-coupled inverters at the output of a pseudo-differential shunt-feedback TIA to create a negative conductance that, when placed in parallel with the output of the inverter, reduces the inverter's output conductance and thus increases its gain. Figure 9.11 shows a pseudo-differential structure driven single-endedly. The dummy photodiode is not illuminated. The cross-coupled inverters are much smaller than the main inverters, denoted as having a width of χW compared to the main inverters having a width of W. Though driven single-endedly, we can still consider the differential half-circuit shown in the top part of Figure 9.11 (b). The voltage-controlled current source (VCCS) at the output is driven by the other output, hence it has a value of $\chi g_m(-v_o)$. The VCCS can be replaced by a conductance of $-\chi g_m$. Hence the overall output resistance has been increased from $R_0 = 1/g_o$ to $1/(g_o - \chi g_m)$. Therefore, the overall gain of the voltage amplifier (without R_F) increases to:

$$A_0 = \frac{g_m}{g_o - \chi g_m}. \tag{9.83}$$

If χ is increased too much, A_0 changes polarity and the TIA can become unstable. Thus χ is selected to increase A_0 but is kept small enough such that $\chi g_m < g_o$.

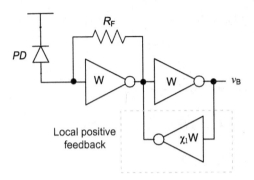

Figure 9.12 Shunt-feedback TIA using local positive feedback.

A fourth technique, similar to the previous one, uses two inverters in a loop to create a negative resistance at the output of the TIA. This is shown in Figure 9.12. A conductance at the output of the TIA is created equal to:

$$g_x = -\chi_1 g_m A, \tag{9.84}$$

where χ_1 is the relative width of the feedback inverter compared to the TIA inverter and A is the voltage gain across the inverter driving node v_B. Note that in general, $A \neq A_0$. In particular, in a commonly used main amplifier topology known as the Cherry-Hooper amplifier, $A \approx 1$, as discussed in Section 9.3. An example of this technique can be found in [57]. The application of this scheme in three locations in a main amplifier was discussed in [68] and is presented in more detail in Figure 9.20.

9.2.3.5 Shunt-Feedback TIA Noise Analysis

Just as the frequency response analysis yields complicated expressions, so too does the noise analysis, despite there being noise sources in only two locations in the circuit. The noise performance of the shunt-feedback TIA is frequently performed by directly writing the input-referred noise density. Here, we start with a consideration of output-referred noise density and then refer the terms to the input.

As shown in Figure 9.13 (a), the TIA has two noise sources, namely the resistor's thermal noise current connected across the feedback resistor and a thermal noise current from the MOSFETs entering the output node. When R_0 models channel-length modulation (and is not a physical load resistor), it does not generate noise.

The source vectors and relevant other aspects of the noise analysis are shown in Table 9.3. For both noise sources, the system of equations is solved for v_o using a source vector (J_i) appropriate to the location of the noise source. In particular, for the feedback resistor we solve the system of equations using $J_i = \begin{bmatrix} i_t \\ -i_t \end{bmatrix}$. Here, v_o due to i_t in the top row of J_i is the transimpedance gain (Z_T) of the TIA, whereas v_o due to i_t in the bottom row is the output impedance of the TIA (Z_O). Hence the overall NTF for the noise of the feedback resistor is the difference between the transimpedance gain and the output impedance. This is further illustrated in Figure 9.13 (b) where the resistor's

Table 9.3 Noise analysis of a shunt-feedback TIA.

Quantity	R_F	MOSFETs
PSD [A^2/Hz]	$\frac{4kT}{R_F}$	$4kT\gamma g_m$
J_i and meaning	$\begin{bmatrix} i_t \\ -i_t \end{bmatrix}, Z_T - Z_O$	$\begin{bmatrix} 0 \\ -i_t \end{bmatrix}, Z_O$
NTF	$-R_0 \frac{g_m R_F + sC_I R_F}{D(s)}$	$R_0 \frac{1 + sR_F(C_F + C_I)}{D(s)}$
O/P contribution [V^2/Hz]	$\frac{4kT}{R_F} \left\| R_0 \frac{g_m R_F + sC_I R_F}{D(s)} \right\|^2$	$4kT\gamma g_m \left\| R_0 \frac{1 + sR_F(C_F + C_I)}{D(s)} \right\|^2$
I/P contribution (s) [A^2/Hz]	$\frac{4kT}{R_F} \left\| \frac{g_m + sC_I}{1/R_F - g_m + sC_F} \right\|^2$	$4kT\gamma g_m \left\| \frac{1/R_F + s(C_F + C_I)}{1/R_F - g_m + sC_F} \right\|^2$
Simplified I/P (f) [A^2/Hz]	$\frac{4kT}{R_F} \left(1 + \left(\frac{2\pi f C_I}{g_m} \right)^2 \right)$	$\frac{4kT\gamma}{g_m} \left(\frac{1}{R_F^2} + (2\pi f (C_F + C_I))^2 \right)$

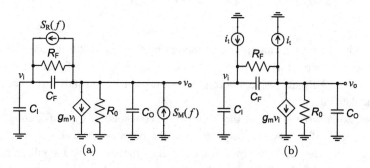

(a)   (b)

Figure 9.13 Noise model of the inverter-based shunt-feedback TIA: (a) general schematic; (b) resistor's NTF as difference between Z_{TIA} and Z_0.

noise current is been split into two separate current sources, one flowing into the input node, v_i, and one flowing out of the output node, v_o. However, approximating the NTF as only the transimpedance gain is close enough. The NTF of the MOSFETs' thermal noise is the output impedance. To simplify expressions in Table 9.3, the denominator of the noise transfer functions is written as $D(s)$ where $D(s)$ is the denominator of (9.55).

Figure 9.14 shows an example of the output noise PSD of an inverter-based shunt-feedback TIA and the contributions from the resistor and the MOSFETs. The contribution from R_F is shown by the black dotted line using the exact NTF given in Table 9.3 ($Z_T - Z_O$) as well as the approximated contribution using only Z_T. The difference between these two is small.

The shape of the noise contribution from the feedback resistor also shows why beyond a certain point additional TIA bandwidth is detrimental to performance. The noise PSD follows the TIA's frequency response. Thus, wider *signal* bandwidth means a wider *noise* bandwidth. If we imagine a TIA with a low enough bandwidth that

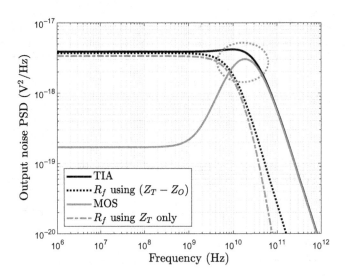

Figure 9.14 Output noise PSDs for an inverter-based TIA.

ISI is introduced, we can argue that increasing the bandwidth reduces ISI and hence improves the signal. However, once increases in noise grow faster than increases in signal, our SNR gets worse. For this reason, we select the bandwidth to be 50 to 70% of the data rate.

We explain the shape of the MOSFETs' contribution by considering the output impedance of the TIA. At low frequency, the output impedance is low ($\approx \frac{1}{g_m}$) due to feedback. Hence, the thermal noise current generates only a small output voltage. However, much below the bandwidth, the output impedance increases due to the zero at $\omega = \frac{1}{R_F(C_F + C_I)}$. In other words, at the frequency where the gain of a resistive TIA of R_F would start to decrease, the output resistance of the shunt-feedback TIA increases due to the fed-back signal being shorted out by the input capacitance. As the loop gain decreases, the output impedance reaches R_0. Ultimately, the output capacitance C_O reduces the output impedance and the noise contribution from the MOSFETs drops. Notice that if the poles of the TIA are real and distinct, there is a range of frequency between the bandwidth of the TIA and the second pole, within which there is signifi-cant noise contribution from the MOSFETs but limited output signal. This region is circled in the figure.

The contributions to input-referred noise of the shunt-feedback TIA are presented in Table 9.3, each found by dividing the output noise contribution by $|Z_T|^2$. Since Z_T and each NTF have the same poles, $D(s)$ divides out when finding the input-referred noise PSDs. Greater insight is found by simplifying the input-referred noise according to two assumptions. The first is that $g_m \gg \frac{1}{R_F}$. The second is that the pole in the I/P contribution created by C_F is higher than the bandwidth to which we integrate and can be ignored.Therefore $1/R_F - g_m + sC_F$ is replaced by $-g_m$.

Both the resistor and the MOSFETs produce components of the input-referred PSD that are flat with respect to frequency and that increase with frequency. This input PSD is shaped by the transfer function of the TIA and whatever additional amplifier stages that are present. The total input noise PSD is:

$$S_i(f) = \frac{4kT}{R_F}\left(1 + \left(\frac{2\pi f C_I}{g_m}\right)^2\right) + \frac{4kT\gamma}{g_m}\left(\frac{1}{R_F^2} + (2\pi f(C_F + C_I))^2\right) \tag{9.85}$$

$$= 4kT\left(\frac{1}{R_F}\left(1 + \frac{1}{g_m R_F}\right) + \frac{(2\pi f(C_F + C_I))^2}{g_m}\left(\gamma + \frac{1}{g_m R_F}\left(\frac{C_I}{C_F + C_I}\right)^2\right)\right) \tag{9.86}$$

$$\approx 4kT\left(\frac{1}{R_F} + \frac{\gamma(2\pi f C_T)^2}{g_m}\right), \tag{9.87}$$

where $C_T = C_I + C_F$. The frequency at which the f^2 term crosses the flat term is found by equating the two terms in (9.87) and solving for f_c, giving:

$$\frac{1}{R_F} = \frac{\gamma(2\pi f_c C_T)^2}{g_m},$$

$$f_c = \frac{1}{2\pi}\sqrt{\frac{g_m}{\gamma C_T}\frac{1}{R_F C_T}}. \tag{9.88}$$

The right side of (9.88) is the geometric mean of two frequencies, the lower of which is $\frac{1}{R_F C_T}$. This term pulls f_c into the band of the TIA. Therefore, the input referred noise increases well before the bandwidth of the TIA.

In looking at the second term of (9.87) we see that the numerator C_T is the sum of the transistor capacitance ($C_{MOS} = C_{GS} + C_{GD}$) and the detector/pad capacitance (C_D) while the denominator is proportional to the transistor capacitance since $g_m = f_T C_{MOS}$. To minimize the f^2 noise term, we first rewrite it in terms of C_{MOS}, giving:

$$\frac{(2\pi f C_T)^2}{g_m} = \frac{(2\pi f(C_D + C_{MOS}))^2}{f_T C_{MOS}} = K\left(\frac{C_D^2}{C_{MOS}} + 2C_D + C_{MOS}\right), \tag{9.89}$$

where $K = \frac{(2\pi f)^2}{f_T}$. Differentiating the right-hand side with respect to C_{MOS} and equating to 0 gives:

$$0 = K\frac{d}{dC_{MOS}}\left(\frac{C_D^2}{C_{MOS}} + 2C_D + C_{MOS}\right) = -\frac{C_D^2}{C_{MOS}^2} + 1. \tag{9.90}$$

To minimize the contribution from the MOSFETs we set $C_{MOS} = C_D$. This is referred to as the "capacitive matching rule" discussed in greater detail in [69]. This relation says that if we increase the input transistors until the TIA's input capacitance is equal to the total of the detector's and other pad parasitic capacitance, we reach a noise minimum. It is important to note, though, that increasing input capacitance of the TIA increases g_m, but at the cost of increased power dissipation. Thus, for many links, smaller transistors are used at the cost of slightly higher noise, but significantly lower power dissipation.

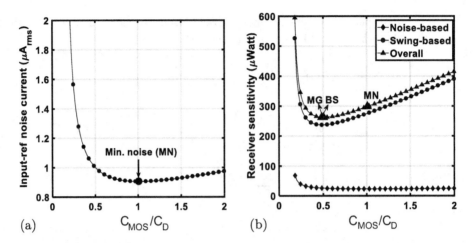

Figure 9.15 Inverter-based TIA noise performance: (a) TIA's input-referred noise current as a function of C_{MOS}/C_D for a fixed 3 dB bandwidth of 8 GHz; (b) receiver sensitivity as a function of C_{MOS}/C_D for a front-end that includes only a TIA. The bandwidth of the TIA is kept constant at 8 GHz through modifications to the feedback resistor. A minimum required output swing of 50 mV is assumed. The bold markers indicate the locations of maximum gain (MG), minimum noise (MN) and best overall sensitivity (BS). Figure from [66].

Figure 9.15 (a) shows the input-referred noise of a shunt-feedback TIA designed for 8 GHz bandwidth as the size of the transistors is varied (C_{MOS}). The noise reaches a minimum when C_{MOS} is equal to C_D. The minimum is relatively shallow, meaning that as the transistors are shrunk to half the size ($C_{MOS}/C_D = 0.5$) the noise increases only slightly, but the power dissipation of the TIA is cut in half. Also, the smaller capacitance allows for a larger R_F. The effect of increased gain is shown in Figure 9.15 (b) where the overall sensitivity is calculated assuming the TIA directly drives a decision circuit with a required input swing of 50 mV. Such a large swing when driven by the TIA output (no additional amplification) means that the best receiver sensitivity, indicated as point BS, falls very close to the receiver with maximum gain (MG). The significant degradation in sensitivity when the decision circuit's requirements are accounted for serves as a motivation for designing a main amplifier to amplify the TIA's output before applying it to a decision circuit. Main amplifier design is presented in Section 9.3.

For a given TIA design the input-referred rms noise current can be found by integrating the output-referred noise PSD, giving a result in units of V^2, square-rooting the result, and then dividing it by the midband gain of the TIA. This is valid when negligible ISI is introduced. Since the output PSD is decreasing with respect to frequency, the result of integration is not very sensitive to the upper limit of integration, provided we integrated beyond the cut-off of the output noise PSD. Consider the noise PSD in Figure 9.14. The circuit has a bandwidth (f_{bw}) of 9.24 GHz, but the output noise only starts to decrease beyond 20 GHz. The rms output noise calculated with various upper limits of integration (f_{max}) is shown in Table 9.4. Increasing the upper limit of

Table 9.4 Integrated
output noise of the PSD
in Figure 9.14.

f_{max} (GHz)	V_{rms} (μV)
f_{bw}	193
$2f_{bw}$	273
$4f_{bw}$	363
70	430
1592	516

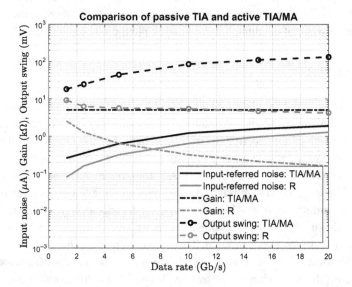

Figure 9.16 Shunt-feedback front-end (TIA/MA) vs passive front-end (R) based on data
from [71].

integration from 70 GHz to 1.6 THz only slightly increases the rms noise. Thus, for a
pessimistic measure of output noise we can integrate to ∞.

Alternatively, one can integrate the input-referred noise PSD, although, as noted
in Appendix B, the integration of input-referred noise PSD must be done carefully.
Notice that since the input-referred noise is increasing with f^2 the result of integration
is sensitive to the upper end of integration. As discussed in the appendix, we have a
definition for the equivalent noise bandwidth of a system when inputted with flat input
noise. In [70] an analogous equivalent noise bandwidth for f^2 input noise is explained,
and this allows the correct integration of input-referred noise PSDs.

It is worth comparing the performance of an inverter-based shunt-feedback front-
end (SFFE) to that of a resistor. This is shown in Figure 9.16. The SFFE was designed
such that its bandwidth was tuneable, while providing constant gain. At each data rate,

both circuits have a bandwidth equal to half the data rate. The solid lines show the input-referred noise of the two systems. Notice that the resistive front-end generates less noise. The dashed lines show the gain of the front-ends. The SFFE has constant gain (by design) of about 5 kΩ. In order to meet the bandwidth specification, the gain of the resistive front-end drops as the data rate increases to less than 200 Ω at 20 Gb/s. Recall that $f_{3dB} = \frac{1}{2\pi RC}$. The dashed lines with the markers show the output swing for the noise-limited input signal. This is the input signal that gives an SNR of 14 at the output of each TIA. The output swing for the resistive TIA is constant, since its mean-squared output noise is constant at kT/C. Although the noise of the SFFE is slightly higher, the significantly larger output swing is due to it having much larger gain. At 20 Gb/s, the SFFE produces an output swing of > 100 mV whereas the resistive TIA's output swing is only 3 mV, meaning that additional gain would be needed to drive latches. Despite the higher noise, active front-ends are needed to provide the necessary gain in the presence of anything but the smallest photodiode capacitance.

9.3   Main Amplifiers

Whatever the selected topology of TIA, it usually will not provide enough gain so that a noise-limited input signal is amplified to a sufficient swing for the latches. The analog sensitivity of the receiver might be excellent, but the overall sensitivity will deteriorate due to the power penalty introduced by the decision circuit, as computed in (9.13). In most low-ISI front-ends the TIA is followed by additional stages of amplification. These amplifiers are referred to as main amplifiers, post amplifiers and, in some cases, limiting amplifiers. Limiting amplifier describes main amplifiers that have sufficient gain such that, during nominal operation, over a range of input amplitudes the amplifier is driven into saturation. This is common when the latches reside on a separate die from the front-end and signals are sent down a short PCB trace. When latches are on the same chip as the front-end, smaller amplitudes are generated and amplifiers will have less gain.

For an NRZ signal, amplifiers that do not "limit" may not have stringent linearity specifications. However, for PAM4 signals or for signals in coherent systems, linearity is important, as described by a total harmonic distortion specification. While we have introduced this section advocating for high gain, if the input signal increases, rather than allow the main amplifier to saturate in the case of a limiting amplifier, we may want to maintain a particular output voltage amplitude and thus some form of automatic gain control (AGC) is needed. This is particularly important when the main amplifier drives an analog-to-digital converter (ADC). To get the best SNR we want the output signal to occupy most of the ADC's input range, but not exceed it. Thus the gain of the main amplifier must adapt to changing input signal levels.

Main amplifier design approaches are heavily shaped by gain-bandwidth trade-offs. There are three aspects of this:

- For a given amplifier topology, designing it for higher gain will reduce its bandwidth.

- When N amplifiers of equal 3 dB bandwidth, f_{3dB}, are cascaded this frequency becomes the $3N$ dB bandwidth. The location of the new 3 dB bandwidth depends on N and the shape of the constituent amplifier's frequency response.
- A cascade of first-order systems has repeated real poles. Feedback can be used to move real poles in the complex plane. Similar to how a high-order filter has poles of high and low Q, a main amplifier can be designed like this to achieve higher bandwidth.

This section begins with a discussion of main amplifiers composed of several first-order sections. This lays out the principles at hand, although better main amplifiers can be designed. These include the Cherry-Hooper stage and several others involving active feedback.

9.3.1   Cascades of First-Order Sections

The ideas introduced here give a foundational perspective on trade-offs in main amplifier design, starting with first-order stages. We want to determine the resulting overall gain and bandwidth when we cascade N stages. We will also determine if there is an optimal number of stages to achieve a particular gain. Practical considerations are also discussed. Consider an amplifier with a transfer function of

$$A(s) = \frac{A_0}{\frac{s}{\omega_0} + 1}, \tag{9.91}$$

where A_0 is the low-frequency gain and ω_0 is its 3 dB bandwidth. For this analysis we assume a constant gain-bandwidth product. Therefore, the gain-bandwidth product of the amplifier is $\omega_{GBW} = A_0\omega_0$. This sort of gain vs bandwidth trade-off is typical in a first-order circuit such as a resistively loaded common-source amplifier or a resistively loaded differential pair. If N of these stages are cascaded the overall amplifier has a "total" transfer function of:

$$A_{tot}(s) = \left(\frac{A_0}{\frac{s}{\omega_0} + 1} \right)^N. \tag{9.92}$$

The low-frequency gain is now A_0^N and the resulting 3 dB bandwidth can be found by solving for the frequency at which the overall magnitude drops by $\sqrt{2}$ from its low-frequency value:

$$\left(\frac{A_0}{\sqrt{1 + \left(\frac{\omega_{3dB}}{\omega_0} \right)^2}} \right)^N = \frac{A_0^N}{\sqrt{2}}, \tag{9.93}$$

$$\omega_{3dB} = \omega_0 \sqrt{\sqrt[N]{2} - 1} \approx \omega_0 \frac{0.9}{\sqrt{N}}, \tag{9.94}$$

where ω_{3dB} is the 3 dB bandwidth of the overall amplifier and the approximation in (9.94) is valid for $N > 2$. As an example, consider the case of a target of 100 V/V

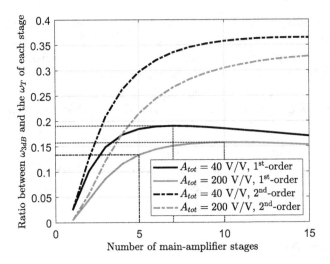

Figure 9.17 Achievable bandwidth as a fraction of GBW (ω_T) for an overall gain of 40 and 200 V/V for first- and second-order stages.

with a target bandwidth of 10 GHz. If implemented with a single stage, this would require a gain-bandwidth product of 100 V/V × 10 GHz or 1 THz, an unachievable target. If this main amplifier were built using a cascade of four first-order stages, each requires a dc gain of only $\sqrt[4]{100} = 3.16$. Solving (9.94) for ω_0 when $N = 4$ and $\omega_{3dB} = 2\pi\,10$ GHz gives a 3 dB bandwidth per stage of 23 GHz. This combined with the required A_0 gives a manageable gain-bandwidth product of 23 x 3.16 = 72.6 GHz. If we rewrite (9.94) to include $\omega_{GBW} = A_0\omega_0$ and $A_0 = \sqrt[N]{A_{tot}}$, we get:

$$\omega_{3dB} = \frac{\omega_{GBW}}{\sqrt[N]{A_{tot}}}\sqrt{\sqrt[N]{2}-1} \approx \frac{\omega_{GBW}}{\sqrt[N]{A_{tot}}}\frac{0.9}{\sqrt{N}}. \tag{9.95}$$

Equation (9.94) can be plotted with respect to N for a given A_{tot}. Up to an optimum N, the achievable bandwidth increases, before dropping. This is shown for the case of $A_{tot} = 40$ and 200 in Figure 9.17. For a given A_{tot}, the optimum N and A for each stage are:

$$N_{opt} = 2\ln A_{tot} \tag{9.96}$$

$$A_{opt} = A_{tot}^{1/N_{opt}} = e^{\frac{1}{2}} = 1.65. \tag{9.97}$$

It is curious that the optimum gain per stage is independent of the total gain. For the example of $A_{tot} = 200$, $N_{opt} = 10$. Equation (9.96) gives 10.6, but 10 is the best integer value. With $N = 10$, $A_{opt} = 1.7$. In Figure 9.17 it can be see that this leads to an overall bandwidth of about 15.8% of the GBW. Reducing the number of stages to 5, increases the required gain per stage to about 2.9 and achieves an overall bandwidth that is 13.3% of the GBW. With fewer stages, this design will dissipate less power and have better noise performance. For these reasons, main amplifiers are usually designed with at most 5 or 6 stages.

The analysis above treats the overall amplifier as one LTI system. Thus, each stage limits the signal bandwidth. However, if there is enough gain that the circuits saturate we may have *less* bandwidth reduction than what we have predicted. Consider the case of logic circuits that switch between 0 and 1. Adding more stages increases the delay but does not further degrade pulses, provided rising and falling delays are matched.

At this point, we must note that first-order sections are only one possible main-amplifier configuration. Higher-order gain stages, such as second- and third-order cells, have advantages over their first-order counterparts, provided their poles are well placed. Before presenting equations, an intuitive argument is made.

If two first-order sections are cascaded, the system is now second-order, but with repeated real poles. The 3 dB bandwidth of each stage becomes the 6 dB bandwidth of the overall system. The 3 dB bandwidth is now 64% of the original bandwidth. The transfer function of this system is:

$$H_{2x1}(s) = A_0^2 \frac{\omega_0^2}{s^2 + 2\omega_0 s + \omega_0^2}.$$
(9.98)

On the other hand, with a slight decrease to the damping factor of the transfer function, we have another second-order transfer function with the same pole magnitude and dc gain:

$$H_2(s) = A_0^2 \frac{\omega_0^2}{s^2 + \sqrt{2}\omega_0 s + \omega_0^2}.$$
(9.99)

Rather than having two repeated poles at $s = -\omega_0$, this second transfer function has poles at $s = -\frac{\omega_0}{\sqrt{2}}(1 \pm j)$ which corresponds to the Butterworth response, also known as the maximally flat response. In this case, the 3 dB bandwidth is now ω_0. Not only has the bandwidth been restored to ω_0, but if we were to cascade two of these second-order sections, rather than the bandwidth dropping to $0.64\omega_0$ as it did in the case of the first-order circuit, it only drops to $0.80\omega_0$. This is remarkable given that the second-order case actually consists of four first-order stages. The 3 dB bandwidth for Butterworth second-order stages is given by:

$$\omega_{3dB} = \frac{\omega_{GBW}}{\sqrt[N]{A_{tot}}}\sqrt[4]{\sqrt[N]{2} - 1}.$$
(9.100)

Figure 9.17 shows the corresponding achievable bandwidth for second-order stages for $A_{tot} = 40$ and 200 V/V. The main takeaways are that a higher percentage of the circuit's GBW is achievable and optimum bandwidth occurs for larger N due to successive second-order stages compressing the bandwidth more gradually. For these reasons, second-order (or even third-order) stages are usually used.

9.3.2 Shunt-Peaked Common-Source Amplifiers

The technique presented in Section 5.2.1 converts a first-order stage to a second-order stage and increases the -3 dB bandwidth of the stage by a factor 1.8. Suppose a gain of 200 V/V is targeted. A cascade of five first-order stages gives a bandwidth of

13.3% of the stage's GBW. Converting these stages to second-order Butterworth stages increases the achievable bandwidth to 21.5% of the stage's GBW, although the GBW will have increased as well by a factor of 1.8, giving significantly higher bandwidth. Therefore, one would expect an overall improvement in bandwidth of $1.8\frac{21.5}{13.3} = 2.9$. However, the zero in the shunt-peaked transfer function alters how cascaded stages interact, even with Butterworth poles, increasing the improvement to 3.3 times. This large BWER occurs with no additional power dissipation, although each inductor adds chip area.

9.3.3   Cherry-Hooper Amplifier

One relatively straightforward alternative to a cascade of first-order stages is the Cherry-Hooper amplifier. An example of an inverter-based design is shown in Figure 9.18. Other configurations based on single-ended and differential common-source amplifiers are also common. There are two advantages offered by the Cherry-Hooper amplifier compared to a cascade of two first-order stages. The first is that it converts two real poles into possibly complex poles, thereby affording the bandwidth advantages described in the previous section.

The second advantage is best seen by considering the Cherry-Hooper amplifier as a cascade of a transconductance amplifier followed by a transresistance amplifier. In a common-source amplifier, such as the one in Figure 5.2 (a), the load resistance, R_L, converts a signal current to an output voltage. The capacitance at the drain sees an equivalent resistance of R_L. On the other hand, a shunt-feedback TIA can have the same current-to-voltage conversion as the resistor while at the same time presenting a resistance in parallel with the capacitance at node v_1 much less than R_F. As discussed in Section 9.2.3, the input resistance of the inverter-based TIA is $\approx \frac{1}{g_m}$.

The analysis of this circuit follows very closely the analysis of the inverter-based shunt-feedback TIA presented in Section 9.2.3. The damping factor is given by:

$$\zeta = \frac{C_X + C_Y + g_{mB}R_FC_{GD2}}{2\sqrt{g_{mB}R_F\left(C_XC_Y + C_{GD2}\left(C_X + C_Y\right)\right)}}. \tag{9.101}$$

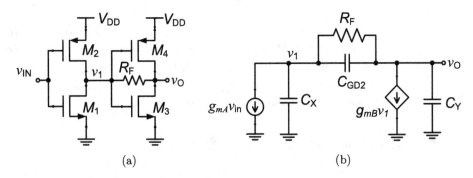

(a)   (b)

Figure 9.18 Inverter-based Cherry-Hooper amplifier: (a) schematic and (b) its small-signal equivalent circuit.

When $\zeta < 1$ the circuit provides us with complex poles. Note that in the analysis above the output resistance of the transistors has been ignored. The impact they have on damping factor is left to the reader. Frequently, a cascade of Cherry-Hooper amplifiers is used where each stage is built from equally sized inverters, leading to $C_X = C_Y = C$ and $g_{mB} = g_{mA} = g_m$. We can also assume that C_{GD} is some fixed percentage of C, denoted by $C_{GD2} = \chi C$. These assumptions simplify (9.101) to:

$$\zeta = \frac{C + C + g_m R_F \chi C}{2\sqrt{g_m R_F \left(C^2 + 2\chi C^2\right)}} \tag{9.102}$$

$$= \frac{1 + g_m R_F \chi / 2}{\sqrt{g_m R_F \left(1 + 2\chi\right)}}. \tag{9.103}$$

Setting $\zeta = 1$ we can write:

$$g_m R_F \left(1 + 2\chi\right) = 1 + g_m R_F \chi + \left(g_m R_F \chi\right)^2 / 4, \tag{9.104}$$

$$\frac{\chi^2}{4} \left(g_m R_F\right)^2 - \left(1 + \chi\right)\left(g_m R_F\right) + 1 = 0, \tag{9.105}$$

$$g_m R_F = \frac{\left(1 + \chi\right) \pm \sqrt{\left(1 + \chi\right)^2 - \chi^2}}{\chi^2 / 2}, \tag{9.106}$$

$$= 2\frac{1 + \chi \pm \sqrt{1 + 2\chi}}{\chi^2}. \tag{9.107}$$

C_X and C_Y are both the sum of a C_{GS} and a C_{DB}. If we assume these MOSFET capacitors are equal to one another ($C_{DB} = C_{GS}$) and that, as an example, $C_{GD} = 0.3 C_{GS}$, then $\chi = 0.15$. The value of $g_m R_F$ that gives $\zeta = 1$ is:

$$g_m R_F = 2\frac{1.15 \pm \sqrt{1.3}}{0.0225} \tag{9.108}$$

$$= 0.873, 204. \tag{9.109}$$

Between these values, $\zeta < 1$. Therefore, over a wide range of R_F, the poles are complex.

An example of an inverter-based optical receiver front-end based on a shunt-feedback TIA and three stages of Cherry-Hooper amplifiers as the main amplifier is shown in Figure 9.19 [72]. The referenced design was in 90 nm CMOS but with four main amplifier stages. All feedback resistors are 245 Ω, resulting in 194 Ω of gain from the TIA and 34 dB gain from the main amplifier. In linear units, this is a gain of 2.66 V/V per Cherry-Hooper stage. This example highlights a straightforward

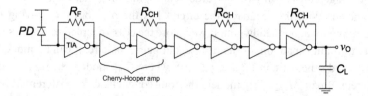

Figure 9.19 Inverter-based optical receiver front-end similar to [72].

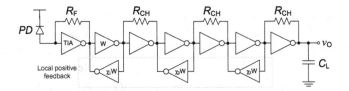

Figure 9.20 Inverter-based optical receiver front-end using local positive feedback in multiple locations.

design approach. By keeping the resistors and transistor widths constant along the signal path, the design reduces to three parameters, namely:

- Feedback resistors, assumed equal, i.e., $R_F = R_{CH}$.
- Width of transistors, taken as equal, W.
- Number of post-amplifier stages, N.

The design approach can be summarized as follows:

- Select the overall bandwidth around 50–70% of the data rate.
- Choose W for $C_I \approx 0.5C_D$, as discussed in Section 9.2.3.
- For a given N, choose the largest R_F to meet the bandwidth target. If the overall gain is insufficient, increment N.

If the load capacitance is smaller than the detector capacitance, more bandwidth can be achieved by decreasing W from input to output.

A variation on the inverter-based receiver in [72] is shown in Figure 9.20 [68]. In this work, the local positive feedback used in [57], presented in Figure 9.12, is applied at the output of the TIA and the output of the first two Cherry-Hooper amplifiers' output nodes.

9.3.4   Other Configurations of Main Amplifiers

This section presents two other approaches for main amplifier design. The first approach uses active feedback to convert first-order stages to second-order stages. An example of this is shown in Figure 9.21 [73]. The second technique is based on third-order stages.

Operation of the circuit in Figure 9.21 (a) is explained as follows, ignoring the effect of the inductors. Without the feedback differential pair formed by $M_{5/6}$, this amplifier stage consists of two first-order stages, meaning that the two poles are real and identical. With the feedback the circuit has the possibility of having complex poles, leading to higher bandwidth. If we ignore the cross-coupled transistors $M_{7/8}$ and the gate-to-drain capacitors of the differential pairs, the circuit can be modelled with the simplified schematic in Figure 9.21 (b), where G_{m1} models the transconductance of differential pair $M_{1/2}$, G_{m2} models the transconductance of differential pair $M_{3/4}$ and G_{mf} models the transconductance of differential pair $M_{5/6}$. The transfer function is of the form:

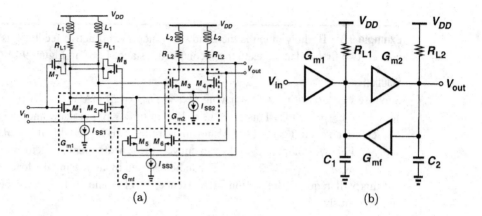

Figure 9.21 Second-order main-amplifier stage using active feedback: (a) schematic; (b) model. ©2003 IEEE. Reprinted, with permission, from [73].

$$H(s) = \frac{A_{dc}\omega_n^2}{s^2 + 2\zeta\omega_n s + \omega_n^2}, \tag{9.110}$$

where

$$A_{dc} = \frac{G_{m1}G_{m2}R_{L1}R_{L2}}{1 + G_{m1}G_{mf}R_{L1}R_{L2}}, \tag{9.111}$$

$$\zeta = \frac{1}{2}\frac{R_{L1}C_1 + R_{L2}C_2}{\sqrt{R_{L1}R_{L2}C_1C_2(1 + G_{mf}G_{m2}R_{L1}R_{L2})}}, \tag{9.112}$$

$$\omega_n^2 = \frac{1 + G_{mf}G_{m2}R_{L1}R_{L2}}{R_{L1}R_{L2}C_1C_2}. \tag{9.113}$$

If the circuit is designed for $\zeta = 1/\sqrt{2}$, then $\omega_{3dB} = \omega_n$. To further investigate this design point, assume $G_{m1} = G_{m2}$, $C_1 = C_2$ and $R_{L1} = R_{L2}$. Therefore $R_{L1}C_1 = R_{L2}C_2 = \tau$ and $G_{m1}R_{L1} = G_{m2}R_{L2} = A_0$. Also, let $G_{mf} = \beta G_{m1}$. With these substitutions,

$$A_{dc} = \frac{A_0^2}{1 + A_0^2\beta}, \tag{9.114}$$

$$\zeta = \frac{1}{2}\frac{2\tau}{\sqrt{\tau^2(1 + \beta A_0^2)}} = \frac{1}{\sqrt{1 + \beta A_0^2}}, \tag{9.115}$$

$$\omega_n^2 = \frac{1 + A_0^2\beta}{\tau^2}. \tag{9.116}$$

Therefore, to achieve the Butterworth design point of $\zeta = 1/\sqrt{2}$, β is selected as:

$$\beta = \frac{1}{A_0^2}. \tag{9.117}$$

Consider a comparison in gain and bandwidth between this circuit and a cascade of two first-order circuits. One can frame this comparison by matching bandwidths and comparing realizable gain or by matching gain and comparing realizable bandwidth.

Example 9.3   By how much is the bandwidth increased if active feedback is used to convert two first-order stages to the second-order stage shown in Figure 9.21?

Solution

Suppose we start with a design having $G_{m1} = G_{m2} = 20$ mA/V, $R_{L1} = R_{L2} = 100\ \Omega$ and $C_1 = C_2 = 160$ fF. This gives us a dc gain of 2 V/V per stage and two poles at 10 GHz for an overall gain of 4 V/V and an overall -3 dB bandwidth of 6.4 GHz. If we apply (9.117) we compute $\beta = 0.25$ to convert the repeated real poles to Butterworth poles. Now, $A_{dc} = \frac{4}{1+1} = 2$. The gain can be restored by increasing the load resistors, but this will require a reduction in β. Hence, (9.114) and (9.117) must be solved simultaneously.

$$A_{dc} = \frac{A_0^2}{1 + \beta A_0^2} = \frac{A_0^2}{2}. \tag{9.118}$$

Therefore, constraining $\zeta = 1/\sqrt{2}$ reduces the overall gain by a factor of two. To preserve the overall gain of 4, we need $A_0^2 = 8 \rightarrow A_0 = 2\sqrt{2}$. This is achieved by increasing the load resistors to 141 Ω. Now the required β is $\frac{1}{8}$, and $G_{mf} = 2.5$ mA/V. Note that if we raised G_m by widening transistors and increasing the bias currents, the load capacitances would scale proportionally. With the new value of load resistors, (9.113) can be evaluated to give:

$$\omega_n^2 = \frac{1 + G_{mf}G_{m2}R_{L1}R_{L2}}{R_{L1}R_{L2}C_1C_2} = \frac{1+1}{R_{L1}R_{L2}C_1C_2} = (2\pi\,10\ GHz)^2. \tag{9.119}$$

Since $\zeta = 1/\sqrt{2}$, $\omega_{3dB} = \omega_n$. By adding active feedback bandwidth has been extended from 6.4 GHz to 10 GHz.

Another approach to wideband main amplifier design involves more complicated feedback networks, such as third-order interleaving feedback [74]. In this scheme, multiple feedback loops allow the design of a sixth-order system with non-identical pole locations. Interleaving feedback was applied to an inverter-based receiver with a Cherry-Hooper main amplifier in [5].

9.4   Automatic Gain Control

In an optical link using 2-level NRZ signalling, the signal swing at the output of the analog front-end must be large enough for the decision circuits even when the input signal is at the sensitivity limit. For larger input signal amplitudes which occur most of the time, the front-end may saturate, operating as a limiting amplifier. In systems where the front-end and decision circuits reside on different chips, the off-chip interface is usually designed for large signals for which a high-gain, limiting front-end is desirable. On the other hand, when the front-end and the decision circuit are on

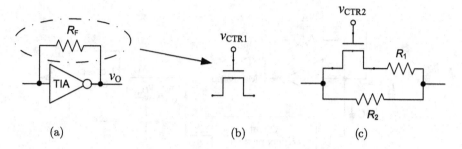

Figure 9.22 Techniques for introducing variable gain to a shunt feedback TIA: (a) original circuit; (b) replacement of R_F by a single MOSFET; and (c) replacement of R_F by a combination of resistors and a MOSFET.

the same chip, the front-end's output swing can be smaller. Front-ends may operate linearly over a wider range of signal inputs. Nevertheless, it is critical that if input signals increase, the main amplifier compresses gracefully without introducing distortion beyond saturation.

Optical links for telecommunications networks using single-mode fibre, coherent detection and complex modulation do not rely on a single decision circuit. Instead, the front-end's output is sampled by an analog-to-digital converter and demodulation is done in the digital domain. The approach requires a linear front-end in order to recover the transmitted bits. While linearity requirements are still less stringent than most wireless systems, care is needed to prevent the amplifier from entering a limiting mode. When input signals are larger, the gain of the main amplifier is reduced, using automatic gain control. The effect of this is that the front-end produces an output of constant amplitude over a wide range of input signal power. A constant output swing also minimizes quantization noise by using the ADC's entire input range.

The simplest way to vary the gain of a TIA or main amplifier is to replace a resistor with a triode region MOSFET. Figure 9.22 shows examples of this. In part (b), R_F has been replaced by a single MOSFET, with its gate connected to v_{CTR1}. Increasing v_{CTR1} will lower the gain. Triode MOSFETs are nonlinear, although signals in the TIA are usually fairly small. In part (c) the MOSFET is combined with two resistors. If we ignore the resistance of the MOSFET when v_{CTR2} is high, the circuit's equivalent resistance ranges from $R_1 \parallel R_2$ when the MOSFET conducts up to R_2 when the MOSFET is off. This scheme can be used in either an analog or digital approach.

9.5   Single-Ended to Differential Conversion and Offset Compensation

In this chapter we have considered TIAs that are connected to only one terminal of the photodiode. The signal from the PD is amplified and ultimately applied to a decision circuit, which is usually differential. For correct operation, the decision circuit is operated in one of two ways:

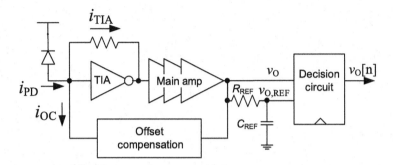

Figure 9.23 Connection between a single-ended front-end and a decision circuit using a low-pass filter for reference generation.

1. A single-ended signal is applied to one input of the decision circuit and a suitably generated reference is applied to the other.
2. A differential signal is generated and applied to the two inputs of the decision circuit.

An additional consideration for an optical receiver is the removal of offsets, which occur due to transistor mismatch and the unipolarity of the optical signal. In Chapter 7, we saw that 1s and 0s of NRZ data are encoded as P_1 and P_0, both greater than 0. This gives rise to a unipolar photo current. With the polarity of PD connection in Figure 9.23, $i_{PD} > 0$. Without the offset compensation block i_{TIA} would be also greater than 0. The TIA has an inverting gain. If we assume a non-inverting main-amplifier gain, v_O will be constrained between V_{REF} and ground as shown in Figure 9.24 (a). In this context, V_{REF} is the output voltage when $i_{TIA} = 0$. The offset compensation block generates a current, i_{OC}, equal to the average value of i_{PD}, resulting in i_{TIA} having zero average. Thus, with offset compensation, v_O is centred around V_{REF} as shown in Figure 9.24 (b).

By centring the output in the middle of the output range, linearity is improved. Figure 9.24 (c) and (d) shows the corresponding situation for a front-end with a differential output. The preceding discussion focuses on the removal of the average value of the photo current. Offset compensation will also reduce the effect of the offsets introduced by the amplifiers themselves.

Figure 9.23 shows a single-ended front-end. The reference for the decision circuit ($v_{O,REF}$) is generated by low-pass filtering the output of the front-end (v_O). The filter is designed such that $R_{REF}C_{REF} \gg NT_b$ where N is the maximum number of consecutive identical digits expected. Therefore, $v_{O,REF}$ is the average value of v_O.

Figure 9.25 shows the connection between a differential front-end and a decision circuit. The offset compensation loop centres the differential output around 0, as shown in Figure 9.24 (d). Any residual, uncompensated offsets in the front-end can be further reduced by offset compensation in the decision circuit.

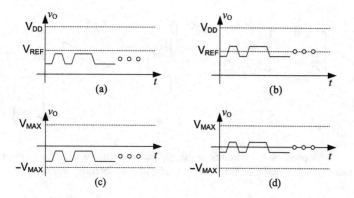

Figure 9.24 Optical receiver front-end output with and without offset compensation: (a) single-ended output, without offset compensation; (b) single-ended output with offset compensation; (c) differential output without offset compensation; (d) differential output with offset compensation.

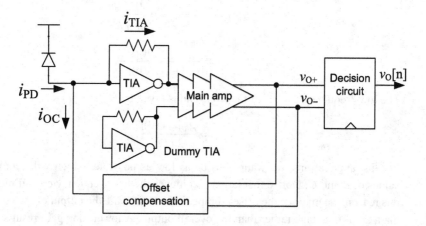

Figure 9.25 Connection between a differential front-end and a decision circuit.

Figure 9.26 (a) shows an offset compensation circuit compatible with the block diagram in Figure 9.23. If we assume infinite voltage gain from the inverter with capacitive feedback, the transfer function from v_O to i_{OC} is given by:

$$H_{OC} = \frac{i_{OC}}{v_O} = -\frac{g_m}{sRC}, \tag{9.120}$$

where g_m is the overall transconductance of the inverter supplying i_{OC}. The offset compensation loop introduces a low-frequency cut-off to the front-end. The closed-loop transfer function Z_{CL} can be found by combining H_{OC} with the front-end transfer function Z_{tot}:

$$Z_{CL} = \frac{Z_{tot}}{1 + Z_{tot}H_{OC}}, \tag{9.121}$$

$$= \frac{Z_{tot}}{1 + Z_{tot}\frac{-g_m}{sRC}}. \tag{9.122}$$

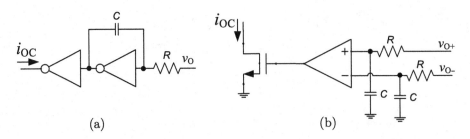

Figure 9.26 Offset-compensation circuits: (a) single-ended circuit suitable for an inverter-based front-end and (b) differential circuit for a front-end with a differential output.

To consider only the low-frequency portion of the closed-loop response, we can assume $Z_{tot} = -|Z_{tot}(0)|$, where $|Z_{tot}(0)|$ is the magnitude of the front-end's dc gain. Therefore:

$$Z_{CL,LF} = \frac{-|Z_{tot}(0)|}{1 - |Z_{tot}(0)|\frac{-g_m}{sRC}} \tag{9.123}$$

$$= -\frac{sZ_{tot}(0)}{s + |Z_{tot}(0)|\frac{g_m}{RC}}. \tag{9.124}$$

The low-frequency cut-off of the front-end is given by:

$$f_L = \frac{g_m|Z_{tot}(0)|}{2\pi RC}. \tag{9.125}$$

In most applications, we want f_L to be as low as possible, which will require large values of R and C. Notice that larger Z_{tot} moves the low-frequency cut-off higher. For this reason, we may see the offset compensation loop take the output of the second last main amplifier stage rather than the overall output. A lower loop gain results in lower cut-off, but leaves the offset of the final main amplifier stage uncompensated.

Considering a finite gain (A_0) in the inverter with capacitive feedback results in finite dc gain and a non-zero pole for H_{OC}, presented as $H_{OC_{NI}}$, where NI denotes non-ideal behaviour.

$$H_{OC_{NI}} = \frac{i_{OC}}{v_O} = -\frac{g_m A_0}{sRCA_0 + 1}. \tag{9.126}$$

Rewriting (9.123) using $H_{OC_{NI}}$ gives:

$$Z_{CL,LF_{NI}} = \frac{-|Z_{tot}(0)|}{1 - |Z_{tot}(0)|\frac{-g_m A_0}{sRCA_0 + 1}} \tag{9.127}$$

$$= -\frac{|Z_{tot}(0)|\,(sRCA_0 + 1)}{sRCA_0 + 1 + |Z_{tot}(0)|g_m A_0}. \tag{9.128}$$

The resulting low-frequency cut-off is now:

$$f_{L_{NI}} = \frac{1 + A_0 g_m|Z_{tot}(0)|}{2\pi A_0 RC} \approx f_L. \tag{9.129}$$

The non-ideal integrator has little effect on f_L. From (9.128) we see that the average value of the photocurrent, considered a dc input ($s = 0$), sees a gain of:

$$\frac{v_O}{i_{avg}} = Z_{CL,LFNI}(s = 0) = \frac{Z_{tot}(0)}{1 + g_m A_0 Z_{tot}(0)} \approx \frac{1}{g_m A_0}, \qquad (9.130)$$

This is relatively small compared to the gain the in-band signal sees.

The preceding discussion argued the need for an offset compensation loop and derived its ac response. What determines the dc node voltage to which it settles? Let's assume each inverter has the same threshold, V_{REF}, meaning that when $V_{IN} = V_{OUT} = V_{REF}$, no current flows in or out of the inverter. When no input is applied to the front-end, all node voltages sit at V_{REF}, and $i_{OC} = 0$. When an input current from the PD is applied to the front-end in Figure 9.23 using the offset-compensation circuit in Figure 9.26 (a), the output (v_O) drops from V_{REF}. Assuming infinite inverter gain, a current is drawn through the resistor of the OC circuit given by:

$$i_{INT} = \frac{V_{REF} - \overline{v_O}}{R}, \qquad (9.131)$$

where $\overline{v_O}$ denotes the average value of the output voltage. The current, i_{INT} is integrated on the capacitor C, leading to a voltage above V_{REF} at the output of the integrator. This is then applied to the OC loop's final transconductor leading to a positive, dc i_{OC} drawn from the photocurrent, reducing the average value of i_{TIA} until the loop settles. Once $\overline{v_O} = V_{REF}$ no further integration occurs. At this point, $i_{OC} = \overline{i_{PD}}$.

Aside from adding a low-frequency cut-off, offset compensation loops introduce two additional issues. The feedback inverter adds additional capacitance at the input node and its thermal-noise current is applied at the input, adding directly to the input-referred noise of the receiver. These issues favour downsizing the final inverter. However, one must be sure that it is large enough to source/sink the required i_{OC}.

Figure 9.26 (b) shows the offset compensation circuitry compatible with the receiver in Figure 9.25. The output of the opamp is the dc value of the differential output. This is converted to a current to be applied to the input. To reduce loading on the input node, the offset signal can be fed back to the dummy TIA (with reversed polarity).

Two capacitors are used for the low-pass filter (decision circuit reference generation, shown in Figure 9.23) and the offset compensation loop detailed in Figure 9.26 (a). For an inverter-based design, without any passive inductors, these capacitors can occupy much more die area than the inverters themselves. The low-pass filter capacitor used for decision circuit reference generation can be eliminated using the connection between the front-end and the decision circuit shown in Figure 9.27 [75]. This figure shows the final Cherry-Hooper amplifier (G_m/TIA) in an inverter-based receiver front-end. With an active offset compensation loop the signals at all nodes are centred around the threshold of the inverters. Although the amplitude of v_O is larger than v_{REF}, since they have opposite phase, the two signals can be used to drive the decision circuits.

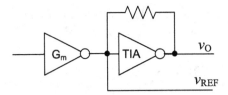

Figure 9.27 Decision circuit reference generation for inverter-based circuits. Voltages v_O and v_{REF} are directly connected to the decision circuit. Inverters G_m and TIA form the receiver's final stage of Cherry-Hooper main amplifier.

9.6   Summary

This chapter presented optical receivers, starting with their metrics and how sensitivity is estimated based on input-referred noise and decision-circuit voltage requirements. Before considering various amplifier topologies the properties of a resistor as a TIA were explored, concluding that they are viable only when capacitance is very low. The common-gate amplifier as a TIA was analyzed for both frequency response and noise performance. Shunt-feedback TIAs were presented next, with an emphasis on inverter-based designs.

To meet overall gain requirements, additional amplifiers are usually needed. Hence, main amplifier design was discussed next, focusing on the gain/bandwidth calculations for cascades of first- and second-order amplifiers. Cherry-Hooper, shunt-peaked and active feedback topologies were considered. Various inverter-based circuit techniques were given. The chapter closed with discussion of offset compensation and the connection between a single-ended front-end and a differential decision circuit.

If the gain of an optical receiver is increased, its bandwidth will decrease, introducing ISI that must be removed. Chapter 10 presents this approach to receiver front-end design.

Problems

9.1   For a total capacitance of 200 fF (photodiode, bond pads) find the largest resistive TIA that can support a data rate of 15 Gb/s while introducing negligible ISI? For a BER of 10^{-12} what peak-to-peak photocurrent is needed? How much peak-to-peak voltage swing will be produced?

9.2   Consider the resistive TIA in Figure 9.28 (b). The optical signal applied to the photodiode (PD) is shown in Figure 9.28 (a). $C_{PD} = 100$ fF, $R_L = 500\ \Omega$.

(a)   Sketch the current $i_{pd}(t)$ and voltage $v_{OUT}(t)$ assuming the responsivity of the photodiode is 0.7 A/W.

(b)   If a decision circuit is connected directly to $v_{OUT}(t)$, what is the worst-case latch sensitivity that can still give 10^{-12} BER?

(c)   What is the maximum data rate this circuit can support with the element values given before non-trivial ISI is introduced?

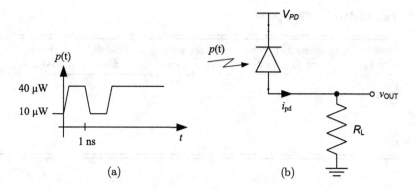

Figure 9.28 Photodiode and resistive TIA: (a) input signal; (b) schematic.

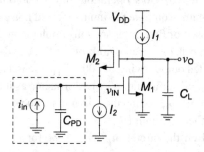

Figure 9.29 Shunt-feedback TIA using source-follower feedback.

9.3 Consider the circuit shown in Figure 9.29. Assume each transistor has an over-drive voltage of 0.25 V and $I_1 = 1$ mA and $I_2 = 0.2$ mA. The current source i_{in} represents the input current coming from a photodiode.

(a) Using an ac small-signal equivalent circuit, derive an expression for the low-frequency transimpedance gain of the circuit ($Z_{TIA} = \frac{v_o}{i_{in}}$) in terms of g_{m1}, r_{o1}, g_{m2}, and r_{o2}.

(b) Find the circuit's input and output resistance.

(c) Write an expression for the low-frequency output noise power-spectral density (i.e., ignore the effect of capacitors). Be sure to give appropriate units. Consider only thermal noise. For this part and the next parts of this question, feel free to simplify assuming $g_m r_o \gg 1$.

(d) Show the steps to convert the output power spectral density to an input-referred rms noise current. Show the units at each step.

(e) If i_{in} is a two-valued signal going between 10 μA and 30 μA when receiving 0s and 1s, sketch the total voltage (dc plus signal) at v_{IN} and v_O over a few UIs of made up data.

9.4 Design a common-gate TIA for 10 Gb/s operation and a photodiode/ESD capacitance of 200 fF. Maximize its gain. Compute the input-referred noise current.

Table 9.5 Main-amplifier options.

Number of stages	Required per-stage bandwidth	Resulting per-stage gain	Overall gain	Power dissipation	FOM
1					
2					
3					
10					
14					

9.5 Design a shunt-feedback TIA for 10 Gb/s operation and a photodiode/ESD capacitance of 200 fF. Maximize its gain. Compute the input-referred noise current.

9.6 Design a main amplifier composed of first-order sections implemented as resistively loaded differential pairs. Design it for a bandwidth of 10 GHz and a gain of 50 V/V. It should support a single-ended output swing of 400 mV$_{pp}$. Generate and plot eye diagrams for various input levels and comment on its performance as it saturates.

9.7 For the assumptions given in Section 9.3.3, find the minimum value of ζ as $g_m R_F$ is varied. What value of $g_m R_F$ gives the minimum ζ?

9.8 Design a main amplifier based on the circuit in Figure 9.21 for the specifications in Problem 9.6. If you *convert* each pair of first-order stages in your design from Problem 9.6 to a second-order stage, how much faster can your new design go for the same gain? How much more gain can your new design have for the same bandwidth?

9.9 Apply shunt inductive peaking to the main amplifier designed in Problem 9.6. Compare two design approaches. In the first design choose the same inductor for each of the identical stages. In the second design, select different inductor values in order to create different amounts of peaking from each stage. Can this second design reach a wider bandwidth and better transient performance?

9.10 The resistively loaded differential pair shown in Figure 1.27 is to be used in a main amplifier (MA) design. Its transistors operate with $g_m = 14$ mA/V, $C_{gs} = 20$ fF and $C_{db} = 12$ fF. You can ignore C_{gd} in this question. A designer is deciding how many stages of MA to use to implement an amplifier with an overall bandwidth of 10 GHz, knowing that the resistor value will have to be selected for each design.

(a) Determine the gain-bandwidth product of the main amplifier stage, assuming it is loaded by another identical stage.

(b) Fill out Table 9.5, assuming a per-stage power dissipation of 4 mW.

(c) For each option, compute the figure of merit: $FOM = \frac{TotalGain}{PowerDissipation}$.

(d) From the point of view of the given FOM, which design is best?

9.11 Design an inverter-based front-end of the topology in Figure 9.19 but with four main-amplifier stages. For a PD/pad capacitance of 150 fF and a load capacitance of 50 fF, what is the highest bandwidth achievable for an overall gain of 74 dBΩ.

(a) Determine its input-referred rms noise current.

(b) What is the front-end bandwidth and, therefore, what data rate can it support?

(c) Add an offset compensation circuit such as that in Figures 9.23 and 9.26 (a) so that the overall low-frequency cut-off of the front-end is no greater 1 MHz. Use inverters in the offset compensation circuit that are the same size as those in the main signal path.

(d) How did the bandwidth and noise performance change as a result of adding the offset compensation loop?

(e) If the final inverter of the offset compensation loop is downsized by a factor of 4, what is the largest dc PD current that can be reasonably compensated? How does the input-referred noise change relative to what was reported in part (d)?

10 Low-Bandwidth (Equalizer-Based) Optical Receivers

This chapter presents optical receivers that intentionally limit the bandwidth to achieve higher gain and better sensitivity. This has the consequence of introducing ISI that must be removed by a suitable equalizer. Important aspects of the noise analysis of these receivers are clarified. Various approaches to equalization in these reduced-bandwidth receivers are presented by way of published examples.

10.1   Introduction

When designing the circuits in Sections 9.2 and 9.3 such that negligible or at least tolerable ISI is introduced, the trade-off between gain and bandwidth often dictates that the TIA's gain is modest and several stages of main amplifiers are needed. Notice that in Figure 9.19 every inverter in the signal path draws the same bias current. One is used for the TIA and, in the case of the design in [72], another eight are used for the main amplifier. With almost 90% of the front-end power dissipated in the main amplifier, its power reduction is an obvious design target. If the gain of the TIA were increased, fewer main amplifier stages would be needed. Also, if the gain of the TIA is larger, the main amplifier contributes even less to input-referred noise. The TIA's input-referred noise PSD is also reduced. However, non-negligible ISI will be introduced that must be addressed or the eye will close. Depending on how the ISI is removed, the relevant noise bandwidth of the TIA might also be significantly reduced compared to the case of a wideband TIA. Overall, a high-gain TIA, fewer main amplifier stages and equalization can lead to a lower power, lower noise design. Recently, as data rates in CMOS are increased to 56 Gb/s and beyond, the techniques in this section are used, not only because ISI is intentionally introduced, but because its introduction has become unavoidable.

To better illustrate the motivation for reducing the bandwidth, we consider the resistive TIA shown in Figure 9.2. At first glance, it may appear that, according to (9.18), by increasing R_L the input-referred noise can be reduced arbitrarily if we tolerate ISI. However, consider the pulse response of a first-order system given by (9.14). With a unit input pulse of duration T_b, the pulse response, $v_{p,out}$ is given by:

$$v_{p,out}(t) = \begin{cases} 0, & \text{for } t < 0 \\ R_L \left(1 - e^{-t/R_L C}\right), & \text{for } 0 \le t < T_b \\ R_L \left(1 - e^{-T_b/R_L C}\right) e^{-(t-T_b)/R_L C} & t > t_b. \end{cases} \quad (10.1)$$

The peak of the first-order pulse reaches a maximum value of $R_L(1 - e^{-T_b/R_L C})$ at $t = T_b$. If T_b is large relative to $R_L C$, the exponential term is negligible and the peak of the pulse response is proportional to R_L. However, as R_L increases, the time constant $R_L C$ increases, meaning that the exponential term is no longer negligible at $t = T_b$. Therefore, the unit pulse response's peak is no longer R_L times the input current magnitude. In fact, as $R_L \to \infty$, $e^{-T_b/R_L C} \to 1 - \frac{T_b}{R_L C}$ and $v_{p,out}(T_b) \to \frac{T_b}{C}$. In other words, the unit input current is integrated on the capacitor C. Of course, if the front-end becomes an ideal integrator, the pulse response does not decay to 0 for $t > T_b$, leading to infinite ISI. Nevertheless, if we remove ISI, the effective gain of the resistive TIA has an upper bound of $\frac{T_b}{C}$. The gain still depends on data rate. If we have a way to remove ISI without adding noise, the minimum peak-to-peak input for BER = 10^{-12} can be computed by finding the input signal that gives an output signal whose main cursor is 14 times the output noise. That is:

$$\overline{v_{n,out}^2} = \frac{kT}{C}, \quad (10.2)$$

$$v_{o,pp,min} = 14v_{n,out} = 14\sqrt{\frac{kT}{C}}, \quad (10.3)$$

$$i_{pp,min} = \frac{v_{p,pp,min}}{Z_{eff}} = 14\sqrt{\frac{kT}{C}}\frac{C}{T_b} = 14f_b\sqrt{kTC}. \quad (10.4)$$

Comparing (10.4) to (9.22) where we prevented significant ISI by fixing $f_b = \frac{1}{\pi R_L C}$ we see that the theoretical sensitivity is improved by $\frac{1}{\pi}$ times. Since the mean-squared output noise is unchanged as we increase R_L, this analysis shows that the gain of the system is only increased by π. Notice as well that (10.4) introduced the notion of the effective gain Z_{eff}. As the resistor value is increased the TIA's *low-frequency gain* is increased to an arbitrarily large value. However, its settling time is lengthened. Within 1 UI (T_b) the output pulse cannot rise to be $i_{in}R_L$. In the limit, the input current is *integrated* on the capacitor.

The ideas presented in the preceding paragraphs regarding the analysis of the resistive TIA extend to the shunt-feedback TIA. Simulation results are shown in Figures 10.1 and 10.3. Figure 10.1 (a) shows the frequency response. As R_f increases from 100 Ω to 100 $k\Omega$, the bandwidth decreases and the low-frequency gain increases. Figure 10.1 (b) shows the pulse response for 10 Gb/s input data. Though the pulse response does increase in magnitude as R_f increases, it reaches a maximum value of less than 1500 Ω, much less than the maximum dc gain of 100 $k\Omega$ shown in Figure 10.1 (a). For lower values of R_f, the pulse response reaches a constant voltage within T_b, showing that the TIA has sufficient bandwidth. Also, 1 UI after the main cursor, there is negligible ISI. For larger R_f, the pulse does not reach a constant voltage and there is ISI at 200 ps. Increases in R_f do not see commensurate increases in the

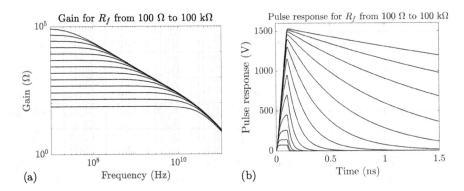

Figure 10.1 Simulation results of a shunt-feedback TIA: (a) frequency response; (b) pulse response for 10 Gb/s data.

height of the pulse response. Hence, input data does not see an *effective gain* of R_f. The largest pulse response has ISI that persists well beyond 10 UI and approaches the pulse response of an integrator.

10.1.1   Effective Gain of an Optical Receiver Front-End

The term effective gain has been used in passing in the previous paragraphs. In this section, we define and illustrate the term. In Section 1.10, expressions for estimating the BER of a receiver are given. For example, (1.32) combines the signal's VEO and the noise level to compute a BER in the presence of ISI. A BER is not calculated from the separation of the constant 1 and constant 0 levels when there is ISI. Further, (1.32) lets us specify a signal-to-noise ratio (SNR) at the decision circuit required for a target BER.

Receiver designers like to give an input-referred noise specification, frequently as a single rms noise current. With this rms input noise, knowing the required SNR for a given BER, a required input current amplitude can be computed. Based on a photodetector's responsivity, the sensitivity of the receiver in terms of OMA can be estimated. For input referral of noise to be useful, we must calculate the same SNR at the input as we measure, or simulate at the output. That is:

$$\frac{VEO_{pp}}{\sigma_{n,o}} = \frac{i_{pp}}{\sigma_{n,i}}, \tag{10.5}$$

where $\sigma_{n,o}$ and $\sigma_{n,i}$ are the output- and input-referred noise, respectively, and i_{pp} is the peak-to-peak input signal assumed to have low ISI, meaning it is also equal to the input VEO. Finally, VEO_{pp} is the output peak-to-peak vertical eye opening.

We define the effective gain, Z_{eff}, as the ratio between input and output as follows:

$$Z_{eff} = \frac{VEO_{pp}}{i_{pp}}. \tag{10.6}$$

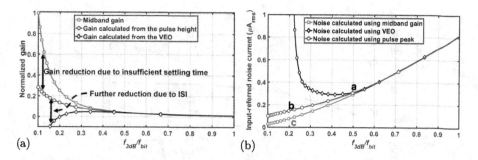

Figure 10.2 Effective gain and impact on noise estimation of a shunt-feedback TIA: (a) three descriptions of gain as a function of bandwidth to data rate ratio; (b) estimation of input-referred noise based on different definitions of gain. ©2019 IEEE. Reprinted, with permission, from [76].

While gain $Z_{FE}(s)$ is a frequency-dependent quantity, Z_{eff} is a scaler that accounts for the aggregate effect of the front-end's frequency response for ideal input data at a particular data rate. If no ISI is *introduced*, $Z_{eff} = Z_{FE}(0)$. This is the case for the three smallest values of R_f in Figure 10.1. We see that no ISI is introduced because the pulse response reaches a constant value before $t = T_b$ and it is close to zero at $t = 2T_b$. This was predictable because the bandwidth of the TIA was at least half of the data rate.

If ISI is introduced, $Z_{eff} < Z_{FE}(0)$ and requires an investigation of VEO either through simulation or a study of the ISI terms. In the case of an ideal equalizer that removes all ISI, leaving only the main cursor, $Z_{eff} = h_0$ where h_0 is the height of the pulse response's main cursor. This is a key point worth repeating. Consider the frequency response and pulse response in Figure 10.1 corresponding to $R_f = 100\ k\Omega$. The pulse response peaks at well under $100\ kV$ as it would if its height were predicted by dc gain. This is because the UI is too short to allow for complete settling. Ideal ISI removal from an equalizer does not restore h_0 to that predicted by dc gain.

With the value of Z_{eff} determined for the appropriate equalization context, the input-referred noise current is computed as:

$$\sigma_{n,i} = \frac{\sigma_{n,o}}{Z_{eff}}. \tag{10.7}$$

The effective gain for different situations is plotted in Figure 10.2 (a) as the bandwidth to data rate ratio is changed. Starting from the right-hand side where the bandwidth of the TIA is high relative to the data rate, and no ISI is introduced, gain estimates from dc gain, pulse height and VEO are equal to one another. Moving toward the left as the bandwidth of the TIA is reduced and ISI is introduced, these gain estimates diverge with the largest estimate being the dc gain, followed by the pulse height, then, lastly, the computed VEO.

Figure 10.2 (b) shows input-referred noise estimates using the three gains from Figure 10.2 (a). Again from the right-hand side to point in Figure 10.2 (b), since the gains are equal, the input-referred noise estimates are also equal. As the bandwidth

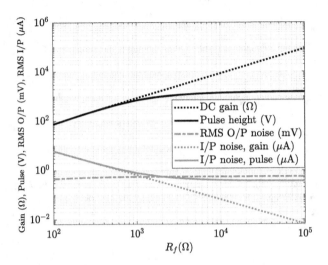

Figure 10.3 Noise simulation results of a shunt-feedback TIA.

decreases, the input-referred noise computed from a VEO-based gain increases dramatically as the eye closes. This is because, as the eye closes, the effective gain tends to zero. As a divisor to output noise when referring noise to the input, this blows up the input noise estimate.

The difference between the remaining two input-referred noise curves is more subtle. If we assume all ISI is removed we use h_0 as the input referral gain. However, this is still more noise than if we use dc gain.

One possible conclusion from Figure 10.2 (b) is that if we can remove all ISI we should arbitrarily reduce the front-end bandwidth. While a DFE can potentially remove all post-cursor ISI, we will see that residual pre-cursor ISI dictates a minimum front-end bandwidth.

Figure 10.3 shows the noise performance of the TIA from several different angles. The dotted black line shows the TIA's dc gain as a function R_f. The solid black line shows the pulse height, which saturates. Pulse height is a data rate-dependent measure of effective gain. For a higher data rate, the dc gain curve would be unchanged, but the pulse height curve would change in two ways. First, it would saturate to a lower value due to reduced integration time. Second, the value of R_f where the pulse height curve breaks off from the dc gain curve would be smaller. The dashed grey curve shows the output noise. Notice that it increases only slightly as R_f increases, but it also saturates. The dotted grey line shows a calculation of input-referred noise using:

$$\overline{i^2_{n,in}} = \frac{\overline{v^2_{n,out}}}{|Z_{TIA}(0)|^2}. \tag{10.8}$$

This is only valid if $Z_{TIA}(0)$ is a reasonable estimate of the data's effective gain. The solid grey line shows the input-referred noise where the dc gain in (10.8) is replaced by the TIA's effective gain:

$$\overline{i^2_{n,in}} = \frac{\overline{v^2_{n,out}}}{\left|Z_{eff}\right|^2},$$

(10.9)

where, in this case, $Z_{eff} = \frac{v_{p,out}(T_b)}{i_{in}} = h_0$. The rest of this chapter draws heavily on [76], where we introduced our notion of effective gain. Before reading the next three sections, the material in Chapter 4 is recommended as background reading.

10.2 DFE-Based Limited-Bandwidth Optical Receivers

An example of a DFE-based optical receiver using one FIR feedback tap is shown in Figure 10.4. The TIA gain must be large enough so that once the DFE cancels the ISI tail, the main cursor of the combined TIA/adder signal path is large enough to be latched. Additional gain between the TIA and summing circuit may be required. Also, depending on the bandwidth of the TIA, taps beyond h_1 may be required. The challenges in this implementation are primarily those related to closing the h_1 tap. Solutions to this are discussed in Section 4.5.2.

A DFE-based optical receiver using IIR feedback is shown in Figure 10.5 (a). Recall from Section 4.5.2 that the 1 UI signal at v_{OUT}, applied to $G(s)$, generates a long tail that can subtract out a long ISI tail of v_{TIA}. Figure 10.5 (b) shows one particular implementation of an IIR DFE [77], which illustrates a critical difference between an electrical link where a DFE is used to compensate for the ISI induced by a distributed electrical channel versus the situation here where the ISI is caused by a spatially

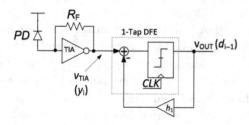

Figure 10.4 FIR DFE-based optical receiver using a one-tap FIR DFE.

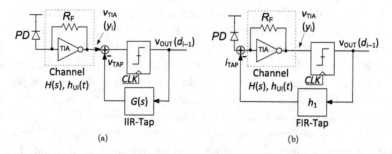

Figure 10.5 IIR DFE-based optical receivers: (a) IIR tap feedback to summer; (b) FIR tap feedback to input of TIA.

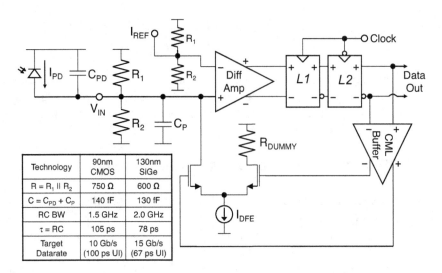

Technology	90nm CMOS	130nm SiGe
$R = R_1 \parallel R_2$	750 Ω	600 Ω
$C = C_{PD} + C_P$	140 fF	130 fF
RC BW	1.5 GHz	2.0 GHz
$\tau = RC$	105 ps	78 ps
Target Datarate	10 Gb/s (100 ps UI)	15 Gb/s (67 ps UI)

Figure 10.6 Implementation of an infinite impulse response decision feedback equalizer-based optical receiver. ©2013 IEEE. Reprinted, with permission, from [77].

proximate reduced-bandwidth circuit. Recall from Section 4.5.2 that a single tap from an IIR DFE can subtract an entire tail of post-cursor ISI if the ISI follows a decaying exponential. When the ISI-inducing part of the system is nearby, the system that introduces the ISI can also be used to generate ISI equal in magnitude but in opposite polarity to the ISI added to the input pulse.

In Figure 10.6, taken from [77], current pulses from the PD are amplified by a high-gain low-bandwidth passive TIA. This has the advantage of a larger main-cursor gain, but with the detriment of a long ISI tail. The input bit is resolved by the latch. A current pulse (I_{DFE}) of < 1 UI duration (i.e., finite duration) is fed back to the TIA's input. When I_{DFE} is chosen with appropriate magnitude, it generates an ISI tail that compensates for the ISI the original input pulse generated. For a delayed pulse to cancel an entire ISI tail, the system generating the ISI must be first-order. Another challenge in this design is that the delay through the TIA, latch, CML buffer and DFE tap must be less than 1 UI, preventing operation at high data rate. This particular design operated at 9 Gb/s in 90 nm CMOS and 15 Gb/s in 130 nm SiGe.

An interesting improvement to address timing closure of the tap is described in [78] and shown in Figure 10.7. In this work the compensatory ISI tail is generated by the RC network coupling two sense-amplifier latches. Using half-rate clocking, two latches are connected to $V_{A\pm}$, the differential output of the analog portion of the front-end. These two sense-amplifier latches share one set of degeneration elements. As shown in Figure 10.7 (a) and (b), the latch's decision in one phase of the clock is stored on the differential voltage $V_{F\pm}$ which subtracts from $V_{A\pm}$. The elements R_F and C_F are programmable, allowing the amplitude and time constant of feedback to be adjusted. Relevant waveforms are shown in Figure 10.8.

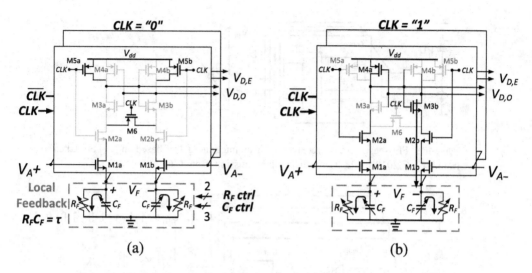

Figure 10.7 Infinite impulse response decision feedback equalizer enabled through latch coupling: (a) operation when CLK = "0" and (b) operation when CLK = "1"; ©2016 IEEE. Reprinted, with permission, from [78].

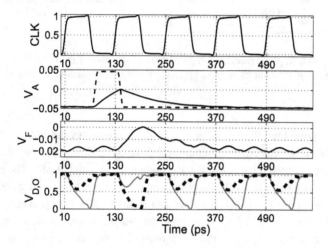

Figure 10.8 Relevant waveforms for an IIR DFE enabled through latch coupling. ©2016 IEEE. Reprinted, with permission, from [78].

10.3 FFE-Based Limited-Bandwidth Optical Receivers

Three examples of FFE-based optical receivers are shown in Figures 10.9 to 10.11. In Figure 10.9 (introduced in [79]) the photodiode's current is integrated on its parasitic output capacitance, generating the voltage v_{IN}. An offset loop sets a current that causes v_{IN} to discharge when the input bit is a 0. This feedback settles to be the average of the 0 and 1 photocurrents assuming equal density of 1s and 0s. In principle this input voltage ramps up and down with input 1s and 0s. By comparing the *change* in v_{IN}

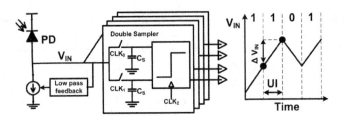

Figure 10.9 FFE-based optical receiver using an integrating TIA based on [79] but taken from [76]. ©2019 IEEE. Reprinted, with permission, from [76].

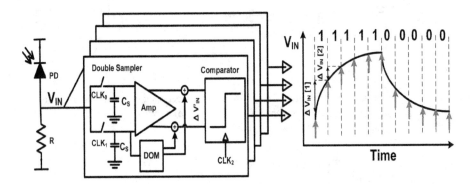

Figure 10.10 FFE-based optical receiver using dynamic offset modulation (DOM) with a resistive TIA based on [30] but taken from [76]. ©2019 IEEE. Reprinted, with permission, from [76].

between consecutive bits, the input bit can be determined. As the bandwidth of the low-pass feedback is reduced, the integration becomes ideal. In this case, the voltage difference for consecutive bits, ΔV_{IN} is independent of the number of CIDs. However, if there is a long run of 1s or 0s the input voltage can saturate to V_{DD} or ground.

The circuit in Figure 10.10 [30] has a resistor R that is chosen to prevent saturation by intentionally making the integration non-ideal. In this case, during a long run of CIDs, ΔV_{IN} decreases. To combat this the threshold of the comparator is adjusted by a process described as dynamic offset modulation.

Discrete-time receivers have added noise due to sampling. Each sample adds a mean-squared noise voltage of kT/C_S to the signal where C_S is the sampling capacitor. Therefore, it is preferred to use larger capacitors to reduce this noise. In the implementations in Figures 10.9 and 10.10, larger capacitors will charge-share with the PD's capacitor, reducing signal swing. The circuit in Figure 10.11 [31] uses a low-bandwidth TIA as a buffer, preventing charge sharing and allowing larger capacitors to be used.

In Figure 10.9, since the voltage applied to the comparator is the difference between consecutive samples, the equalization operation has the z-domain transfer function of:

$$H(z) = 1 - z^{-1}. \tag{10.10}$$

In the case of Figures 10.10 and 10.11, the equalization operation can be shown to be:

$$H(z) = 1 - \alpha z^{-1}. \tag{10.11}$$

Optimum equalization is found using $\alpha = e^{-T_b/\tau}$ where τ is the dominant time constant of the front-end. In Figure 10.10, $\tau = RC_{PD}$ assuming $C_S \ll C_{PD}$, whereas in Figure 10.11, τ is the reciprocal of the TIA's dominant pole. As the time constant of the front-end increases, the front-end's operation approaches ideal integration and $\alpha \to 1$.

10.4 CTLE-Based Limited-Bandwidth Optical Receivers

A representative block diagram of a CTLE-based optical receiver is shown in Figure 10.12. The CTLE block can add additional low-frequency gain similar to the main amplifier in a conventional receiver, but with the added feature of some high-frequency boosting to compensate for the high-gain but low-bandwidth TIA.

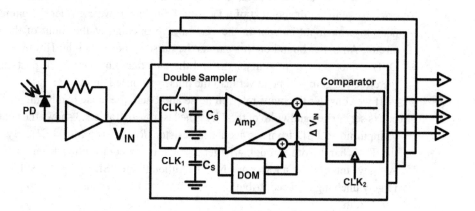

Figure 10.11 FFE-based optical receiver using DOM with a shunt-feedback TIA based on [31] but taken from [76]. ©2019 IEEE. Reprinted, with permission, from [76].

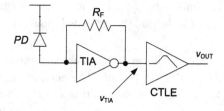

Figure 10.12 CTLE-based optical receiver block diagram.

10.5   Noise Analysis of Reduced Bandwidth Front-Ends

The best way to estimate the noise performance of the front-end is to calculate the integrated noise at the input of the decision circuit (i.e., output of the front-end) and compare it to the vertical eye opening once the introduced ISI and its partial cancellation have been taken into account. In doing this, the input signal amplitude needed for a vertical eye opening 14x the rms noise gives an estimate of the input sensitivity for BER = 10^{-12} assuming an ideal decision circuit. This *analog* sensitivity is increased by the requirements of the decision circuits by adding to it $\frac{V_{MIN}}{gain}$, where *gain* is the gain of the front-end. In the case of a reduced-bandwidth front-end the ratio between the input amplitude and the vertical eye opening will not be the dc gain of the front-end. Rather, the input signal will see an effective gain that is less than the dc gain. If the goal is to estimate an input-referred rms noise, the designer can either:

- Compute an output-referred rms noise and refer it to the input by dividing it by the effective gain of the front-end; or
- Compute an input-referred noise PSD and integrate its flat and f^2 components up to appropriate bandwidths.

In the case of DFE-based receivers, TIA noise appears at the input of the DFE. If we assume that the DFE taps contribute very little noise, we integrate the TIA's output noise only. To refer the output noise to the input we divide the output noise by the effective gain. In the context of a DFE-based receiver, we consider the gain from the peak-to-peak input current to the vertical eye opening at the input of the decision circuit. An upper bound on this gain is the pulse response height. However, since the DFE cannot remove pre-cursor ISI and the taps may not remove all post-cursor ISI, the effective gain will be lower than the pulse response height.

If we consider the input-referred noise approach, the noise bandwidth of the front-end is set by the bandwidth of the TIA. The last row of Table 9.3 shows the contribution to input noise PSD from the feedback resistor, R_F and the MOSFETs. By reducing the TIA's bandwidth through a larger R_F, the noise contribution from R_F decreases. Also, we integrate to a lower bandwidth. Although the PSD of the MOSFETs' noise is largely unchanged, its contribution to noise is reduced because we integrate to a lower bandwidth.

As presented in [76] the typical input-referred noise integration approach will lead to erroneous results. Specifically, the computed input-referred noise will only be accurate if the input signal sees a gain of $Z_{TIA}(0)$. That is, it is accurate when we have sufficient bandwidth. Thus, in a limited-bandwidth context, after integrating the input-referred PSD to the appropriate bandwidth, the estimated input-referred noise must be scaled by $\frac{Z_{TIA}(0)}{Z_{eff}}$.

For CTLE- and FFE-based receivers we integrate the noise at the output of the equalizer. Not only does the equalizer add noise, but the TIA's noise is shaped by the transfer function of the equalizer. For a CTLE it is obvious that we use its continuous-time (CT) (s-domain) transfer function. However, in the case of a discrete-time FFE, we can still use its CT transfer function. The justification can be seen in Figure 10.13

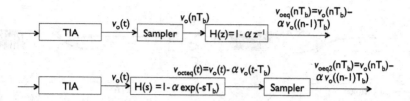

Figure 10.13 Justification for swapping the order of the sampler and equalizer for discrete-time FFE noise analysis.

where we see that the overall outputs of the two systems are identical. Thus, from the point of view of the output signal of the TIA, the sampler and FFE can be swapped. This allows the analysis of TIA noise to be done in continuous time. The noise analysis proceeds as follows:

- The discrete-time (z-domain) FFE transfer function is converted to continuous time by replacing z^{-1} by $\exp(-sT_b)$.
- The PSD of the output noise of the TIA is filtered by the CT FFE transfer function.
- We find the rms output noise at the input of the sampler by integrating the FFE's output noise to ∞.
- This rms output noise can be referred to the input by dividing it by the effective gain. This gain is the ratio between the vertical eye opening at the input of the sampler and the peak-to-peak input current of the TIA.

Past work on FFE equalization of low-bandwidth TIAs is based on a TIA response that is reasonably well described by a first-order transfer function. Figure 10.14 shows TIA and equalizer output noise spectra for TIAs with different spacing between the dominant and non-dominant poles. For both (a) and (b), the noise contribution of the resistor is shown by the dashed line while the contribution of the MOSFETs is shown by the solid grey line. Notice that in Figure 10.14 (a), the TIA's second pole is much higher than the first pole and hence the noise contribution of the MOSFETs extends to much higher frequency than in Figure 10.14 (b). The total TIA noise is plotted by the solid black line. In spite of the reduced non-dominant pole, both TIAs have similar bandwidth and pulse response. The dashed black line shows the transfer function of the FFE. At low frequency, it has a magnitude of $1 - \alpha$ whereas it peaks at $1 + \alpha$ at $f_b/2$. Since the transfer function is based on exponentials, it is periodic. It appears compressed at high frequency because of the logarithmic horizontal axis. The FFE suppresses TIA noise at low frequency, but boosts signal and noise at high frequency. In Figure 10.14 (a), the out-of-band MOSFET noise is amplified. The dashed grey trace shows the total noise, which is reduced when the non-dominant pole is reduced.

The pulse response of the TIAs for both non-dominant pole scenarios is shown in Figure 10.15. The change in C_L has a small effect on the pulse response of the TIA and the ability of the FFE to equalize it.

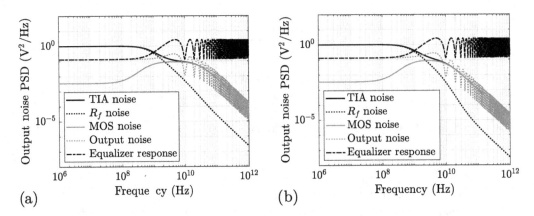

Figure 10.14 Effect of non-dominant pole on FFE-based low-bandwidth front-end noise: (a) smaller C_L and hence higher frequency non-dominant pole and (b) larger C_L and hence lower frequency non-dominant pole.

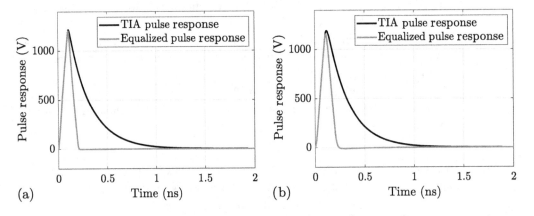

Figure 10.15 Effect of non-dominant pole on FFE-based low-bandwidth front-end pulse response: (a) smaller C_L and hence higher frequency non-dominant pole and (b) larger C_L and hence lower frequency non-dominant pole.

Discrete-time FFEs also have kT/C noise due to sampling. Unlike the samples of the TIA's noise which may be correlated from one sample to the next, kT/C is uncorrelated from one sample to the next.

If we take an input-referred noise perspective on the CTLE- or FFE-based receiver we see that the equalizer has little impact on the input-referred noise PSD's shape. However, because of equalization, the effective bandwidth of integration is extended. The relevant bandwidth is no longer that of the TIA, but is now the bandwidth of the combined TIA and equalizer.

An in-depth comparison of the various equalizer-based low-bandwidth front-ends was done in [76]. The paper considered a low-bandwidth shunt-feedback TIA followed by each class of equalizers. For a fair comparison, the bandwidth of the TIA

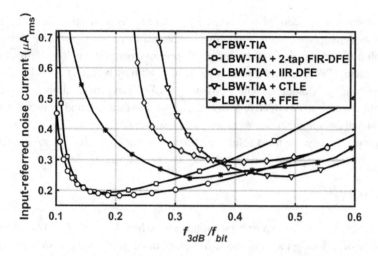

Figure 10.16 The best-case noise performance for each receiver architecture at $f_b = 30$ Gb/s. The horizontal axis represents the ratio of the bandwidth of the overall front-end to the data rate. The optimum value of the TIA's pole Q is found to be 0.707 for the IIR DFE and CTLE-based front-ends. The optimum values of the TIA's pole Q of the two-tap FIR DFE and FFE-based front-ends are found to be 0.577 and 0.3 (and $\alpha = 0.7$), respectively. ©2019 IEEE. Reprinted, with permission, from [76].

and its pole Q were optimized to get the lowest input-referred noise. The results are summarized in Figure 10.16. The conclusions are as follows:

- When no equalizer is used, denoted as full bandwidth (FBW), increasing gain and decreasing bandwidth below 50 to 70% achieves lower input-referred noise. The best bandwidth from an SNR point of view is 40% of the data rate. Also, the best Q of the second-order TIA was found to be 0.707. These results are consistent with other work.
- The best architecture from a noise point of view is the IIR DFE, since it was assumed to be capable of cancelling all post-cursor ISI, leaving only pre-cursor ISI. The presence of pre-cursor ISI is the reason why the bandwidth should not be reduced below 21% of the data rate.
- When a two-tap FIR DFE is used, the optimum scenario was found to be a bandwidth of 18% of the data rate and a lower pole Q of 0.577. Ultimately, this architecture has similar noise performance to the IIR DFE.
- Of the CTLEs considered in the paper, a CTLE with one real zero and complex poles with $Q = 0.707$ improved noise the most. This CTLE extends the bandwidth of the TIA by a factor of 1.4. This means that the bandwidth of the front-end, consisting of TIA and CTLE, is 49% of the data rate with the TIA's bandwidth at 35%, slightly lower than the 40% design point when no equalizer is used.
- When an FFE is used, the best case ratio of non-dominant to dominant pole was found to be 9. In this case the bandwidth of the TIA was only 6% of the data rate, but the FFE extended the bandwidth to about 32% of the data rate.

This discussion of reduced-bandwidth front-ends along with a portion of the published work underemphasizes the need for a front-end to provide sufficient gain, without which a power penalty will be incurred. Improvements in analog sensitivity gained by reducing TIA noise by lowering its bandwidth may be offset by a decision-circuit power penalty. Although a reduced-bandwidth TIA can provide more gain than a full-bandwidth design, its effective gain at the target data rate is bound by its pulse response height and does not increase as much as its low-frequency gain. Finally, when estimating power penalty, we must refer the decision circuit's requirement to the input using the front-end's effective gain, rather than its low-frequency gain.

10.6   Summary

This chapter began by exploring what happens when the gain of a TIA is increased at the cost of reducing its bandwidth. Increases in dc gain gave rise to increases in the pulse response's main cursor, but were limited to that which an ideal integrator would yield. The need for equalization was established. Recent work that used DFEs, FFEs and CTLEs to remove ISI were presented.

To correctly estimate the sensitivity of equalizer-based optical receivers, we introduced the notion of effective gain. This is critical when the ratio between input and output eye height is much smaller than the circuit's dc gain. This approach enabled the noise analysis of the three types of receivers, leading to the conclusion that, from the noise point of view, a DFE-based approach gives the best sensitivity.

Problems

10.1   The results in Figure 10.16 were generated assuming the Q of a shunt-feedback TIA can be easily adjusted, independently from its gain.

(a)   Find the optimum bandwidth as a percentage of data rate for a shunt-feedback TIA, designed in your technology of choice. For this part, assume no follow-on equalization.

(b)   Then, increase its gain, thereby shrinking its bandwidth, and find its optimum design point assuming all post-cursor ISI is cancelled (IIR DFE assumptions) and eye closure is only the result of pre-cursor ISI.

(c)   How do these results differ from Figure 10.16?

(d)   Report the TIA's low-frequency gain, the ratio between pulse response height and input, and the ratio between VEO and input pulse height.

10.2   Similar to Problem 10.1, find the best bandwidth to data rate ratio assuming a one-tap DFE follows the TIA.

10.3   Determine the best bandwidth to data rate ratio for a common-gate TIA when no equalizer is used and when an equalizer of your choice is used.

10.4   Revisit the circuit designed in Problem 9.11. If the assumption of an ideal DFE is made, redesign the circuit for the same gain, but with possibly fewer main amplifier stages. Compare the estimated sensitivity and power dissipation of the two designs. To limit the complexity of this problem, ignore offset compensation circuitry for both designs.

11 Advanced Topics in Electrical and Optical Links

The previous chapters have introduced the fundamentals of electrical and optical links. This chapter presents advanced topics that draw on these fundamentals. Section 11.1 explains how implementing PAM4 signalling impacts equalization and circuit design. In Chapter 5 we have seen how a transmitter can be segmented into slices to support a transmitter-side feed-forward equalizer (FFE) and impedance control. This idea can be extended to build a DAC-based transmitter. Similarly, to detect a PAM4 signal we need three decision thresholds. This too can be extended into 32 or 64 levels, thereby implementing an ADC of five or six bits. A brief overview of DAC/ADC-based links, also known as DSP-based links, is presented in Section 11.2.

A consequence of PAM4 signalling is a smaller vertical eye opening. Designing for a BER of 10^{-12} is not always viable in higher-loss channels. Instead, forward error correction (FEC) can be used to convert a raw BER in the range of 10^{-3} to a post-FEC BER of 10^{-12}. This is facilitated with an ADC-based receiver. The power dissipation of the DSP is shrinking as CMOS processes shrink. However, these algorithms introduce latency much longer than the "time-of-flight" of the signal down the channel, which may have negative performance consequences for the overall system. As alluded to in Chapter 4, ISI can be removed digitally using techniques known as "maximum likelihood sequence estimation," discussed very briefly in Section 11.4.

11.1 PAM4 Links

Transmitters and receivers must both be modified when we move from a two-level signal (PAM2) to a four-level signal (PAM4). This section outlines this as well as presenting two metrics for linearity, namely ratio level mismatch and transmitter and dispersion eye closure quaternary.

11.1.1 Overview of Transceiver Modifications to Support PAM4

Starting with a transceiver for a PAM2 (NRZ) link, the following modifications are needed to support PAM4:

1. Ignoring equalization, the PAM4 transmitter must generate four signal levels instead of two in the case of PAM2.

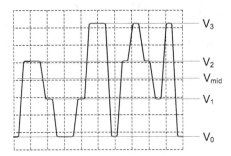

Figure 11.1 Signal level definitions for ratio level mismatch (RLM).

2. The PAM4 receiver must slice the signal with three decision thresholds instead of one in the case of PAM2.
3. The transceiver must be linear enough.

11.1.2   PAM4 Linearity Metrics

Two metrics are commonly used to describe PAM4 transmitter performance, namely:

1. Ratio level mismatch (RLM); and
2. Transmitter and dispersion eye closure quaternary (TDECQ).

11.1.2.1   Ratio Level Mismatch

Figure 11.1 shows a PAM4 signal. Levels V_0 through to V_3 are the four signal levels, when enough CIDs are transmitted that the signal reaches a steady state. That is to say, there is negligible ISI. The following signal differences are computed:

$$V_{mid} = \frac{V_3 + V_0}{2}, \tag{11.1}$$

$$ES_1 = \frac{V_1 - V_{mid}}{V_0 - V_{mid}}, \tag{11.2}$$

$$ES_2 = \frac{V_2 - V_{mid}}{V_3 - V_{mid}}, \tag{11.3}$$

giving V_{mid} as the middle of the overall signal and ES_1 and ES_2 are normalized signal levels for 01 and 10 symbols, ideally equal to 1/3. RLM is computed as:

$$R_{LM} = \min\left(3ES_1, 3ES_2, (2 - 3ES_1), (2 - 3ES_2)\right). \tag{11.4}$$

If $ES_1 = ES_2 = 1/3$, then $R_{LM} = 1$.

Example 11.1   Compute the RLM for the following signal levels:

$$V_0 = -0.5 \text{ V}, \tag{11.5}$$

$$V_1 = -0.17 \text{ V}, \tag{11.6}$$

$$V_2 = 0.2 \text{ V}, \tag{11.7}$$
$$V_3 = 0.55 \text{ V}. \tag{11.8}$$

Solution

The calculations are as follows:

$$V_{mid} = \frac{V_3 + V_0}{2} = \frac{0.55 - 0.5}{2} = 0.025, \tag{11.9}$$

$$ES_1 = \frac{V_1 - V_{mid}}{V_0 - V_{mid}} = \frac{-0.17 - 0.025}{-0.5 - 0.025} = 0.371, \tag{11.10}$$

$$ES_2 = \frac{V_2 - V_{mid}}{V_3 - V_{mid}} = \frac{0.20 - 0.025}{0.55 - 0.025} = 0.333. \tag{11.11}$$

With these values, RLM is computed as follows:

$$R_{LM} = \min\left(3ES_1, 3ES_2, (2 - 3ES_1), (2 - 3ES_2)\right), \tag{11.12}$$
$$= \min\left(1.11, 1, 0.885, 1\right) = 0.885. \tag{11.13}$$

Acceptable values of RLM are in the range of 0.95. This metric, though, leaves out several important aspects for signal integrity such as ISI and timing skew between the sub-eyes.

11.1.2.2 Transmitter and Dispersion Eye Closure Quaternary

This metric describes the performance of a PAM4 transmitter for optical links. Unlike RLM, which only considered settled values for the different signal levels, TDECQ takes into account ISI. More precisely, TDECQ considers the signal after a five-tap receiver-side FFE, taking into account the FFE's effect on ISI and noise. The output of a TDECQ measurement is a value in dB which quantifies the optical power penalty of the transmitter relative to an ideal transmitter at the same OMA. A detailed discussion of the metric, its measurement and how different impairments affect the measurement can be found here [80]. The next paragraphs give an overview.

Figure 11.2 [80] shows a block diagram of a measurement setup for measuring TDECQ. The transmitter under test drives the worst-case channel for a given optical communication standard. The output is received by a reference receiver consisting of a fourth-order Bessel–Thomson filter with a bandwidth of 50% of the symbol rate. Noise with a standard deviation of σ_G is added to the signal before it passes through a five-tap FFE. The noise is increased (while optimizing the FFE and sampling position in the eye) until the symbol error ratio rises to the standard's target value, frequently taken as: SER $= 4.8 \times 10^{-4}$. TDECQ is defined as:

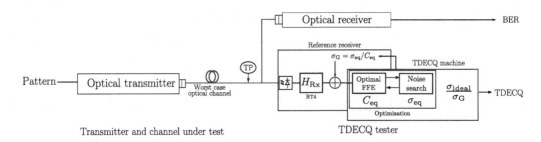

Figure 11.2 Measurement setup for TDECQ. ©2019 IEEE. Reprinted, with permission, from [80].

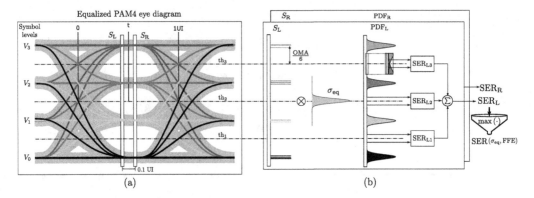

Figure 11.3 Waveforms showing calculation of TDECQ. ©2019 IEEE. Reprinted, with permission, from [80].

$$TDECQ = 10 \log_{10} \frac{\sigma_{ideal}}{\sigma_G}, \tag{11.14}$$

where σ_{ideal} is the tolerable noise when an ideal signal is received. Due to residual ISI and noise amplification from the FFE, $\sigma_{ideal} > \sigma_G$.

Figure 11.3 (a) [80] shows a PAM4 eye diagram. To measure TDECQ, a sampling scope generates slices of the eye at two locations 0.1 UI apart, denoted as S_L (left slice) and S_R (right slice). For real channels, we only see an eye this nice at the output of the five-tap FFE. Figure 11.3 (b) shows histograms at S_L. Due to ISI, each PAM4 level has multiple crossing points, degrading SER. Noise (σ_{eq}) is added to this histogram until the computed SER reaches the target dictated by the standard. The noise term (σ_G) in (11.14) results from the input-referral of σ_{eq} computed as:

$$\sigma_G = \frac{\sigma_{eq}}{C_{eq}}, \tag{11.15}$$

where C_{eq} is computed from the coefficients (c_i) of the FFE as:

$$C_{eq} = \sqrt{\sum_{i=0}^{4} c_i^2}. \tag{11.16}$$

Finally, for an ideal signal with no ISI, σ_{ideal} is given by:

$$\sigma_{ideal} = \frac{OMA}{6Q_t}, \tag{11.17}$$

where OMA is the optical modulation amplitude between settled first and fourth levels of the PAM4 signal and Q_t is the required SNR at the target symbol error ratio. When $SER = 4.8 \times 10^{-4}$, $Q_t = 3.414$.

Although an ideal signal gives rise to a TDECQ of 0 dB, an ideal transmitter will not. This is because even the fourth-order filter introduces some ISI, resulting in a TDECQ of approximately 0.8 dB, according to [80]. Common specifications of TDECQ of 3.4 and 4.5 dB for single-mode and multi-mode channels, respectively, allow about 2.6 and 3.7 dB of transmitter and channel impairments.

11.1.3   PAM4 Transmitters

The circuit techniques for generating multiple signal levels have already been discussed in Sections 5.4 and 5.6.4, where implementation of an FFE was presented for CML and SST circuits, respectively. Recall that a simple FFE implements:

$$y[n] = d_i[n] - \alpha d_i[n-1], \tag{11.18}$$

where d_i has values $+1$ and -1. Here, the signal takes on values $1+\alpha$, $1-\alpha$, $-1+\alpha$, and $-1-\alpha$, based on two consecutive bits. For evenly spaced PAM4 levels, we need signals of 1, 1/3, $-1/3$, and -1. Setting:

$$\frac{1-\alpha}{1+\alpha} = \frac{1}{3}, \tag{11.19}$$

$$\alpha = \frac{1}{2}. \tag{11.20}$$

Therefore, a PAM4 transmitter implements:

$$y[n] = d_{msb}[n] + 0.5 d_{lsb}[n], \tag{11.21}$$

where d_{msb} and d_{lsb} are the most significant and least significant bit of the two bits encoded each symbol. In (11.21) bits are ± 1. The segment generating the MSB must be exactly twice as large as the segment generating the LSB, otherwise RLM degrades.

Although PAM4 is used to lower the baud rate in an effort to reduce the effect of channel loss, PAM4 links will nevertheless use equalization at both ends of the link. For an FFE with N coefficients (α_i) the overall transmitter response is given by:

$$y[n] = \sum_{i=0}^{N-1} \alpha_i d_{msb}[n-i] + 0.5 \sum_{i=0}^{N-1} \alpha_i d_{lsb}[n-1]. \tag{11.22}$$

It would be mathematically equivalent to apply the equalizer difference equation after adding the MSB and LSB data streams. This would mean applying a four-valued signal to each FFE tap. However, both CML and SST FFEs work best by fully switching each tap. Thus we want to apply digital signals to the coefficients. Overall, this scheme requires $2N$ sub-drivers. Further segmentation may be required to ensure matching.

In general, $y[n]$ in (11.22) can take on 4^N values and requires a clocked delay line for both LSB and MSB. An alternative approach, presented in Section 11.2, views a driver as a DAC. The $2N$ bits needed to compute each $y[n]$ are mapped to a required output level, often through a look-up table, and applied to the DAC.

11.1.4   Considerations for PAM4 Receiver Circuits

PAM4 receivers must have superior linearity compared to their NRZ counterparts. An NRZ optical link where no receiver-side equalization is performed can use a highly nonlinear receiver consisting of a TIA followed by a limiting (saturating) amplifier. This scenario is less common in electrical links because most electrical links introduce sufficient ISI to require receiver-side equalization, and hence reasonable linearity. If we assume equal peak-to-peak amplitude in the received signal, the PAM4 link must resolve smaller eyes, allowing the introduction of smaller errors, hence requiring better linearity.

In Chapter 6 we considered two implementations of an inverter-based CTLE, shown in Figure 6.23. The upper signal path in the additive equalizer in Figure 6.23 (a) required a gain of $2A - 1$ while the lower path in the subtractive equalizer in Figure 6.23 (b) only needed a gain of A. The signal-handling capabilities of these circuits is dictated by output swing. Thus a larger gain translates to a smaller input range. In general, PAM4 receivers must be designed with linearity in mind.

11.2   DAC/ADC-Based Links

As data rates cross the 100 Gb/s threshold, we see the more frequent deployment of DAC/ADC-based links. These are transceivers in which the transmitter is segmented into a sufficient number of slices that it can be viewed as a DAC and the receiver employs an ADC to quantize the signal. In such a link, most of the equalization is performed in the digital domain.

11.2.1   Motivation for DSP-Based Links

There are several technology trends at play that motivate DAC-based transmitters and ADC-based receivers. These include:

- Shrinking CMOS technology reduces the energy dissipation of digital computation needed in a DSP-based equalizer.
- DSP can be parallelized obviating the need to have 1 UI computation.
- DSP makes reconfiguration of Rx equalization easier.
- Power dissipation of high-speed ADCs is improving.

In fact, DAC-based transmitters are a no-brainer as Tx outputs were already subdivided into slices to perform impedance control, implement FFEs and produce PAM4 signals. The same total width of slices is reorganized into generally binary weighted slices to produce a DAC-based Tx. There still needs to be impedance trimming, but FFE operation translates to setting an M-bit code. Therefore, the complexity of a long Tx FFE is reduced. Output signal values are taken from look-up tables.

11.3 Digital Equalization in DAC/ADC-Based Links

The performance of high-speed digital-to-analog converters (DACs) and analog-to-digital converters (ADCs) has steadily improved, driven in part by the needs of optical links for long-haul transmission. The modulation formats used favour signal generation using a DAC and detection using an ADC. Equalization has also been implemented in the digital domain. This approach can be used in electrical links and short-reach optical links, although, until recently, the energy use of the data converters was too high for this approach to be practical. Nowadays top-tier electrical and short-reach optical links implement digital-domain equalization and employ DACs and ADCs.

Most high-speed ADCs are built using a high-speed sampler and a large number of time-interleaved ADCs operating in the range of 1 GS/s. Success-approximation-register ADCs are the most prevalent. If a 32 GS/s ADC is implemented using 32 1 GS/s sub-ADCs, equalization operations are parallelized and the digital clock frequency is lowered to 1 GHz. This potentially relaxes the performance required from digital logic blocks, allowing implementation using standard cells. However, there are still critical timing loops. For example, the 32 lanes of computation depend on each other. For timing closure of the DFE, 32 operations must take place sequentially, within 1 ns.

A conventional mixed-signal electrical-link transceiver suitable for a high-loss channel typically uses:

1. Two to five taps of Tx-side FFE, programmed based on the target application of the transceiver.
2. Rx-side CTLE: this could be one CTLE that introduces multiple pole-zero pairs or multiple CTLEs, with different zero frequencies.
3. Rx-side DFE having a combination of fixed location taps and some programmable taps.

DSP-based electrical-link transceivers suitable for high-loss channels are using:

1. Two to five of Tx-side FFE, programmed based on the target application of the transceiver as was the case for the mixed-signal approach.
2. Rx-side CTLE as in the case of the mixed-signal approach. In choosing the transfer function of the CTLE, the objective is to reduce the required ADC resolution. This is done by increasing the main cursor's size relative to the peak-to-peak signal.
3. A high-speed ADC implemented using a time-interleaved approach, taking one sample per symbol. This has consequences on the design of the clock and data recovery system, dictating the use of a "baud-rate" CDR.

4. An Rx-side *digital* FFE, using a comparatively large number of taps. For example, a 112 Gb/s PAM4 transceiver (i.e., 56 Gbaud/s) that can tolerate 37.5 dB loss at 28 GHz implemented in 7 nm FinFET technology [81] uses a 31-tap FFE.
5. Rx-side DFE, frequently having only one-tap [81] [82].

The choices made in a DSP-based receiver appear counterintuitive. Firstly, in Section 4.5.2 the stringent timing constraint for the first tap of the DFE was explained. Thus one might wonder why, if a DFE is used, it would be used to cancel the first post-cursor. One might also wonder if, with such an extensive FFE, we need a DFE at all?

Following the ADC, the signal $x[k]$ has noise, due to noise in the channel and circuits as well as quantization noise from the conversion from an analog signal to a discrete-valued signal. If this noise $x_n[k]$ has a variance of $\bar{x}_n^2$, the noise at the output of the jth FFE tap of weight α_j is given by:

$$\bar{x}_j^2 = \alpha_j^2 \bar{x}_n^2. \tag{11.23}$$

The FFE implements the following difference equation:

$$y[k] = \sum_{j=0}^{N} \alpha_j x[k - j], \tag{11.24}$$

where N is the number of delays. Since (11.24) is expressed in a causal fashion, the tap associated with the main cursor is $j = m$, $0 < m < N$. Thus tap α_m will be the largest. If $x_n[k]$ is assumed to be a white noise process, successive samples of $x_n[k - j]$ are uncorrelated. Thus the noise associated with $y[k]$ is given by:

$$\bar{y}_n^2[k] = \bar{x}_n^2 \sum_{j=0}^{N} \alpha_j^2. \tag{11.25}$$

Although the noise grows with the number of FFE taps, (11.25) also shows that if eliminating the first post-cursor requires a large tap weight, this tap will add more noise than smaller tap weights used to eliminate smaller ISI terms.

Example 11.2   Revisit the pulse response in Table 1.2 and apply an FFE to equalize the channel with this pulse response, similar to what was done in Example 4.3, but consider the noise at the output of the FFE in terms of $\bar{x}_n^2$. Consider an alternative in which a DFE is used to remove the effect of h_1 and thus the FFE eliminates all other terms.

Solution
We can find the effect of an FFE on a channel with a given pulse response (h_k) by applying the pulse response to the FFE and finding the resulting pulse response of the combined channel/equalizer. This is summarized in Table 11.1. The final row of the table computes the sum of the squares of the FFE contribution, which is a proxy for the noise at the output of the FFE. Cases 1 to 4, repeated from Example 4.3, show the

Table 11.1 Resulting pulse response after an FFE for a channel with $h_k = 0.02, 1, 0.5, 0.2$.

FFE weights i	Case 1 0, 1, 0, 0 h_{rx_1}	Case 2 0, 1, −0.5, 0 h_{rx_2}	Case 3 0, 1, −0.5, 0.05 h_{rx_3}	Case 4 −0.02, 1, −0.5, 0.05 h_{rx_4}	Case 5 −0.02, 1, 0, −0.2, 0.1 h_{rx_5}
−24	0	0	0	−0.0004	−0.0004
−1	0.02	0.02	0.02	0	0
0	1	0.99	0.99	0.98	0.99
1	0.5	0	0.001	−0.003	0*
2	0.2	0.05	0	0	0.002
3	0	−0.1	−0.075	−0.075	0
4	0	0	0.01	0.01	0.01
5	0	0	0	0	0.02
ISI:	0.72	0.17	0.106	0.0884	0.0324
VEO:	0.28	0.82	0.884	0.8916	0.9576
$\sum_{j=0}^{N} \alpha_j^2$	1	1.25	1.2525	1.2529	1.0504

* Assumed to be cancelled by a DFE.

effect of adding more taps to the FFE. In comparison to Case 1 which had only an h_0 tap, the h_1 tap of the FFE contributed most of the additional noise in Cases 2 to 4. The last column (Case 5) shows a scenario in which the first post-cursor of the pulse response at the output of the FFE is not eliminated, but is eliminated by a subsequent DFE. This results in lower noise since the largest FFE tap is eliminated.

The results in Example 11.2 were for a pulse response that an FFE could equalize relatively straightforwardly. This was because the post-cursor tail was almost a geometric series. If, on the other hand, the pulse response has $h_1 = h_0$, but small ISI beyond h_1, an FFE is ill-suited. With the relatively noiseless ISI elimination of the DFE, one might want to use a longer digital DFE. Timing closure challenges have steered designers away from this option.

In designing a DAC/ADC-based link, several decisions must be made. These include:

- Voltage- or current-mode transmitter architecture.
- Amount of transmitter-side FFE.
- Amount and complexity of pre-ADC equalization in the receiver. Usually some is done to reduce the dynamic range of the input signal and therefore the required ADC resolution.
- Number of bits of resolution and type of ADC. Most are in the 5–8 bit range using time-interleaved SAR sub-ADCs. A large number of ADCs, say 32, share a small number of track-and-hold circuits, which operate quickly. Each SAR can take 32 UI to compute its N-bit output. From the system designer's perspective, resolution is the most important decision.
- Type and length of receiver-side digital equalization. Most recently published links are using long FFEs (> 20 taps) and short DFEs (one-tap).

There are different ways of looking at these trade-offs. In some cases, more Rx pre-equalization can relax ADC resolution for a given BER. Which approach is better depends on power dissipation. A higher resolution ADC uses more power, as will computation with higher-resolution signals; however, the latter is likely the smaller issue.

A good introduction to link budget analysis, and in particular how ADC resolution influences BER, is [42]. When modelling a link, our goal is to estimate its BER. In Chapter 1 simple expressions were presented taking into account signal amplitudes and noise levels. The argument of the $Q(\cdot)$ function contains the standard deviation of noise, although it is its Gaussian shape that gives rise to the exponential functions in (1.27) and (1.28). In analyzing an ADC-based receiver, quantization error must be accurately modelled. Since it is bounded to $\pm 0.5 LSB$ it contributes less to BER than to Gaussian noise of equal variance. Even a uniform distribution may not be accurate when the number of bits of ADC resolution is very small.

11.4 Maximum-Likelihood Sequence Detectors

Rather than invert the channel as an FFE or CTLE does, or subtract away ISI like a DFE, maximum-likelihood sequence detectors process a stream of received bits and determine what was the most likely stream of transmitted bits [83]. This notion is applicable to equalization as well as error detection, where errors are the result of additive noise. Several algorithms have been demonstrated. Using an ADC in the receiver allows for a multi-level representation of the received signal. With this available, designers can consider a whole range of options, including digital equalization and MLSE.

Despite the power reduction possible with small CMOS transistors, implementation of MLSE dissipates too much power for use in short-reach electrical links. However, longer-reach links at 112 Gb/s make use of MLSE. Since this text is focused on the analog and mixed-signal IC design aspects of wireline communication, a detailed treatment of MLSE is beyond our scope.

11.5 Summary

This chapter began by introducing how a PAM2 (NRZ) link would need to be modified to support PAM4 signalling. Aside from generating four levels, the link needs to be linear. Linearity metrics for electrical-link transmitters (RLM) and optical-link transmitters (TDECQ) were defined.

Segmentation of drivers into slices and the establishing of more decision thresholds in receivers naturally leads to DAC- and ADC-based transmitters and receivers, respectively, giving rise to more equalization in the digital domain. The typical equalizer configurations for mixed-signal links and DSP-based links were summarized. The chapter ended by introducing the notion of maximum-likelihood sequence estimation.

Problems

11.1 Explain the timing constraints in DSP-based FFEs and DFEs.

11.2 Compute the RLM for a PAM4 signal with:

$$V_0 = 0.20 \text{ V}, \tag{11.26}$$
$$V_1 = 0.38 \text{ V}, \tag{11.27}$$
$$V_2 = 0.61 \text{V}, \tag{11.28}$$
$$V_3 = 0.81 \text{ V}. \tag{11.29}$$

11.3 Consider a channel with the pulse response from Table 1.2. Assume a receiver-side FFE implements the coefficients in Case 4 of Table 11.1. If the received signal before the equalizer has a peak-to-peak signal of 200 mV, what is the maximum possible peak-to-peak VEO assuming ideal equalization? That is, how large is $\pm h_0$? How large is the peak-to-peak VEO using the coefficients in Case 4 of Table 11.1?

11.4 Here, we extend Problem 11.3 to PAM4. Assume a PAM4 signal has the same peak-to-peak signal of 200 mV and then passes through the same channel. What is the maximum possible VEO of the sub-eyes assuming ideal equalization? How large is the VEO of the sub-eyes using the coefficients in Case 4 of Table 11.1?

11.5 Assume that before the FFE the receiver has a weakly nonlinear amplifier, modelled as:

$$v_o = v_s \tanh(v_i/v_s), \qquad (11.30)$$

where v_s is a voltage normalization coefficient for the nonlinearity. As v_s increases the amplifier becomes more linear. Find values for v_s such that the eye openings in Problems 11.3 and 11.4 are each reduced by 20%.

11.6 Find the value of v_s in (11.30) that gives rise to an RLM of 0.95, for an input amplitude of ± 1. What is the THD for an input sinusoid with an amplitude of 1 V?

12 Overview of Synchronization Approaches

This chapter, along with the next three, covers the general topic of clock generation and distribution as well as clock and data synchronization. Clocking circuitry, including clock and data recovery systems, can dissipate 30–50% of total transceiver power. Jitter degrades BER as much as amplitude noise. Therefore wireline designers must pay attention to the jitter performance of clock generation circuitry and clock distribution circuitry as much as they focus on the amplitude noise behaviour of the receiver's signal path. Although clock generation can dissipate non-trivial power, the centralized generation of a clean reference clock allows the amortization of its power dissipation across multiple transceiver lanes, although its distribution is also challenging. Therefore, a thorough treatment of all aspects of clocking is important. This chapter gives an overview of the principal synchronization approaches used in wireline systems as well as clock distribution circuitry.

To introduce the main challenges in Tx/Rx synchronization, consider the scenario shown in Figure 12.1 in which a driver is connected to a receiver via a channel. Both the driver and the slicer are activated by the same clock signal, *Clk*. This scenario, referred to as a *synchronous* system, occurs within a CPU or GPU where a globally distributed clock can feed all flip-flops. In the processor example, the channel might in fact be a cascade of logic circuits adding a delay but, nevertheless, a delay of less than one UI. The system functions correctly so long as the driver's output, launched by one edge of the clock, arrives at the slicer (another DFF) before the next clock edge. Therefore DFFs that are consecutive in the data path are physically close to one another and they receive clocks that have little difference in clock phase relative to the clock that launched the data.

The block denoted by "x N" is a clock multiplication unit (CMU) that generates a high-frequency clock N times the frequency of its *RefClk* input. Ideally, every N^{th} edge of *Clk* is aligned to an edge of *RefClk*. Thus jitter in the reference clock signal is mapped onto *Clk*.

The synchronous approach runs into trouble when the physical distance between driver and slicer increases. We can no longer assume that the clock phase at the two ends of the link is synchronized. Also the delay along the channel may exceed one UI, meaning that greater care must be taken to sample the received signal in the middle of the eye.

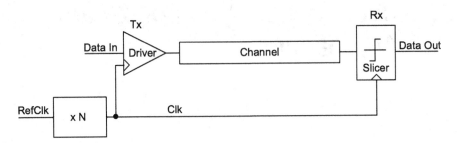

Figure 12.1 Simplified block diagram of a synchronous link.

12.1   Clock-Forwarded (Source-Synchronous) Links

The next simplest form of synchronization uses an additional channel to forward a clock from the transmitter to the receiver. An example of this is shown in Figure 12.2. The forwarded clock is received by *ClkRx*, amplified and possibly advanced/delayed to best align it to the received data. A feature of this scheme is that the clock and data will have highly correlated jitter, meaning that although each signal has jitter relative to an ideal jitter-free clock, their relative timing can be well matched. In fact, this property allows this configuration to tolerate a higher jitter CMU.

Since the clock at the receiver (CRx) is a delayed version of the clock from the Tx (CTx), both clocks have the same frequency, but have different phase. This fits the general definition of a mesochronous link, discussed in Section 12.2.1. However, many mesochronous links are deployed with a common low-frequency reference clock requiring CMUs in both Tx and Rx. Source-synchronous links, though a special case of mesochronous links, are thus treated separately in this section.

A published example of a clock-forwarded link is shown in Figure 12.3 [84]. A total of 47 data lanes connect die A and die B with 19 sending data from die A to die B and 28 from die B to die A. Its operation along with its advantages and disadvantages are as follows:

- Each die has a clock generation circuit denoted by IL-VCO. The clock generated in die A is *forwarded* to die B as fclk_A. It is also used to launch the data sent from txbundle_A[1] and txbundle_A[0]. Therefore, jitter from the IL-VCO in die A is superimposed onto the data stream transmitted from die A to B. The best clock signal for clocking the receiver will have edges that move back and forth with the jitter on the incoming data stream (i.e., fclk_A). The clock receiver can track this jitter up to some frequency and supply a receive clock with highly correlated jitter. This feature can relax the transmit clock's jitter specifications.
- The extra lane for the clock will dissipate approximately as much power as a data lane. However, the power overhead of two clock lanes is amortized over 47 data lanes.
- In die B, f_clkA is used to clock rxbundle_B[1] and rxbundle_B[0]. If there is a small delay between the clock arriving at txbundle_A[0] relative to txbundle_A[1],

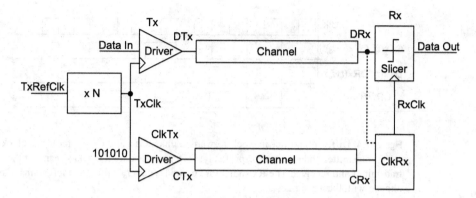

Figure 12.2 Simplified block diagram of a clock-forwarded link.

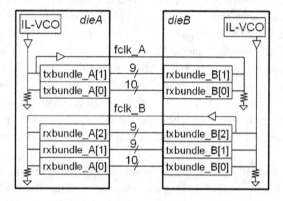

Figure 12.3 Block diagram of a bidirectional clock-forwarded link. ©2010 IEEE. Reprinted, with permission, from [84].

this delay is approximately duplicated at the receiver between rxbundle_B[0] and rxbundle_B[1].

- A similar approach is used to launch data from die B to die A.
- Notice that although each die has a locally generated clock signal, the clock used to capture data in its receivers is the clock *forwarded* from the other die. This is done because power-supply noise and other causes of jitter will affect the data stream and the forwarded clock in a similar fashion. If the receiver side clock were used, its jitter would be uncorrelated with the data stream's jitter.
- In many cases the forwarded clock is a signal at half of the data rate. To maximize jitter tracking, it is often generated by using a full-rate clock to clock a transmitter configured to send a 1010 data stream. For example if the data rate is 10 Gb/s, a 1010 pattern has a fundamental period of 2 UI and is therefore a 5 GHz clock.
- There will be a timing offset between the clock edges and the middle of the data eye. The circuitry in Figure 15.23 from Section 15.6 is used to align the clock and track slowly varying changes in this timing offset.

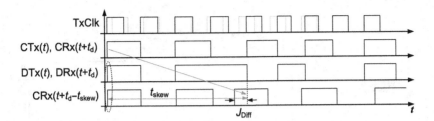

Figure 12.4 Differential jitter in clock-forwarded links: CTx and CRx refer to the clock signals at the transmitter and receiver, respectively; DTx and DRx are the data streams at the transmitter and receiver. The Rx signals are plotted with a t_d shift to the left so that they are aligned with the Tx signals.

Although the circuitry shown in Figure 15.23 is not trivial, it is still simpler than what is required in the absence of a forwarded clock. It is reasonable to ask when is a source-synchronous link *not* the best approach. The first criterion is the amortization of the clock lane over multiple data lanes. We see clock-forwarded links in parallel interfaces such as memory interfaces and other chip-to-chip links. When an interface consists of a single data lane, a separate lane for a forwarded clock doubles the required I/Os and power dissipation. This is more burdensome than building a clock and data recovery unit in the receiver. The other criterion relates to the premise of jitter correlation and specifies the maximum link length over which jitter correlation provides an advantage.

A thorough discussion of jitter in clock-forwarded links is found in [85]. Key aspects are presented below. Before delving into the mathematics, the fundamentals are presented in Figure 12.4 and summarized as follows:

- A jitter-free TxClk is plotted in grey; the real TxClk, with jitter, is shown in black.
- Near $t = 0$, jitter has advanced the clock, whereas after four clock periods, jitter has delayed the clock relative to the ideal clock.
- TxClk jitter is mapped to DTx and CTx, the latter of which is generated via a "1010" data pattern applied to an identical Tx.
- Edges of DTx and CTx are aligned.
- If the time delay (t_d) between Tx and Rx are equal, edges of DRx and CRx are also aligned to one another. The plots of the DRx and CRx are shifted left by t_d so they fall on top of DTx and CTx.
- When both clock/data see the same time delay, their edges move back and forth, relative to the ideal clock, together, meaning that at the sampling point, their jitter is perfectly correlated. This requires that both signals see the same time delay.
- If there is a skew, t_{skew}, between DRx and CRx, clock and data edges are no longer aligned at the receiver. This is shown in the last trace in Figure 12.4 where a skew of about 4 UI is present. Within the dashed ellipse, DRx and CRx are aligned. However, a few periods later the clock is early by an amount J_{Diff}. The edge near $t = 0$ was early, relative to the ideal clock. Due to a skew of about four clock periods, this early edge is used to capture the data UI that was launched by a late clock.

Jitter between clock and data is highly correlated if the same clock edge that was used to launch a given UI is also used to capture that UI. Skew between clock and data degrades jitter correlation.

Referring back to Figure 12.2, we assume at the transmitter end of the link, TxClk has sinusoidal jitter $J_{C,Tx}$ at a particular frequency f_j and amplitude A_j given by:

$$J_{C,Tx} = A_j \sin 2\pi f_j t. \tag{12.1}$$

Since both the clock signal and the data signal are launched by the same driver topology, we assume both have the same jitter. Thus, $J_{D,Tx} = J_{C,Tx}$. Assume a time of flight between Tx and Rx for the data of t_d and $t_d - t_{skew}$ for the clock. Clock and data jitter at the receiver become:

$$J_{D,Rx} = A_j \sin \left(2\pi f_j (t - t_d) \right), \tag{12.2}$$

$$J_{C,Rx} = A_j \sin \left(2\pi f_j (t - t_d + t_{skew}) \right) = A_j \sin \left(2\pi f_j (t - t_d) + 2\pi f_j t_{skew} \right). \tag{12.3}$$

At the receiver, we are concerned with the relative timing of the clock and data signals. Thus, we define the differential jitter as:

$$J_{Diff} = J_{D,Rx} - J_{C,Rx} = A_j \sin \left(2\pi f_j (t - t_d) \right)$$
$$- A_j \sin \left(2\pi f_j (t - t_d) + 2\pi f_j t_{skew} \right) \tag{12.4}$$
$$= 2A_j \sin \left(-\pi f_j t_{skew} \right) \cos \left(2\pi f_j (t - t_d) + \pi f_j t_{skew} \right). \tag{12.5}$$

The cos term gives the time-varying jitter, while the sin term, along with A_j gives the amplitude of differential jitter. The worst-case J_{Diff} is given by:

$$A_{Diff} = 2A_j \sin \left(\pi f_j t_{skew} \right). \tag{12.6}$$

As expected, if $t_{skew} = 0$, there is no differential jitter. Consider the example that follows for $t_{skew} \neq 0$.

Example 12.1   A given clock signal has jitter at 100 MHz with an amplitude of 10 UI for a 10 Gb/s link. What is the longest path-length difference (in s) between clock and data for which the differential jitter (J_{Diff}) is less than 0.5 UI?

Solution
Rearranging (12.6) we get:

$$t_{skew} = \frac{1}{\pi f_j} \sin^{-1} \frac{A_{Diff}}{2A_j}, \tag{12.7}$$

$$= \frac{1}{\pi 10^8} \sin^{-1} \frac{0.5}{20} = 80 \text{ ps}. \tag{12.8}$$

This means that a path-length *difference* of 80 ps converts 10 UI of peak per-lane jitter to a differential jitter of 0.5 UI.

The example above considered jitter at lower frequency. Higher jitter frequencies require correspondingly shorter path-length differences. Note as well that in the worst case, when $t_{skew} = \frac{1}{2f_j}$, $A_{Diff} = 2A_j$. This means that per-lane jitter is amplified by a factor of two.

Returning to the question of when clock-forwarding no longer makes sense due to path-length differences, consider that data in transmission lines travels at approximately 15–20 cm/ns. For a 1 m backplane cable to present 80 ps of mismatch requires 1.6% mismatch, a level that is possible. Optical links, which see longer time of flight, are more sensitive to path-length mismatch. Thus for longer channels, we want the receive clock to be extracted from incoming data so that the data's jitter can be tracked.

12.2   Non-Clock-Forwarded Links

Without a forwarded clock, per-lane clock and data recovery will be needed. In this section we consider two scenarios, namely, the mesochronous system, where all clocks are derived from a common reference clock, and the plesiochronous system, where transmit and receive clocks are derived from physically distinct devices, leading to (ideally) small frequency offsets.

12.2.1   Common Reference Clock: Mesochronous

Figure 12.5 shows a link without a forwarded clock. Both the transmitter and receiver generate clock signals using separate CMUs. However, CMUs are fed by the same reference clock. Characteristics of this type of link are the following:

• TxClk and RxClk will have the same frequency since they are derived from the same physical signal.
• RxClkRec is a phase-adjusted version of RxClk, generated using a clock-recovery (CR) circuit. It is aligned to sample the data in the middle of the eye.
• This system, with differing phases but matched frequency between Tx and Rx is referred to as mesochronous.
• The details of the CR circuit determine if RxClkRec can track the jitter on the incoming data. This is discussed in more detail in Chapter 15.

One important aspect of mesochronous links is the impact of reference clock jitter. The CMUs used in the transmitter and receiver, usually implemented as phase-locked loops (PLLs, Chapter 14), will track reference clock jitter up to each PLL's bandwidth. If the PLLs have the same bandwidth, reference clock jitter appears equally in the Tx and Rx clocks, behaving as correlated jitter, similar to the clock-forwarded case. However, the invariable bandwidth difference between Tx and Rx PLLs leads to frequencies at which reference clock jitter appears in only one of the clock signals. Consider Example 12.2:

Example 12.2   A wireline link generates Tx and Rx clocks from PLLs sharing a 200 MHz reference clock. Although designed to be identical, the bandwidth of the

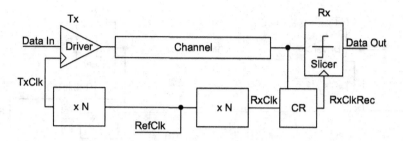

Figure 12.5 Simplified block diagram of a mesochronous link.

PLL can range between 5 and 15 MHz. What is the worst-case scenario for PLL bandwidth from a jitter tracking point of view?

Solution
If the Tx PLL bandwidth is 5 MHz, reference clock jitter above 5 MHz is rejected. For this example, we assume a brick-wall PLL response. If the Rx PLL bandwidth is 15 MHz, jitter above 15 MHz is rejected. Both PLL outputs have correlated jitter up to 5 MHz. Reference clock jitter between 5 and 15 MHz only appears on the Rx clock. If the Rx CR cannot track data jitter within this frequency range, high reference clock jitter at these frequencies will lead to bit errors.

This effect must be taken into account, ultimately dictating that PLL bandwidths are relatively well controlled. A clean reference is also required so that the jitter that propagates to only one clock signal is low enough.

Mesochronous links are practical when the Tx and Rx are close to one another such as on a single board. Beyond this scale, distributing the RefClk is no longer practical and plesiochronous links are designed.

12.2.2   Different Reference Clocks: Plesiochronous

As the distance between Tx and Rx increases, distributing the reference clock is no longer practical and clocks are generated off different reference signals, generated from different crystal oscillators. Although the oscillation frequency of crystal oscillators can be well matched, they are not identical, with differences in frequency in the range of 10s of parts per million (ppm). An example of such a system is shown in Figure 12.6. Here, the TxClk and RxClk frequencies, both being multiples of their respective reference clock signals, will also differ by the same ppm. Unlike the mesochronous system in which synchronizing the Rx clock to the incoming data requires only a phase offset that tracks jitter, in a plesiochronous system, the phase used to sample the data must be continuously updated so as to match the incoming data stream's phase and frequency.

To illustrate this point, consider the following example in which 10 Gb/s data is transmitted from a Tx clock at 10.00000000 GHz and is to be sampled by an Rx clock

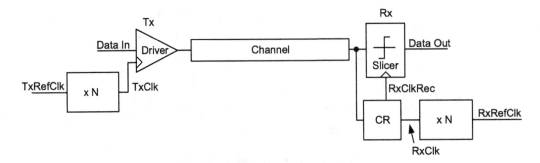

Figure 12.6 Simplified block diagram of a plesiochronous link.

at 9.99900000 GHz clock. The Tx clock is 100 ppm higher in frequency. The data UI is exactly 100 ps, whereas the Rx clock period is 100.01 ps. If a 100.01 ps UI sampling clock were used, these extra 10 fs would accumulate each clock period. If the Rx clock is aligned to the middle of the eye at $t = 0$ s, after 5,000 cycles, it has shifted by 50 ps, and is now sampling the edges. At 10 Gb/s, 5,000 cycles occur in 0.5 μs. The drift of the sampling point from the middle of the eye can be detected using a phase detector, which can adjust the phase of the sampling clock. This happens continuously so that the phase of the Rx clock steadily accumulates a phase offset relative to its reference.

The block diagram in Figure 12.6 includes a reference clock at the receive side. The CR block generates sampling clocks (RxClkRec) that have accumulating phase offsets from RxClk using phase rotators/phase interpolators. Other clock recovery systems directly infer the required clock frequency from the data stream, so-call "referenceless" clock recovery. These approaches are presented in detail in Chapter 15.

12.3   Clock Distribution

Chapter 13 presents the design of oscillator circuits used to generate clock signals while Chapter 14 explains how oscillators are controlled to produce signals with specific frequency and phase relationships relative to reference clocks. Once a clock is generated, it needs to be distributed across a chip to multiple receivers or transmitters. We often argue that we can afford to dissipate power when generating low-jitter clocks since this power is amortized across multiple data lanes. We must not ignore the task of distributing this signal and the possible degradation that can occur.

A critical building block in clock distribution is the clock buffer. This can be either a CML differential pair or a static CMOS inverter. The differential pair is designed to preserve a given clock signal swing while having a dc gain of at about 2 V/V. In many applications the clock is a fraction of the data rate. For example, consider the case of 1/4-rate sampling. The clock frequency is 1/4 of the data rate, but four clock phases are needed. By using differential signals, we immediately have two clock phases by switching the polarity of the signal. To have four phases we need two differential signals, separated by 90°. One clock is referred to as the in-phase clock while the signal that is delayed by 1/4 of the period is referred to as the quadrature clock. These terms are often abbreviated to I and Q. We should view clock distribution as not merely the distribution of one signal but as the distribution of four signals. The network must preserve the duty cycle of each signal and the quadrature relationship between them.

The general requirements of clock distribution circuits when considering a single clock phase are:

1. Provide matched rising and falling delay.
2. Low delay sensitivity to supply voltage.
3. Add minimal random jitter.
4. Preserve 50% duty cycle.

Matching rising and fall delay helps to preserve 50% duty cycle. Supply voltage fluctuations can modulate the delay of circuits, introducing deterministic jitter. This source of jitter is said to be deterministic because it can be predicted if we know the supply voltage waveforms. This is opposite to thermal noise in circuits which gives rise to random jitter. A 50% duty cycle is important if a given clock is used on its rising and falling edge, as noted above. However, even if sampling occurs on only one phase, since clock periods are only a few gate delays long, we must ensure that these short pulses are not eroded further, possibly being swallowed completely.

When multiphase clocks are considered, the following is also required:

1. If two separate signals are used for CLK and $\overline{CLK}$, ensure they are $180°$ out of phase.
2. Ensure a $90°$ relationship among quadrature clocks.

12.3.1   Matched Rise and Fall Delay

When mismatch between input transistors is ignored, CML circuits, due to their symmetry, show matched rising and falling delays. On the other hand, CMOS circuits need to be designed carefully. Assuming equal threshold voltage between NMOS and PMOS transistors, the shortest average delay is achieved when $\frac{W_P}{W_N} = \sqrt{\frac{\mu_N}{\mu_P}}$, although to have matched delays, which will be longer due to larger loading, we select:

$$\frac{W_P}{W_N} = \frac{\mu_N}{\mu_P}. \tag{12.9}$$

The inverters in a typical standard cell library are usually designed for shortest average delay, unless they have been specifically designed as clock buffers, in which case they will adopt the sizing criterion in (12.9). This sizing criterion also places the switching threshold of the inverter at $V_{DD}/2$, assuming the PMOS/NMOS threshold voltages have the same magnitude (i.e., $V_{tn} = |V_{tp}|$). As discussed below in Section 12.3.4, deviations in the switching threshold from $V_{DD}/2$ alter the clock's duty cycle.

12.3.2   Low Delay Sensitivity to Supply Voltage

Thanks to the tail current source, CML buffers have delay that is relatively insensitive to V_{DD} fluctuations, provided the current source stays in saturation. On the other hand, the delay of CMOS inverters shows a reciprocal relationship between delay and V_{DD} assuming a square-law transistor model. Velocity saturation reduces this sensitivity but does not eliminate it. Current starving can also be used to reduce this sensitivity.

12.3.3   Single-Ended to Differential Conversion: Maintaining 180° Phase Shift

Cross-coupled inverters, denoted by W_C in Figure 12.7, can be used to force *CLK* and *CLKB* to be 180° out of phase. The circuit works as follows:

- The single-ended clock signal *CLKIN* generates an inverted version *CLKB*1 through an inverter.
- *CLKIN* also drives a transmission gate with a delay similar to that of the first inverter. The transmission gate is sized with M_P and M_N equal to the PMOS and NMOS in the W_1 inverters.
- *CLK*1 and *CLKB*1 will be nearly 180° out of phase.
- The next inverters loaded by cross-coupled inverters bring the phase difference closer to 180°.
- Additional stages of W_1 and W_C inverters can be used.

The W labels capture the relative size of the cross-coupled inverters compared to those in the main signal path. If the cross-coupled inverters are too large they form a latch and the main-path inverters cannot flip their state. All inverters are sized for equal rise and fall delay, following (12.9). If the delay to *CLK*1 and *CLKB*1 is unequal, the output signals can be restored to being 180° out of phase, but the outputs may not necessarily have 50% duty cycle. Wherever true/complement clock signals are distributed using CMOS circuits, cross-coupled inverters can be used.

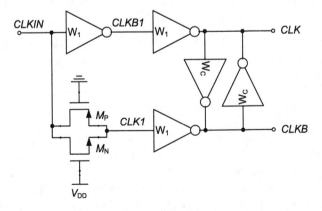

Figure 12.7 Inverter-based circuit for generating *CLK* and *CLKB*.

12.3.4   Duty-Cycle Detection and Correction

The duty cycle of a signal is defined as the duration of the high interval as a percentage of the whole period, usually measured between $V_{DD}/2$ crossings in the case of a static CMOS circuit or between zero crossings for the differential output of a CML circuit. Deviations from an ideal duty cycle of 50% cause two problems:

1. A chain of circuits that erode duty cycle can ultimately collapse clock pulses altogether.
2. In a subrate system, one latch is activated on the rising edge of the clock while another on the falling edge. Deviations from 50% mean that if one UI is sampled at the middle of the eye, the UI sampled from the opposite edge is not sampled in the middle of the eye.

To correct the duty cycle of a signal we can adjust the switching threshold of buffers in the signal path. Since input signals have non-zero rise and fall times, increasing the buffer's threshold delays its response to a rising edge and vice versa.

An example of such a threshold-adjustable inverter is shown in Figure 12.8 [86]. The ac-coupling capacitor C separates the dc operating point of the input signal from the inverter's input. Without M_P, M_N and R_{DAC}, the feedback resistor R_F biases the inverter at its switching threshold. If $CLKIN$ has a duty cycle of χ, CLK has a duty cycle of $1 - \chi$ since it is an inversion of $CLKIN$. By adding M_P and M_N, bias voltage V_{DAC} can be programmed to be above or below the nominal voltage at $INVI$. This can raise (or lower) the dc bias at node $INVI$, which in turn means that the effective switching threshold of the inverter is lowered (or raised). Note that the change to the threshold is opposite in direction to the change to $INVI$.

Consider the waveforms in Figure 12.9. A 50% duty-cycle input $CLKIN$ is applied to an inverter with three different switching thresholds denoted by the three horizontal lines. The output signals corresponding to the three scenarios for inverter threshold are shown with no delay relative to the input. In practice, the three outputs would be shifted to the right. Consider the middle threshold drawn with the solid line. When the input signal crosses the threshold, the output is also equal to the threshold. When the

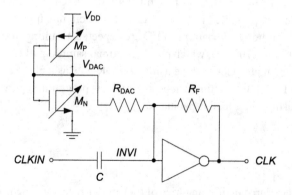

Figure 12.8 Circuit for controlling the duty cycle of a clock [86].

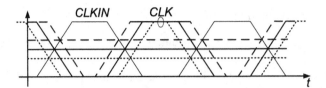

Figure 12.9 Waveforms for the circuit in Figure 12.8 with varying values for V_{DAC}. Horizontal lines show the resulting threshold of the inverter. Solid line: $V_{DAC} = 0.5V_{DD}$, threshold is at $0.5V_{DD}$; dotted line: $V_{DAC} > 0.5V_{DD}$, threshold is below $0.5V_{DD}$; dashed line: $V_{DAC} < 0.5V_{DD}$, threshold is above $0.5V_{DD}$.

threshold is lowered to the dotted line, the threshold-crossing for a rising edge moves to the left while the crossing for a falling edge moves to the right. If the gate were a non-inverting buffer, its output duty cycle would be increased. Since it is an inverter, its output duty cycle is decreased. The higher threshold, drawn with the dashed line, has the opposite effect on the output duty cycle.

Transistors M_P and M_N represent digitally programmable elements. By activating more transistor width for M_P and deactivating width of M_N, V_{DAC} can be raised. If $R_{DAC} = R_F$, the voltage at $INVI$ will increase by:

$$\Delta INVI \approx \frac{1}{A_0} \Delta V_{DAC}, \qquad (12.10)$$

where A_0 is the intrinsic gain of the inverter.

A low-pass filter can be used to measure the average value of a signal, which is a reasonable measure of its duty cycle. In the case of a static CMOS circuit, we expect a signal with a duty cycle of $\chi\%$ to have an average value of $\chi V_{DD}/100$. Using negative feedback, we can compare the actual average value to $0.5V_{DD}$, detecting and correcting non-50% duty cycle.

12.3.5   I/Q Clock Generation, Detection and Correction

Clock signals $I, Q, \bar{I}$ and $\bar{Q}$ should nominally have a phase relationship of $0°, 90°, 180°$ and $270°$. This allows correct sampling in a 1/4 rate system. Such clock signals can be generated by differential ring oscillators of even length, discussed in Section 13.2, or quadrature LC oscillators, presented in Section 13.3. These circuits nominally generate quadrature signals, although deviations in I/Q are expected, resulting in something like $0°, 90° + \phi, 180°$ and $270° + \phi$, which prevent optimal sampling of every second UI. Detection of I/Q mismatch can be done using an XOR-gate as a phase detector, as presented in Section 14.1.3. Correction can be as simple as programming the delay of clock buffers by adjusting their load capacitance.

12.4   Summary

This chapter introduced three main categories of transmitter/receiver synchronization, namely: synchronous, mesochronous and plesiochronous. Plesiochronous links have small frequency offsets in their reference clocks, requiring continuous adjustment of the clock phase used to capture data in the receiver. The jitter tolerance of a source-synchronous (clock-forwarded) link, which is a special case of a mesochronous link, was presented in detail. It was observed that even when identical jitter is added to transmitted clock and data signals, a timing skew along their channels can convert correlated jitter to differential jitter. This can increase the link's BER. This motivated the conclusion that a source-synchronous link would be suitable over a shorter distance and longer links would operate without a forwarded clock signal.

Challenges of distributing clock signals around an IC were discussed, noting that clock buffers need matched rising/falling delay, complementary outputs and the ability to correct duty-cycle and quadrature errors. Circuit examples were given for single-ended to differential clock conversion and duty-cycle correction.

Problems

12.1 A source-synchronous approach is being considered for an optical link operating at 10 Gb/s. If the time-of-flight in the fibre can vary by at most 1%, find the longest link possible while keeping clock/data skew to less than $\frac{T_b}{6}$.

12.2 Find the link standard with the largest separation between Tx and Rx that uses a common reference clock.

12.3 Design a clock correction circuit that corrects duty-cycle and quadrature error of clock signals produced by static CMOS circuits.

12.4 Design a series of differential clock buffers that can distribute a clock signal along a 2-mm-long path. Assume per-unit-length capacitance of 0.1 fF/μm and resistance of 0.2 Ω/μm. Assume a terminal load of 100 fF and a maximum input capacitance of the first buffer of 30 fF.

13 Oscillators

Throughout this book we have seen widespread use of clock signals used to activate decision circuits in receivers, DFFs in transmitter FFEs, as well as multiplexers and demultiplexers. In this chapter we introduce circuits that are used to generate clocks. In Chapter 14 we present techniques for phase-locking clock signals to other reference clocks and in Chapter 15 we discuss locking clock signals to data streams. Before we consider the control loops wrapped around clock-generation circuits, we need to understand the following concepts:

- What is an oscillator?
- What determines if a circuit oscillates?
- What are the metrics of a good oscillator?
- What are the advantages and disadvantages of ring oscillators relative to oscillators that use LC tanks?
- How do we implement each type of oscillator?
- How do we make oscillators voltage controlled?

After an introduction to the metrics of oscillators, this chapter introduces the main categories of oscillators, and how they are made to be voltage controlled. The important trade-offs between power, phase-noise and tuning range are explained. Methods for converting a voltage-controlled oscillator (VCO) to a digitally controlled oscillator are included in Section 14.4.

13.1 Oscillator Metrics

An oscillator is a circuit that generates a periodic output without a periodic input. Under ideal operating conditions, this amounts to:

$$v_{OUT}(t) = v_{OUT}(t - T),\qquad(13.1)$$

where T is the period of oscillation. In some cases, particularly the case of oscillators built using LC tanks, $v_{OUT}(t)$ has a sinusoidal component, such that:

$$v_{OUT}(t) = A_0 \sin(2\pi f_0 t + \theta) + V_{OUT},\qquad(13.2)$$

where A_0 is the amplitude of oscillation, θ is a phase offset, V_{OUT} is the output's dc offset and f_0 is the frequency of oscillation. Therefore $T = 1/f_0$. In the case of an

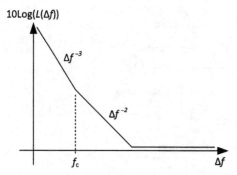

Figure 13.1 Typical oscillator phase noise power spectral density as a function of frequency offset from the carrier. Both axes are logarithmic.

oscillator whose output is a differential signal, $V_{OUT} = 0$. If $V_{OUT} = 0$, $v_{OUT}(t)$ crosses 0 at equally spaced intervals of time, $T/2$. We use the term "zero crossings" loosely to refer to the times at which the oscillatory portion of the signal crosses zero, noting that when $V_{OUT} > A_0$, v_{OUT} does not actually cross zero.

Ring oscillators, on the other hand, produce non-sinusoidal outputs, although we can still refer to a period, amplitude and dc offset.

A real oscillator will not have zero crossings that are uniformly spaced. These deviations in its zero crossings from the ideal, uniformly spaced crossings, known as jitter (presented in Section 1.11 and Figure 1.26), can be expressed mathematically as:

$$v_{OUT}(t) = A_0 \sin\left(2\pi f_0 t + \theta + \phi_n(t)\right) + V_{OUT}, \tag{13.3}$$

where $\phi_n(t)$ is a random process known as phase noise, arising from the thermal- and flicker-noise sources of the circuit elements. Note that we can also have jitter induced by power supply noise and substrate coupling. Phase noise is characterized by its power spectral density, an example of which is shown in Figure 13.1. The horizontal axis is the frequency offset (Δf) from the carrier, f_0. At smaller offsets, the PSD follows a $1/f^3$ relationship up to the flicker-noise corner (f_c), beyond which it follows a $1/f^2$ dependence until a noise floor is reached. In general, lower phase noise can be achieved by dissipating more power in the oscillator. A figure of merit (FOM) that can be used to compare oscillators is given by:

$$FOM_1 = 10\log\left(\frac{f_0^2}{\Delta f^2 L(\Delta f) P_{diss}}\right), \tag{13.4}$$

where P_{diss} is the power dissipation of the oscillator, usually in mW, and $L(\Delta f)$ is the PSD of phase noise at a frequency offset Δf. Note that for a given oscillator in the $1/f^2$ part of the phase noise PSD, $\Delta f^2 L(\Delta f)$ is a constant. Hence, choosing a different value for Δf will not change FOM_1.

FOM_1 does not take into account tuning range. Most useful oscillators are voltage controlled. An example tuning curve is shown in Figure 13.2 where over some range of V_{Tune}, f_0 increases. Even if the oscillator is used to generate only one frequency set by

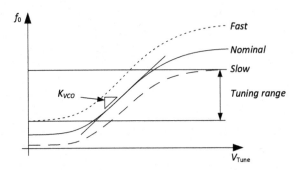

Figure 13.2 VCO tuning curve.

a standard, sufficient tuning range is needed to rein in the oscillation frequency when it is pushed from its nominal value due to process, voltage and temperature variations.

Example 13.1   Consider a situation in which an oscillator is needed to generate a 5 GHz clock signal. As such, it was designed so that for nominal process, voltage and temperature, it oscillates at 5 GHz. Suppose at low temperature, high supply voltage and fast process parameters, the nominal oscillation frequency increases to 7 GHz while at high temperature, low supply voltage and slow process parameters, the frequency of oscillation drops to 3.5 GHz. How much tuning range is needed? Is this design centred?

Solution
The fast oscillator needs to be slowed down by 2 GHz, which is 28% of 7 GHz. The slow oscillator needs to be sped up by 1.5 GHz, which is 43% of 3.5 GHz. The overall tuning range that is required is 71%. Since the positive tuning range is larger than the negative tuning range, the design is not centred. It is not necessarily the case that the best approach is to centre the nominal design to operate at the target frequency under nominal process conditions. Rather, the nominal design point should be set such that the tuning range is balanced when PVT variation is considered.

The tuning curve in Figure 13.2 also illustrates the effect of PVT variation on the available tuning range. The overall tuning range is the difference between the lower end of the tuning range for the oscillator operating under fast PVT conditions and the upper end of the range for slow PVT conditions. This range can be considerably smaller than the range for nominal PVT conditions.

When we model the behaviour of a phase-locked loop or clock and data recovery unit, we use the slope of the oscillator's tuning curve to model its voltage-to-frequency conversion. Referred to as the VCO gain, we define the voltage-to-frequency conversion as:

$$K_{VCO} = \frac{d\omega_{osc}}{dV_{tune}} = 2\pi \frac{df_{osc}}{dV_{tune}}. \qquad (13.5)$$

In addition to tuning range and low phase noise, other specifications of an oscillator include:

- Guaranteed start-up: across all anticipated PVT variation, the oscillator must start oscillating. This requirement can be mapped to a minimum gain of the gain stages used.
- Minimum output voltage swing.
- Low deterministic jitter due to power supply noise.
- Low die area.
- Constant VCO gain: as we will see in Chapter 14, the dynamics of a PLL depend on the VCO gain. Hence we may want the VCO gain to be constant over the tuning range or in the presence of PVT variation.

13.2   Ring Oscillators

Ring oscillators are a class of oscillators built with chains of inverting gain stages, connected in a feedback loop. This section presents their fundamentals, criteria for oscillation and both CML and CMOS implementations. Making both classes voltage controlled is also discussed.

13.2.1   Fundamentals of Ring Oscillators

Consider the three-stage ring shown in Figure 13.3 (a). Each inverter has the static input-output voltage transfer characteristic shown in Figure 13.3 (b). If a single inverter is put in feedback, $v_O = v_I = v_{th}$. Note that v_{th} denotes the switching threshold of the inverter and not the threshold voltage of any transistor. For the three-stage ring oscillator, $v_3 = v_2 = v_1 = v_{th}$ is a valid dc operating point of the circuit as computed by a SPICE-style simulator. These voltages satisfy the dc analysis of the circuit. However, this operating point is not stable. Around this operating point, each inverter has an inverting gain. Consider what happens when noise increases node v_1.

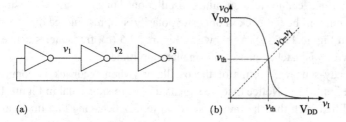

(a)
(b)

Figure 13.3 Three-stage ring oscillator: (a) schematic; (b) per-stage voltage transfer characteristic.

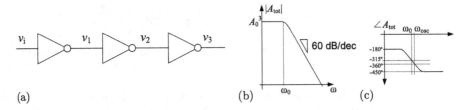

Figure 13.4 Open-loop transfer function of a three-stage ring oscillator: (a) schematic; (b) magnitude response; (c) phase response.

With some delay, v_2 decreases. Then v_3 increases, which will reduce v_1. This may appear to compensate for the initial increase in v_1 (i.e., be negative feedback). However, since this compensation occurs three inverter delays after the initial disturbance, this can *reinforce* a signal for which three inverter delays is half of its period. To better illustrate this, and to put this discussion on a mathematical footing, consider the transfer function of the circuit around the $v_I = v_O = v_{th}$ operating point. The transfer function of one inverter is approximated by:

$$A(s) = \frac{V_o(s)}{V_i(s)} = -\frac{A_0}{\frac{s}{\omega_0} + 1},$$ (13.6)

where A_0 is the dc gain of the inverter and ω_0 is the 3 dB bandwidth of the inverter when used as an amplifier. If we break the feedback loop, splitting v_3 into v_3 and v_i, as shown in Figure 13.4 (a), the overall transfer function ($A_{tot}(s)$) is the cube of $A(s)$:

$$A_{tot}(s) = \frac{v_3}{v_i} = -\frac{A_0^3}{\left(\frac{s}{\omega_0} + 1\right)^3}.$$ (13.7)

The magnitude and phase of (13.7) are plotted in Figure 13.4 (b) and Figure 13.4 (c), respectively. The dc gain is A_0^3. The phase starts at $-180°$ and extends to $-450°$, crossing $-360°$ at a frequency denoted as ω_{osc}. This frequency, calculated below, is slightly higher than ω_0. Although this is beyond the bandwidth of the inverters, provided A_0 is high enough, the gain at ω_{osc} may still be larger than one. Put another way, a signal at this frequency is amplified and is in phase with the input as shown in Figure 13.5 (a). When we connect v_3 to v_i this signal is further amplified around the ring, leading to a sustained oscillation. For $\omega \ll \omega_{osc}$ a disturbance is amplified, but will be fed back in opposite polarity, suppressing the disturbance. This is shown in Figure 13.5 (b). As shown in Figure 13.5 (c), frequencies above ω_{osc} are returned with mismatched phase and smaller amplitude.

The steady-state waveforms of the oscillator (when feedback is connected) are shown in Figure 13.6. Notice that the signals that were sinusoidal in Figure 13.5 have been amplified such that the inverters are no longer behaving like linear amplifiers. The LTI system descriptions are convenient for discussing the onset of oscillation but large-signal oscillations are better described using a "logic gate" perspective. In this

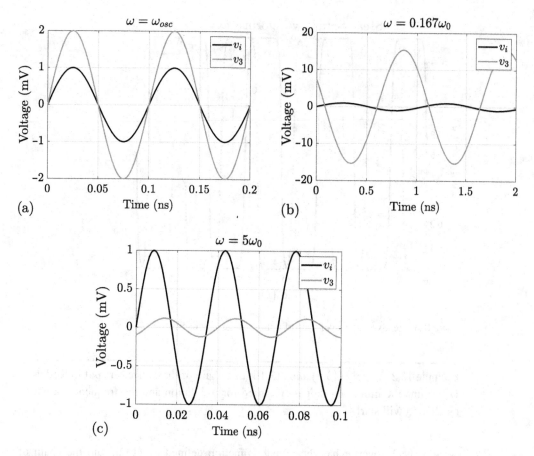

Figure 13.5 Open-loop transient response of a three-stage ring oscillator: (a) $\omega = \omega_{osc}$; (b) $\omega = \omega_0/6$; (c) $\omega = 5\omega_0$.

context, each gate has a delay of τ_D. The overall period of the three-stage ring will be $6\tau_D$. For an N-stage ring, the large signal frequency of oscillation is given by:

$$f_{osc,LS} = \frac{1}{2N\tau_D}. \tag{13.8}$$

When considering whether a candidate topology of ring oscillator has the potential to oscillate, we can apply the Barkhausen criterion. It states that for a system described by the transfer function $H(s)$, put in unity-gain positive feedback the system may oscillate if the following two conditions are met:

$$\angle H(j\omega_c) = n360°, \tag{13.9}$$

$$|H(j\omega_c)| > 1. \tag{13.10}$$

Note that if we consider a system put in negative feedback, (13.9) becomes:

$$\angle H(j\omega_c) = 180° + n360°. \tag{13.11}$$

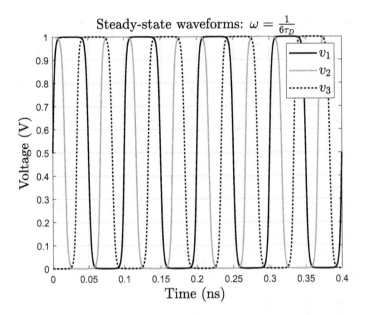

Figure 13.6 Steady-state waveforms of a three-stage ring oscillator.

Example 13.2   Apply (13.9) and (13.10) to a chain of three inverters put in feedback. Determine the minimum dc gain required for oscillation and the frequency at which oscillation will start.

Solution
Assume each inverter has the transfer function defined by (13.6) and the chain of three has the transfer function given in (13.7). The total phase starts at $-180°$ at dc and increases to $-450°$ as $\omega \to \infty$. To determine the potential oscillation frequency we find ω_{osc} such that $\angle A_{tot}(j\omega_{osc}) = -360°$.

$$\angle \left(-\frac{A_0^3}{\left(\frac{s}{\omega_0}+1\right)^3} \right) = -360°, \tag{13.12}$$

$$\angle \left(\frac{A_0^3}{\left(\frac{s}{\omega_0}+1\right)^3} \right) = -180°, \tag{13.13}$$

$$\tan^{-1}\left(\frac{\omega_{osc}}{\omega_0} \right) = 60°, \tag{13.14}$$

$$\omega_{osc} = \sqrt{3}\omega_0. \tag{13.15}$$

The gain at dc is A_0^3 and decreases as ω increases. To oscillate, we need the gain of each inverter at ω_{osc} to be greater than 1. Using (13.15) we can solve (13.10) for the minimum A_0 that gives sufficient gain:

$$\left|\frac{A_0}{\frac{j\omega_{osc}}{\omega_0} + 1}\right| > 1, \tag{13.16}$$

$$\sqrt{\frac{A_0^2}{\sqrt{3}^2 + 1^2}} > 1, \tag{13.17}$$

$$A_0 > 2. \tag{13.18}$$

Therefore, if the inverters behave as first-order amplifiers, with a minimum dc gain of 2, the three-stage ring will start oscillating at $\omega_{osc} = \sqrt{3}\omega_0$.

Example 13.3 Apply (13.9) and (13.10) to a chain of four inverters put in feedback.

Solution
Assume each inverter has the transfer function defined by (13.6). With four, the total phase starts at $0°$ at dc and increases to $-360°$ as $\omega \to \infty$. The gain at dc is A_0^4 which is greater than 1. Therefore, the chain of inverters "oscillates" at 0 Hz. We refer to this condition as latching up. Output voltages are either V_{DD} or gnd. As $\omega \to \infty$, $|A(j\omega)| \to 0$. Therefore there is no other candidate frequency for oscillation.

Example 13.4 Can a six-stage inverter-based ring oscillator oscillate?

Solution
Following a similar approach as Example 13.3, assume each inverter has the transfer function defined by (13.6). With six, the total phase starts at $0°$ at dc and increases to $-540°$ as $\omega \to \infty$. The gain at dc is A_0^6 which is greater than 1. Therefore, the chain of inverters can "oscillate" at 0 Hz. However, $\angle A_{tot}(j\omega_c) = -360°$ when $\angle A(j\omega_c) = -180° - 60°$. This occurs when $\omega_c = \sqrt{3}\omega_0$. So long as $A_0 > 2$ the Barkhausen criterion is met for this frequency. Does the circuit oscillate at $\sqrt{3}\omega_0$ or latch up? Since the gain at dc is higher, the circuit will latch up and not oscillate. However, upon start-up it may exhibit quasi-oscillatory behaviour until it latches up.

To generate quadrature clock signals, we need to have some number of inverters in the ring provide a delay of $T_{osc}/4$. Since the delay through all the inverters in the chain is $T_{osc}/2$, a delay of $T_{osc}/4$ occurs half-way down the chain. Thus, quadrature clock generation using a ring oscillator requires an even number of stages. However, we need the dc loop gain to be $-180°$ instead of zero to avoid latch-up. Figure 13.7 shows conceptually how this is done. The loop has five inversions, leading to negative feedback at dc. To satisfy (13.9) we need the additional phase at ω_{osc} from the four inverters to be $-180°$, meaning each inverter contributes $-45°$. The "zero-delay inversion" is accomplished using differential circuits and cross-coupling the output of the last inverter when feeding it back to the input of the first.

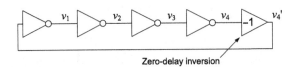

Figure 13.7 Four-stage ring oscillator shown with zero-delay inversion.

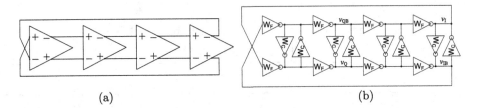

(a)   (b)

Figure 13.8 Four-stage ring oscillators implemented by cross-coupling the signal path using: (a) differential (CML) inverters; (b) pseudo-differential static CMOS inverters.

Two implementations of a four-stage ring are shown in Figure 13.8. Part (a) shows a four-stage ring implemented by differential pairs (CML inverters). The output from the final stage is cross-coupled when fed back to the input of the first inverter. Part (b) shows a four-stage pseudo-differential ring implemented by coupling two four-stage rings. The cross-coupled inverters, labelled as W_C, at the output of each stage are required. Without them the eight inverters, labelled as W_F, form one large eight-stage ring that will latch up. In this situation, the adjacent outputs in Figure 13.8 (b) would be equal. Cross-coupled inverters prevent this, allowing the circuit to oscillate as a differential four-stage ring.

The cross-coupled inverters form latches that, while ensuring anti-phase outputs of the forward path, need a minimum current to toggle. They must be sized to be strong enough to prevent latch-up, but not too strong, otherwise the forward path of inverters will not be able to toggle their state.

Like any ring, the steady-stage oscillation frequency is set by large signal delays (13.22). Nevertheless, outputs taken two stages apart will be separated by $T_{osc}/4$. Care must be taken to ensure that the differential inverters in Figure 13.8 (a) have common-mode rejection. The common-mode circuit is a four-stage ring *without* an extra inversion. Therefore, the circuit can latch up if the common-mode gain is greater than 1.

13.2.1.1 Summary of Criteria for Oscillation

The following summarizes relationships for ring oscillators assuming each stage has a first-order response as given by (13.6). If we assume the necessary signal inversion when the number of stages is even, oscillations start when each stage adds an additional phase of:

$$\phi = \frac{180°}{N}.$$ (13.19)

This occurs at ω_{osc} given by:

$$\omega_{osc} = \tan\left(\frac{180°}{N}\right)\omega_0.$$ (13.20)

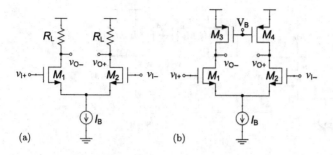

Figure 13.9 CML implementation of each inverter: (a) Resistors as loads; (b) triode-region MOSFETs.

The minimum required dc gain for oscillation is:

$$A_{0,min} = \sqrt{1 + \tan^2 \left(\frac{180°}{N} \right)} = \sqrt{1 + \left(\frac{\omega_{osc}}{\omega_0} \right)^2}. \tag{13.21}$$

Once oscillations build, the large signal frequency of oscillation ($f_{osc,LS}$) is set by gate delays:

$$f_{osc,LS} = \frac{1}{2N\tau_D}. \tag{13.22}$$

13.2.2 CML Implementation of Ring Oscillators

Figures 13.9 and 13.10 show common implementations of each differential inverter in a ring oscillator such as the one shown in Figure 13.8 (a). In the case of Figure 13.9 (a), $|A_0| = g_m R$, ignoring channel-length modulation (r_o). Figure 13.9 (b) uses triode-region MOSFETs. Here, the conductance of them is controlled by the bias voltage V_B. Assuming $M_{3/4}$ operate in deep triode, the small-signal gain is given by:

$$|A_0| = g_{m1} r_{ds3}, \tag{13.23}$$

$$|A_0| = \frac{\mu_n \frac{W_n}{L_n} V_{OV_n}}{\mu_p \frac{W_p}{L_p} V_{OV_p}}. \tag{13.24}$$

If the loads leave deep triode, the operation is discussed where the symmetrical load in Figure 13.10 (b) is considered. The bias current will determine V_{OV_n} while V_{OV_p} is set by an externally applied bias voltage. Hence (13.24) is bias-current dependent.

Figure 13.10 (a) uses diode-connected loads. Here, the gain is the ratio of the transconductance of the NMOS and PMOS, which when operated in strong inversion is given by:

$$|A_0| = \frac{g_{m1}}{g_{m3}} = \sqrt{\frac{\mu_n \frac{W_n}{L_n}}{\mu_p \frac{W_p}{L_p}}}. \tag{13.25}$$

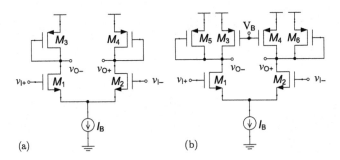

Figure 13.10 CML implementation of each inverter: (a) diode-connected MOSFETs; (b) symmetrical load.

Notice that the gain in (13.25), unlike that in (13.24), is independent of bias current, so long as the transistors are in strong inversion and depends only on the mobility ratio and dimensions. If NMOS diode-connected loads are used, the gain depends only on dimensions in strong inversion.

Thus far, we have examined the loads from the point of view of the circuit's small-signal gain so as to establish limits to ensure start-up. Once the circuit starts oscillating, we must consider the large-signal behaviour. When considering small-signal gain, care must be taken to ensure that across process, voltage and temperature variation the criterion in (13.21) is met. This can lead to a nominal design with 1.5x the minimum gain.

13.2.2.1   CML Load Symmetry and Jitter

The common-mode rejection of differential circuits relies on symmetry. For example, if both sides of a differential circuit see an increase in current of Δi, we have only a common-mode output if the Δi on each side sees the same impedance. As an example, resistively loaded differential pairs have this property since the loads are linear. Transistors, on the other hand, are nonlinear. The triode-region loads of Figure 13.9 (b) have an impedance that increases with current. For any non-zero differential output voltage, common-mode disturbances in current are converted to differential voltages. This makes the circuit susceptible to power-supply induced jitter.

A symmetrical load, though nonlinear, leads to symmetrical oscillation and symmetrical load impedance, leading to superior supply induced jitter [87][88]. Consider the dc operating point of each output, V_O. For an oscillator with a single-ended peak-to-peak swing of V_{pp} a symmetrical load will swing between:

$$V_O - V_{pp}/2 < v_O < V_O + V_{pp}/2. \tag{13.26}$$

For this to occur, the load must be linear or have a symmetrical nonlinearity around V_O. At the dc operating point, $I_B/2$ flows through each load. The differential pair is nearly fully switched when oscillating with a larger output, leading to a current through the load of

$$- \Delta I + I_B/2 < i_L < \Delta I + I_B/2. \tag{13.27}$$

Equation (13.26) holds for a resistive load since the voltage drop across it depends linearly on its current. If the triode-region MOSFET load remains in deep triode, it is nearly linear. However, ensuring deep triode requires a large overdrive voltage. The load shown in Figure 13.10 (b) uses triode-region MOSFETs ($M_{3/4}$) and diode-connected MOSFETs ($M_{5/6}$). Using the notation in Figure 13.11, the total current through the load, i_{Tot} is given by:

$$i_{Tot} = i_D + i_T,\tag{13.28}$$

$$= \frac{1}{2}\mu_p C_{OX}\frac{W_5}{L_5}(v_L - V_t)^2 + \ldots$$

$$\mu_p C_{OX}\frac{W_3}{L_3}((V_{DD} - V_B - V_t)v_L - 0.5v_L^2),\tag{13.29}$$

where i_D is the drain current of M_5 and i_T is the drain current of M_3. Notice that i_D has a positive quadratic term whereas i_T has a negative quadratic term. Expanding (13.29) further we get:

$$i_{Tot} = \mu_p C_{OX}\left(\frac{1}{2}v_L^2\left(\frac{W_5}{L_5} - \frac{W_3}{L_3}\right)\ldots\right.$$

$$+ v_L\left(V_{OV_3}\frac{W_3}{L_3} - V_t\frac{W_5}{L_5}\right)\ldots$$

$$\left.+ \frac{1}{2}\frac{W_5}{L_5}V_t^2\right).\tag{13.30}$$

If we choose equal dimensions for the four load transistors, denoted as $\frac{W}{L}$, the v_L^2 drops out and (13.30) simplifies to:

$$i_{Tot} = \mu_p C_{OX}\frac{W}{L}(v_L(V_{OV_3} - V_t) + V_t^2).\tag{13.31}$$

With no quadratic term, this combined load is linear with respect to v_L. Interestingly this only requires the matching of the geometry and is valid even as V_{OV_3} is adjusted. This is important because, as we will see later, it will be through the control of V_B that the load resistance and oscillation frequency can be controlled. The conductance of the load is found by differentiating (13.31) with respect to v_L:

$$g_L = \frac{di_{Tot}}{dv_L} = \mu_p C_{OX}\frac{W}{L}(V_{OV_3} - V_t),\tag{13.32}$$

and the small-signal gain of the circuit is:

$$|A_0| = \frac{g_{m1}}{g_L} = \frac{\mu_n\frac{W_1}{L_1}V_{OV_1}}{\mu_p\frac{W}{L}(V_{OV_3} - V_t)}.\tag{13.33}$$

Square-law operation of the MOSFETs was assumed in the analysis above. How well this technique works with scaled CMOS devices is left to the reader to explore. Since the behaviour of the load captured in (13.29) was derived assuming M_3 is in triode and M_5 was not in cut-off, linearity requires M_3 stays in triode, but not deep triode.

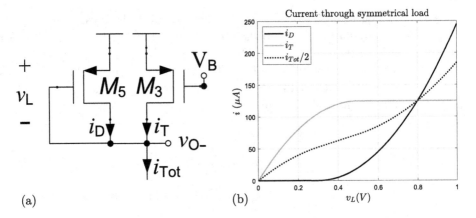

Figure 13.11 Operation of symmetrical load: (a) schematic; (b) I–V characteristic.

Figure 13.11 (b) shows i_D, i_T and the average current, $i_{Tot}/2$ assuming $V_{OV_3} = 0.5$ V and $V_t = 0.3$ V. Here, M_5 is on and M_3 is in triode only for $0.3 < v_L < 0.5$ V. Below, expressions for total current and load conductance are shown over a wider range of v_L:

$$\frac{i_{Tot}}{\mu_p C_{OX} \frac{W}{L}} = \begin{cases} V_{OV_3} v_L - 0.5 v_L^2, & 0 \le v_L < V_t \\ v_L (V_{OV_3} - V_t) + V_t^2, & V_t \le v_L < V_{OV_3}, \\ 0.5(v_L - V_t)^2 + 0.5 V_{OV_3}^2 & v_L \ge V_{OV_3} \end{cases} \qquad (13.34)$$

$$\frac{g_L}{\mu_p C_{OX} \frac{W}{L}} = \begin{cases} V_{OV_3} - v_L, & 0 \le v_L < V_t \\ V_{OV_3} - V_t, & V_t \le v_L < V_{OV_3}. \\ v_L - V_t & v_L \ge V_{OV_3} \end{cases} \qquad (13.35)$$

If we consider $\frac{V_{OV_3}+V_t}{2}$ as the mid point of the voltage range, the conductance increases symmetrically around this voltage. Thus the load is linear between V_t and V_{OV_3} but symmetric from 0 to $V_t + V_{OV_3}$. In [87][88] $V_{CTRL} = V_t + V_{OV_3}$ and the circuits are biased such that $V_{CTRL}/2$ appears across each load when zero differential input is applied to the inverters.

To illustrate the benefit of symmetry, consider the curve in Figure 13.11 (b) where for our numbers, the voltage swings around $v_L = 400$ mV. If at a particular instant in time, one load sees 200 mV and the other sees 600 mV across it, the loads are operating in the nonlinear regions. However, the slope (conductance) of i_{tot} is the same at 200 and 600 mV, meaning a supply-noise injected current pulse will disturb both sides by the same Δv_L. The disturbance will remain common mode. This would also occur if the loads were resistors, but not stand-alone triode-region or diode-connect loads.

13.2.2.2 Adding Voltage Control to CML Ring Oscillators

All oscillators must be tuneable in frequency, either through a tuning current or voltage. In the case of CML ring oscillators, we usually have tractable tuning relationships with respect to bias current. However, the systems in which we use ring oscillators often generate control voltages, thereby requiring a V-to-I conversion circuit, often referred to as a V-to-I converter, or transconductor.

Example 13.5   If the bias current of the triode-loaded CML circuit in Figure 13.10 is cut in half, what percentage of the old oscillation frequency will the new oscillation frequency be?

Solution
It can be helpful to consider oscillation frequency from two points of the view. The first is the ac small-signal estimation of ω_{OSC}. The second is from a large-signal delay model. In this example, we consider both. For the CML circuits presented, the transfer function has a dc gain and a pole ω_0. This pole is equal to $\frac{1}{RC}$ where R is the small-signal resistance of the load and C is the load capacitance. In a CML ring oscillator, the load capacitance comes from the capacitances of the MOSFETs connected to the output such as those in the following stage and any output buffers added to each stage. There is also a small wiring capacitance. The MOSFET capacitances have weak dependence on bias conditions, but for the purpose of VCO tuning we can assume they stay relatively constant. The small-signal resistance of the triode load will change slightly as the bias current changes as the transistor moves closer to or (in this example) farther from saturation. If we have designed the circuit to have a relatively linear load, we have implicitly designed it so that the operating point stays away from saturation. Hence we expect the small-signal resistance to also stay relatively constant. Therefore, changing the bias current does not on its own have a large impact on the RC time constant of each stage, keeping the oscillation frequency fairly constant.

But logic circuit delay depends on current? If we view the CML circuit as an inverter, the peak current available to charge the load capacitance is I_B. If we ignored the current flowing through the load resistance, we can write:

$$\tau_D = \frac{\Delta V C_L}{I_B}, \tag{13.36}$$

where ΔV is a portion of output swing needed to consider the gate having switched. From this it *appears* that cutting the bias current I_B in half will double the delay. However, if done on its own, the swing (ΔV) of the CML gate is also halved. Since the required ΔV is also halved, the delay predicted by (13.36) is unchanged. Therefore, from both points of view, bias current adjustment is not sufficient to control oscillation frequency in a CML ring oscillator.

As noted in the example above the output swing depends on bias current. This is another problem associated with only changing the bias current. Too small of an output swing and the circuits the VCO drives can perform poorly. Low swing makes the clock chain more susceptible to noise. We need a mechanism for controlling output swing as the bias current is changed.

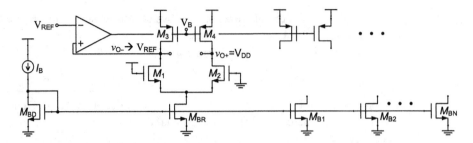

Figure 13.12 Replica bias for a CML VCO.

13.2.2.3   Replica Biasing of CML Ring Oscillators

Figure 13.12 shows a replica-bias scheme that fixes VCO swing while bias current is adjusted. The current I_B is set by a transconductor, not shown. This current is mirrored by M_{BD} and M_{BR} to bias the replica cell as well as to $M_{B1,2,3...N}$ which bias the CML stages in the ring oscillator. The replica cell has the same transistor dimensions as the delay cells in the ring. It is biased with one input at V_{DD} and the other grounded. This fully switches the differential pair, with all of I_B flowing through M_1. The amplifier adjusts V_B so that the drain of M_1 is equal to V_{REF}. We reach this conclusion by assuming that there is high-gain negative feedback around the loop consisting of the amplifier and the load. Under these assumptions the amplifier's two inputs become equal. Therefore, V_B settles to the bias voltage that results in the desired voltage drop when the differential pair is fully switched. We then distribute V_B to the loads of the cells in the ring. This sets the peak-to-peak differential swing to be approximately $2(V_{DD} - V_{REF})$. Though triode-region loads are shown, any load with tuneable resistance can be used such as the symmetrical triode/diode load.

13.2.2.4   Delay Interpolation in Ring Oscillators

Figure 13.13 shows another mechanism for varying the delay of circuit blocks which can be applied to CML-based and static-CMOS VCOs. A three-stage ring VCO is shown in part (a) where each stage provides a delay of τ_D. This delay needs to be tuneable. As shown in part (b) each stage is implemented using an interpolator structure where the output, v_O, is a weighted sum of the interpolator's two inputs, v_{IN} and v_D. The stage inverts the signal, thus we consider the idealized interpolation operation generating $\overline{v_O}$. One input to the interpolator is the block's overall input v_{IN} whereas the other input to the interpolator is a version of the input v_{IN}, delayed by an additional delay, τ_A, denoted as v_D. The interpolator takes a weighted average of these two inputs and produces an output, subject to an additional delay denoted by τ_{min}. This is shown in part (c). When $x = 1$ the output is equal to v_{IN} delayed by τ_{min}. As x is tuned from 1 to 0, the delayed signal v_D takes a greater weight in the output and the output is delayed further. The maximum delay is $\tau_{min} + \tau_A$. What is critical to the operation of this circuit is that the rise/fall time of v_D and v_{IN} is as long as the delay difference, otherwise, the output waveforms become distorted. This is considered in Problem 13.7.

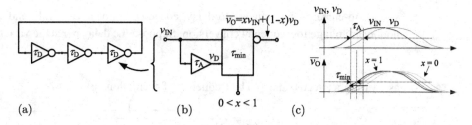

Figure 13.13 Conceptual view of a delay interpolator: (a) three-stage VCO; (b) block diagram of an inverting delay interpolator; and (c) waveforms for various values of x.

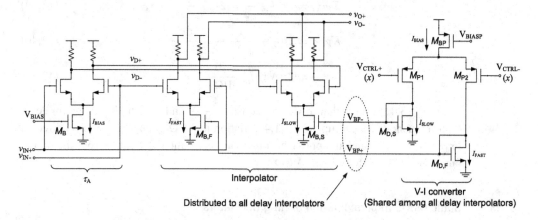

Figure 13.14 Implementation of a CML-based delay interpolator.

A CML-based implementation of a delay interpolator is shown in Figure 13.14. In this case, the delay cells are resistively loaded differential pairs, but other types of loads can be used. The outputs of two are tied together to produce the overall output v_O. The bias voltages for the tail current sources are generated by the V-to-I converter circuit on the right. Based on the differential input $V_{CTRL+/-}$ (representing x), the bias current I_B is steered between two diode-connected transistors, $M_{D,F}$ and $M_{D,S}$, each of which will mirror the current to the delay cells via the voltages v_{BP+} and v_{BP-}. The sum of the currents in the output stages tied together is I_B. Therefore, the output swing will not change as the circuit is tuned. We can see that if all of I_B is steered through $M_{D,F}$ the signal path in the interpolator reduces to the "fast path" consisting of the single stage connecting v_{IN} to v_O. If all of I_B is steered through $M_{D,S}$ the signal path in the interpolator reduces to the "slow path" consisting of two CML stages, the first of which implements τ_A. If each circuit has a delay of τ_A, the overall delay ranges from τ_A to $2\tau_A$. A ring oscillator can be created by replacing one or all of the inverters with delay interpolators. If all are replaced, the total delay can have a 1:2 range, leading to a 1:2 range in frequency.

The V-to-I converter in Figure 13.14 can be modified to use degeneration resistors to linearize its voltage to current conversion. Although we can make the voltage-to-current, and current-to-delay conversion linear, we still end up with nonlinear voltage to frequency conversion. Suppose the delay per stage varies as:

$$\tau_{Stage} = \tau_A(2 - x),$$ (13.37)

as x varies between 0 and 1. The frequency of oscillation is:

$$f_{osc} = \frac{1}{2N\tau_{Stage}} = \frac{1}{2N\tau_A(2 - x)}$$ (13.38)

Since $0 < x < 1$, f_{osc} has a 1:2 range. The VCO gain is:

$$K_{VCO} = 2\pi \frac{df_{osc}}{dV_{tune}} = 2\pi \frac{df_{osc}}{dx} \frac{dx}{dV_{tune}},$$ (13.39)

$$K_{VCO} = 2\pi \frac{1}{2N\tau_A(2 - x)^2} \frac{dx}{dV_{tune}},$$ (13.40)

$$K_{VCO} = 2\pi N\tau_A f_{osc}^2 \frac{dx}{dV_{tune}}.$$ (13.41)

From (13.41) we see that if we have constant $\frac{dx}{dV_{tune}}$ and a 1:2 tuning range for delay (and frequency) we end up with a 1:4 variation in VCO gain. The reader is directed to [89] for details and an approach to maintain constant VCO gain across the tuning range for a CML interpolating VCO.

13.2.3   Static CMOS Implementation of Ring Oscillators

The fundamentals of static CMOS ring oscillators were introduced at the beginning of this chapter. Pseudo-differential static CMOS implementation was presented in Figure 13.8 (b). What remains in designing a CMOS ring is adding voltage control. This is discussed next.

13.2.3.1   Adding Voltage Control to Static CMOS Ring Oscillators

The delay of an inverter depends on the peak current flow during switching transitions. If this current can be controlled, delay and therefore frequency can be tuned. This is frequently done through an approach known as current-starving.

Figure 13.15 shows a current-starved inverter to which an extra PMOS and NMOS are added that limit (starve) the current during switching transitions. The amount of current-starving depends on the dimensions of $M_{P,S}$ and $M_{N,S}$ and the bias voltages V_{SP} and V_{SN}. As an example assume a regular inverter is sized for equal rising and falling propagation delay. If we add current-starving devices to it of equal width, i.e., $W_{P,S} = W_P$ and $W_{N,S} = W_N$, and connect their gates such that $V_{SP} = 0$ V and $V_{SN} = V_{DD}$, we expect the delay of the current-starved inverter to be approximately twice that of the original inverter. This is because the pull-down network that was previously a single NMOS of $\frac{W}{L}$ has been replaced by two equally sized transistors in series giving an equivalent *strength* of $\frac{W}{2L}$ which leads to approximately half of the current when the output falls, leading to twice the delay. This discussion is approximate due to short-channel effects and increased capacitance due to the addition of the extra

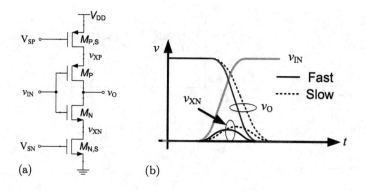

Figure 13.15 Current-starving in a static CMOS ring oscillator: (a) inverter schematic; (b) waveforms.

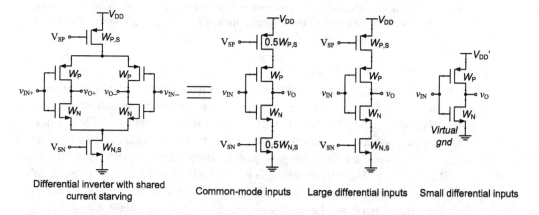

Differential inverter with shared current starving   Common-mode inputs   Large differential inputs   Small differential inputs

Figure 13.16 Current-starving in a differential static CMOS ring oscillator.

transistor. However, the example illustrates one key design point. Adding frequency control through current-starving always reduces the frequency of the ring relative to the maximum possible frequency. As $W_{P,S}$ and $W_{N,S} \to \infty$, the oscillation frequency approaches the frequency of the oscillator formed by regular inverters.

In the preceding discussion in which the bias voltages were connected to extreme values we expect that the current-starving MOSFETs will be in triode and we can consider equivalent pull-up and pull-down resistances of the gates. However, as V_{SN} is reduced, we may find that $M_{N,S}$ enters saturation. Example waveforms for two scenarios denoted as "fast" and "slow" are included in Figure 13.15 (b). These correspond to a higher and lower bias for V_{SN}, respectively. Notice that for higher V_{SN} the delay to v_O is shorter and the internal node v_{XN} rises less than for the "slow" case.

In general, V_{SN} and V_{SP} are tuned such that the maximum current sourced and sunk by the inverter are equal. If $\frac{W_{P,S}}{W_P} = \frac{W_{N,S}}{W_N}$ and $\frac{W_P}{W_N} = \frac{\mu_n}{\mu_p}$, the common-mode voltage of V_{SN} and V_{SP} is near $V_{DD}/2$.

Figure 13.16 shows an approach to current-starving in a differential CMOS ring oscillator such as that in Figure 13.8 (b). By sharing a single pair of current-starving

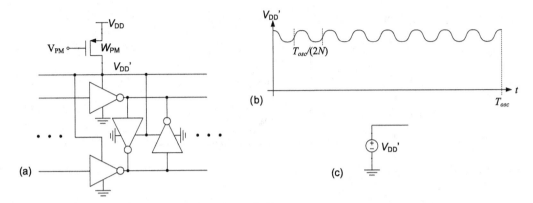

Figure 13.17 Single-FET current-starving in a static CMOS ring oscillator. One large PMOS controls the supply current to the ring, generating an internal supply voltage V'_{DD}: (*a) schematic highlighting connections to one stage. Main inverters and cross-coupled inverters are current-starved; (b) internal supply voltage ripple at twice f_{osc}; and (c) approximation of biasing condition. Voltage source of V'_{DD}.

transistors between the two, coupled inverters, differential oscillation is favoured over common-mode oscillation. We can see this by examining the common-mode and differential half circuits. The common-mode circuit is straightforward. If both sides of the inverter switch in the same direction, the current-starving transistors are shared between the two sides. Each side gets the current-conducting capacity of a MOSFET of half the width. The differential-mode is a bit trickier. In the analysis of differential pairs, we usually consider a differential half circuit for *small-signal* operation. Here, we have large-signal oscillation. Therefore, the middle half circuit, labelled "Large differential signals," is more appropriate. If only one output rises, it sees the capacity of the entire $W_{P,S}$. The falling output sees the capacity of W_{N_B}. Therefore we can put the entire width of the current-starving MOSFETs into the differential-mode half circuit. If we consider the start-up conditions of the circuit from an operating point in which the two inputs v_{IN+} and v_{IN-} are equal, for small signals around this operating point, nodes v_{X_P} and v_{X_N} can be considered as virtual-ground nodes as in the case of a differential pair. In this case, the differential half circuit does not see the current-starving devices at all. However, the strength of the remaining inverter transistors is set by the voltages present at the internal nodes denoted as V'_{DD} and the virtual ground.

As a final note, current-starving is also applied to the cross-coupled inverters so that their strength relative to the forward path is preserved.

Instead of current-starving each stage of the ring oscillator separately, we can employ the approach shown in Figure 13.17 where a single current-starving PMOS is used. Notice that it not only starves the inverters in the coupled rings, but also the cross-coupled inverters. With switching events of consecutive stages overlapping, this technique approximates supply regulation. As the on-resistance of the PMOS is increased, the effective supply voltage applied to the ring oscillator, V'_{DD} decreases. A single-ended N-stage ring has N rising edges per period, during which current is drawn from the supply. The differential ring draws supply current $2N$ times per period, giving rise to supply ripple with a period of $T_{osc}/2N$. Given the capacitance at node V'_{DD} that reduces ripple, we can view the ring as being biased by a regulated supply of V'_{DD}. Since the swing of the ring oscillator is now V'_{DD} the VCO's output must be buffered by inverters supplied by the nominal supply V_{DD}.

13.2.4   Phase Noise in Ring Oscillators

For a given power dissipation, ring oscillators will have higher phase noise than LC oscillators. However, LC oscillators occupy much larger chip area. Ring oscillators are compatible with a digital design flow since there is no need to interrupt the chip's power and ground distribution to make way for an inductor. For reasons discussed in Chapter 14, many wireline systems can use one clean, larger LC-based PLL that feeds several, compact ring-based PLLs or ILOs. If designed carefully, this combination can meet jitter specifications.

To design a low-noise ring oscillator, the universal knob that the designer can turn is power dissipation. For any topology, the phase noise $L(\Delta f)$ depends on the power dissipation as:

$$L(\Delta f) \propto \frac{1}{P_{diss}} \frac{f_{osc}^2}{\Delta f^2}. \tag{13.42}$$

The phase noise can be lowered by scaling the oscillator by widening transistors, reducing load resistors and increasing bias currents, following the approach described in Appendix Section B.5 [90].

13.3   LC Oscillators

LC oscillators are used in both wireless and wireline communication systems. This section gives a brief introduction and discusses basic implementation issues. Improved phase noise of LC oscillators comes from the narrow-band response of the LC tank. Higher quality factor leads to better phase noise. Therefore, predicting phase noise requires accurate inductor models, not only for estimating its nominal inductance value but also for its parasitic resistance and capacitance. Some technologies supply well-modelled inductors of either specific values or reasonable parameterized models. Without such models, oscillator designers model inductors using separate electromagnetic simulators such as HFSS from Ansys.

A simplified model of a parallel LC circuit (tank) is introduced in Figure 13.18. The resistor models losses in the inductor and is not intentionally added. Although such losses might be more intuitively captured by a resistor in series with the inductor, the series model can be transformed to the parallel model. The impedance of the tank, $Z(s)$ in Figure 13.18 (a) is given by:

$$Z(s) = \frac{1}{sC + \frac{1}{sL} + \frac{1}{R}} = \frac{s/C}{s^2 + \frac{s}{RC} + \frac{1}{LC}}. \tag{13.43}$$

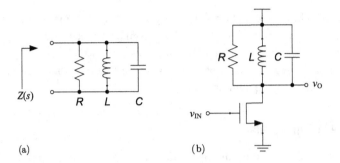

Figure 13.18 Parallel LC tank circuits: (a) stand-alone tank and (b) common-source amplifier loaded by an LC tank.

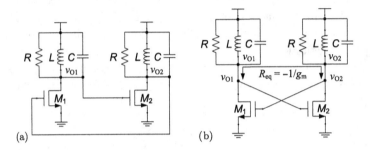

Figure 13.19 LC oscillator based on two tank-loaded common-source amplifiers: (a) drawn as a cascade of two stages and (b) drawn as a symmetrical circuit.

The maximum magnitude of $Z(j\omega)$ occurs at $\omega_{osc} = \frac{1}{\sqrt{LC}}$. At this frequency, the inductor and capacitor resonate out and we are left with the resistor. Therefore $Z(j\omega_{osc}) = R$. Below this frequency, $Z(j\omega)$ is inductive whereas at higher frequency, $Z(j\omega)$ is capacitive. If the tank loads a common-source amplifier as in Figure 13.18 (b), the small-signal gain becomes:

$$A(s) = -g_m \frac{s/C}{s^2 + \frac{s}{RC} + \frac{1}{LC}}. \tag{13.44}$$

The gain is $-g_m R$ at ω_{osc}. At dc and as $\omega \to \infty$, the gain is 0. If the magnitude of the gain exceeds unity at ω_{osc}, putting two of these amplifiers in a loop as shown in Figure 13.19 (a) leads to a loop gain greater than one with a phase of 360°. In Figure 13.19 (b) the circuit is redraw to show identical tanks loaded by the cross-coupled MOSFETs. These produce a negative resistance of $-1/g_m$, as discussed in Chapter 3, which is in parallel with the tank resistance R. Provided $g_m > 1/R$ the real part of the overall tank impedance is negative and the system oscillates. Although the LC oscillator is composed of two inverting amplifiers in a loop, it does not latch up at dc because the dc gain of each stage is 0.

To reduce an LC oscillator's sensitivity to power-supply voltage fluctuations, it is usually made differential by adding a biasing current source I_{BIAS} as shown in Figure 13.20. Since at dc the inductor is a short circuit, the dc operating point of

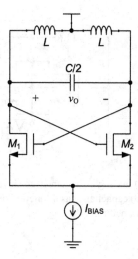

Figure 13.20 Differential LC oscillator biased with a current source.

the gates of the MOSFETs is V_{DD}. The outputs swing above and below V_{DD}. The maximum single-ended peak swing will occur if the differential pair is fully switched. It is limited by:

$$V_{DD} - I_{BIAS}R/2 < v_{O+} < V_{DD} + I_{BIAS}R/2. \tag{13.45}$$

However, full switching only occurs if $g_m R \gg 1$. Nevertheless, care must be taken so as to not exceed the safe voltage range for the MOSFETs.

13.3.1 Adding Voltage Control to an LC Oscillator

Since the frequency of oscillation is $1/\sqrt{LC}$ one of L or C must be tuneable to convert the oscillator to a VCO. A MOSFET with source, drain and body terminals grounded as shown in Figure 13.21 (a) has a capacitance vs gate voltage curve as shown in Figure 13.21 (b). In the vicinity of $V_G = V_t$ the small-signal capacitance increases from a minimum value to a maximum value. Recall from the operation of a MOSFET that for $V_G > V_t$ an inversion layer is formed under the gate. This means that the gate-to-channel structure is a parallel plate, separated by the oxide. Hence the capacitance is:

$$C_{max} = C_{OX} = C_{OX,A}WL = \frac{\epsilon_{OX}WL}{t_{OX}}, \tag{13.46}$$

where $C_{OX,A}$ is the per-unit-area oxide capacitance, t_{OX} is the thickness of the gate oxide and ϵ_{OX} is the permittivity of the gate oxide. For smaller gate voltages, in the absence of an inversion layer, the capacitive structure is the series combination of the oxide capacitance and the capacitance across the depletion region (C_{dep}). Therefore:

$$C_{min}^{-1} = C_{OX}^{-1} + C_{dep}^{-1} = \frac{C_{OX} + C_{dep}}{C_{OX}C_{dep}}. \tag{13.47}$$

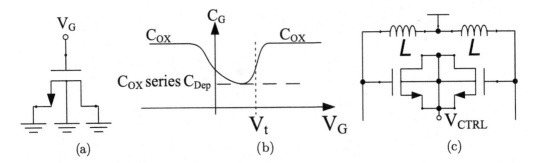

Figure 13.21 MOSFET as a varactor in an LC VCO: (a) biasing arrangement for investigating MOS gate capacitance as a function of gate voltage; (b) capacitance as a function of gate voltage; and (c) portion of an LC VCO using a MOS varactor.

Therefore,

$$C_{min} = \frac{C_{OX}C_{dep}}{C_{OX} + C_{dep}}. \tag{13.48}$$

Figure 13.21 (c) shows how NMOS transistors, used as variable capacitors (a.k.a. varactors) can be added to an LC VCO. To consider the tuning range of this approach, consider the ratio of maximum to minimum capacitance:

$$C_{range} = \frac{C_{max}}{C_{min}} = 1 + \frac{C_{OX}}{C_{dep}}. \tag{13.49}$$

Suppose as an example, $C_{range} = 2$. A 1:2 variation in capacitance leads to a 1:$\sqrt{2}$ variation in tuning range. Any additional parasitic capacitance at the drains of the MOSFETs, such as their C_{db}, wiring capacitance, or parasitic capacitance of the inductor will add to both C_{max} and C_{min} further reducing C_{range}.

The capacitance of the MOSFET is a nonlinear function of the voltage across it. Thus far we have considered dc voltages. When used in an LC VCO, the voltage across the varactor is varying, meaning the output signal sees a range of capacitance value. Approximately speaking, the frequency of oscillation is set by an effective capacitance across the voltage range of the output nodes.

A second issue with the circuit in Figure 13.21 (c) is that giving the tuning voltage control of the entire tuning range leads to a high VCO gain. One solution is to split the capacitance into smaller varactors. By driving most of them with binary voltages we switch in capacitance in increments, leading to a "band selection" operation. The remaining tuning range is allocated to a smaller amount of capacitance that is continuously controlled. This approach reduces the VCO gain on the analog control port and also can keep most of the capacitance biased away from the highly nonlinear region where capacitance varies abruptly.

13.3.2   Phase Noise in LC Oscillators

Phase noise in LC oscillators is due to thermal noise from the cross-coupled transistors as well as the up conversion of flicker noise from the tail current source.

13.4   Summary

This chapter introduced oscillators by first presenting typical metrics for oscillators and typical phase-noise trends (Power spectral density vs frequency offset). Two important aspects of a VCO's tuning characteristic were presented, starting with tuning range (and the effect of process variation) and VCO gain (K_{VCO}).

Ring oscillators were presented in detail. Criteria to start oscillation were derived using a first-order gain model for the inverters. Steady-state oscillation frequency was expressed based on the inverters' logic circuit delay. Differential rings, which can support even numbers of stages, were explained, giving rise to rings that can generate quadrature clock signals. The implementation of CML and CMOS inverter stages was explored. In the case of CML stages, various loads, including symmetrical loads, were analyzed.

Replica biasing, which jointly adjusts the load resistance and bias current of a CML inverter, can be used to make a ring oscillator voltage controlled. Delay interpolation was also investigated. For CMOS oscillators, various current-starving techniques were considered.

The brief treatment of LO oscillators explained their operation by considering the active circuitry as adding a negative resistance in parallel with an LC tank. An alternative intuition of putting two bandpass transfer functions in a loop was explained. Tuning of LC oscillators is usually done using MOS varactors, controlled with both digital and analog signals.

An LC oscillator will have lower phase noise than a ring oscillator for a given power dissipation. Usually an I/O interface's global clock will be generated with an LC oscillator. Other, local clocks can be generated by ring techniques, either in a phase-locked loop or as an injection-locked oscillator. Ring oscillators have the benefit of occupying less chip area. When used with a wide bandwidth PLL or ILO, acceptable jitter performance is possible.

Problems

13.1   In the technology of your choice, determine the maximum oscillation frequency of a three-stage static CMOS inverter-based ring oscillator, when biased at nominal supply. You should investigate the choice of $W_P : W_N$ and see how it determines the duty cycle of the outputs. Denote this frequency as $f_{osc,max}$. Compare the oscillation frequency to the frequency at which the small-signal frequency response of each stage provides 60° of phase. To investigate the ac response of the inverter, bias it at the point where $v_{IN} = v_O$. Compare the oscillation frequency to that predicted by (13.22). Confirm that the frequency of oscillation is independent of the choice of widths, so long as the selected $W_P : W_N$ ratio is maintained.

13.2   In the technology of your choice, determine the maximum oscillation frequency of a three-stage CML ring oscillator, when biased at nominal supply voltage. Use a minimum peak-to-peak differential swing of $V_{DD}/2$. Compare the oscillation frequency to the frequency at which the small-signal frequency response of each stage provides $60°$ of phase. Compare the oscillation frequency to that predicted by (13.22). How much does the frequency of oscillation change if the swing is increased to $2V_{DD}/3$?

13.3   The oscillators designed in Problems 13.1 and 13.2 do not have any output-buffering. Investigate the impact on oscillation frequency of adding a buffer to each phase that is the same as the implementation of each delay stage.

13.4   The oscillators designed in Problems 13.1 and 13.2 were investigated for nominal operating conditions. Investigate the impact on oscillation frequency of process corners, temperature variation and supply voltage variation. Consider $\pm10\%$ variation of supply voltage, temperature of 0 to $85°$ and the foundry-supplied process corners. Find the individual and aggregate affects of these variations. What is the worst-case spread relative to the nominal oscillation frequency?

13.5   Design a four-stage, differential, static-CMOS ring oscillator. Use the $W_P : W_N$ ratio found in Problem 13.1. Cross-coupled inverters must be added to prevent latch-up. The ratio of the width of their transistors to those in the "main path" is denoted by r_4. Determine the acceptable range of size ratio, r_4, between the cross-coupled inverters and those in the "main path." Produce a sketch of $f_{osc,4}$ vs. r_4. Repeat this process for a five-stage oscillator. On the same axes, plot $f_{osc,5}$ vs. r_5.

13.6   Add frequency control to a ring oscillator such as that previously designed in Problems 13.1 or 13.5. Use the approach in Figure 13.17. Choose the size of the PMOS current-starving transistor such that $V'_{DD} = 0.9V_{DD}$ when $V_{CTRL} = 0$ V. For nominal PVT, find the tuning range as V_{CTRL} is swept from V_{DD} to 0 V. Repeat for the variation outlined in Problem 13.4. With this variation, find the upper and lower end of the tuning range changes. Define the tuning range as $f_{max} - f_{min}$ where f_{max} is the lowest of the upper ends and f_{min} is the largest of the lower ends. How much smaller is the tuning range when PVT is included?

13.7   Sketch the output waveform of a delay interpolator if the rise time of the input signal is equal to the added delay, that is $\tau_r = \tau_A$. Repeat for the case where $\tau_A = 5\tau_r$. Consider $x = [0, 0.25, 0.5, 0.75, 1]$.

13.8   Modify (13.25) for weak inversion operation. Is it possible to get the circuit to oscillate if all transistors are in weak inversion?

13.9   Explore the symmetrical load in a sub-micro process. How symmetrical is it?

13.10   Design an LC oscillator. How much tuning range can you achieve if all of the capacitance is connected to a control voltage?

14 Phase-Locked Loops and Injection-Locked Oscillators

This chapter presents systems that use a voltage-controlled oscillator (VCO) in a feedback loop to lock its phase to that of a reference clock. These systems, called phase-locked loops (PLLs), generate the signals used to clock decision circuits and MUX/DEMUX circuits. In this chapter the VCO is locked to a reference clock signal. In Chapter 15 we consider how to lock a clock signal to an incoming data stream.

An introduction to PLLs and the notion of phase comparison starts this chapter. The typical type II analog PLL is analyzed. Split tuning and details of frequency division are presented. Digital PLLs are now commonplace, necessitating an overview. Injection-locked oscillators (ILOs) play an important role as clock buffers and multi-phase generators This chapter gives an overview of ILO dynamics covering topics of jitter-tracking bandwidth, lock range and injection strength.

14.1 PLL Introduction and Phase Detection

In this section we consider the relationship between the quantities of phase and frequency as well as why we compare phase even if the objective is to control the frequency of an oscillator. To start, consider a VCO's output $v_{OSC}(t)$ as:

$$v_{OSC}(t) = A_0\cos(\omega_o t + \theta). \tag{14.1}$$

We often refer to ω_o as the angular frequency and θ as the phase. However, we can rewrite (14.1) as:

$$v_{OSC}(t) = A_0\cos(\phi(t)), \tag{14.2}$$

where we lump the entire argument together and call it the total phase, $\phi(t)$. To avoid confusion, we will call θ the phase offset and ϕ the total phase. Frequency is the derivative of total phase, allowing us to write:

$$\omega = \frac{d\phi}{dt} = \frac{d(\omega_o t + \theta)}{dt} = \omega_o, \tag{14.3}$$

$$\phi = \int_0^t \omega_o d\tau = \omega_o t + \theta. \tag{14.4}$$

Consider the block diagram shown in Figure 14.1 (a). Here, the frequency of the VCO is divided down and compared to the frequency of a reference clock. Ideally, this produces an output frequency of:

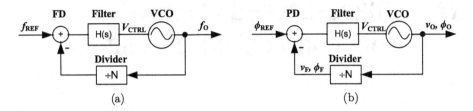

Figure 14.1 Frequency-locked loop (FLL) vs phase-locked loop (PLL): (a) FLL and (b) PLL. FD denotes frequency detector and PD denotes phase detector.

$$fo = Nf_{REF}. \tag{14.5}$$

Assuming at least one integrator in $H(s)$ and no offsets in any block, this scheme will work. However, integrators always have finite gain and there will be offsets in various blocks, so we will actually have:

$$fo = Nf_{REF} + \Delta f. \tag{14.6}$$

If, on the other hand, we compare phase, we have the block diagram shown in Figure 14.1 (b). Even considering errors such as finite gain of integrators and offsets, we have an input/output relationship of:

$$\phi_O = N\phi_{REF} + \Delta\phi. \tag{14.7}$$

Though we are left with a possible static phase error $\Delta\phi$, we will have no static frequency error. Hence we will generate clock signals using phase comparison rather than frequency comparison.

14.1.1   Overview of PLL Operation

Figure 14.1 (b) shows a generic block diagram of a PLL. A voltage-controlled oscillator generates the output signal, $v_O(t)$, having a total phase ϕ_O. The output signal is divided by a divider circuit which produces an edge in v_F every N edges of v_O. The relative phase of the reference and fed-back signals is compared to produce a signal representing the phase difference. This error signal is filtered by a low-pass filter $H(s)$ to generate a control voltage V_{CTRL} at the input of the VCO. The next subsections present different approaches for comparing the phase of two signals.

14.1.2   Nonlinear Phase Detector

A single DFF, as shown in Figure 14.2 (a), can be used as a phase detector. Consider the waveforms shown in Figure 14.2 (b) and (c). In (b), the reference clock arrives before the fed-back signal. Since V_F drives the clock port of the DFF, a logic "1" is repeatedly sampled and the output, V_O, is high. If the rising edge of V_F arrives before the edge of V_{REF}, as in part (c), a logic "0" is sampled and the output is low. Figure 14.2 (d) shows the transfer characteristic between time offset, ΔT, and output voltage. In general, transfer characteristics of phase detectors will be interpreted as

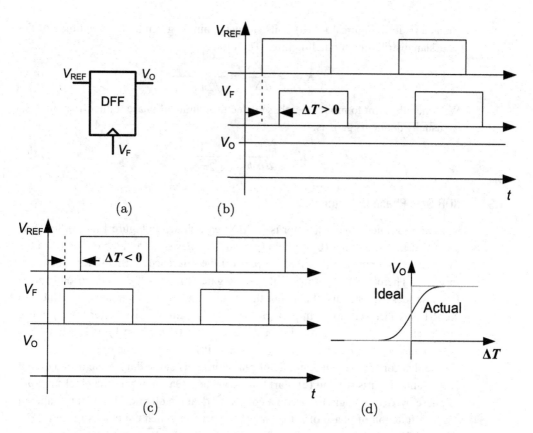

Figure 14.2 Early-late phase detection using a D flip-flop: (a) schematic; (b) waveforms when reference clock, V_{REF}, arrives first; (c) waveforms when fed-back clock, V_F, arrives first; (d) voltage transfer characteristic as a function of ΔT.

showing the average output voltage with respect to phase or time offset over a clock period or several clock periods. Considering the DFF as ideal: the output is high for any ΔT or $\Delta \phi > 0$ and low for ΔT or $\Delta \phi < 0$. We expect a step function at $\Delta T = 0$. However, clock signals have jitter. Also, for time offsets close to zero the DFF can exhibit metastability, a condition in which the output state changes very slowly, and with random outcomes. The actual behaviour of the phase detector is the convolution of its ideal response with the jitter and metastability probability density functions, leading to a smooth transition from low to high. This is an important point because when we model a system such as a PLL, we need to specify the gain of each block. Step functions are not differentiable and complicate system modelling, but on the other hand the actual behaviour simplifies modelling, provided we have a reasonable estimate of the voltage transfer characteristic's slope at $\Delta \phi = 0$. Although we linearize its characteristic, because the ideal operation is a step function, we refer to this phase detector as a nonlinear phase detector or a bang-bang phase detector.

If we assume a root-mean-squared differential jitter between V_{REF} and V_F of $T_{j,rms}$ in units of time or ϕ_{rms} in units of phase, the slope of the actual voltage transfer characteristic in Figure 14.2 (d) at its steepest point is given by the amplitude of the Gaussian distribution modelling jitter. Therefore,

$$K_{\Delta T} = \frac{dV_O}{d\Delta T} = \frac{1}{\sqrt{2\pi}\, T_{j,rms}}. \tag{14.8}$$

We generally want to model the phase detector in units of phase, rather than time, so we usually define its gain as:

$$K_\phi = \frac{dV_O}{d\Delta\phi} = \frac{1}{\sqrt{2\pi}\,\phi_{rms}}. \tag{14.9}$$

14.1.3   XOR Gate Phase Detector

The simplest linear phase detector is an XOR gate shown in Figure 14.3 (a). Its operation is depicted in parts (b) through (g). Part (b) shows its truth table. The output is "1" when the inputs are different. In part (c) the reference and fed-back clocks are in phase. Therefore the output of the XOR gate is always "0." Though our aim is to phase lock two signals, this is not the steady-state phase relationship to which an XOR-based PLL will lock. In part (d) the fed-back clock arrives after the reference clock. The XOR gate outputs "1" for the interval of time where V_F is low but V_{REF} is high and where V_F is high and V_{REF} is low. Part (e) shows the opposite situation. Notice that for the same magnitude of phase difference, the duty cycle of the output is the same in parts (d) and (e). Part (f) shows the situation where the clocks are out of phase by $\Delta\phi = \pi$, giving rise to a constant "1" at the output. The average value of V_O as a function of phase offset is shown in part (g). When the clocks reach a 180° phase difference, V_O is always high. Between 0 and 180°, $\overline{V_o}$ rises from 0 with a slope of $\frac{1}{\pi}$ to reach 1. However, since the XOR gate is a combinational logic circuit (does not store state), it cannot distinguish the polarity of the phase difference. Hence, the transfer characteristic of the phase detector has even symmetry. PLLs that use an XOR gate will be designed to lock to $\theta = \frac{\pi}{2}$.

The primary advantage of an XOR gate phase detector is that it is fast. High-speed CML XOR gates can operate well into the GHz range. In these cases, the XOR gate can be viewed as a multiplier, where the time-varying component of its output is given by:

$$v_O(t) = V_F(t)V_{REF}(t), \tag{14.10}$$

$$= \cos(\omega_f t + \theta_{(f)})\cos(\omega_{ref} t + \theta_{ref}), \tag{14.11}$$

$$= \frac{1}{2}\cos((\omega_f - \omega_{ref})t + \theta_f - \theta_{ref}) +$$

$$\frac{1}{2}\cos((\omega_f + \omega_{ref})t + \theta_f + \theta_{ref}). \tag{14.12}$$

Here we assume a unit amplitude of the two inputs. If we consider the case where the frequency of the signals is the same, the expression for the output can be simplified to:

$$v_O(t) = \frac{1}{2}\cos(\theta_f - \theta_{ref}) + \frac{1}{2}\cos(2\omega_{ref} t + \theta_f + \theta_{ref}), \tag{14.13}$$

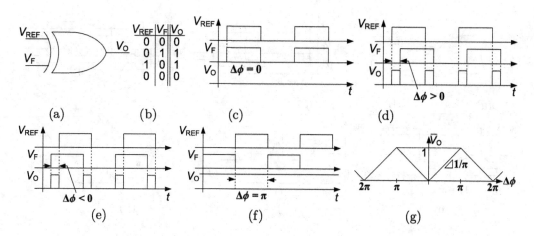

Figure 14.3 XOR gate as a phase detector: (a) circuit symbol; (b) truth table; (c) waveforms for $\Delta\phi = 0$; (d) waveforms for $\Delta\phi > 0$; (e) waveforms for $\Delta\phi < 0$; (f) waveforms for $\Delta\phi = \pi$; (g) $\overline{V_O}$ as a function of $\Delta\phi$.

giving rise to a dc term and a term at twice the reference frequency. Considering the average value of the output we have:

$$\overline{v_O(t)} = \frac{1}{2}\cos(\theta_f - \theta_{ref}). \tag{14.14}$$

This average value will be driven to zero by the PLL, leading to:

$$0 = \frac{1}{2}\cos(\theta_f - \theta_{ref}), \tag{14.15}$$

$$\theta_f - \theta_{ref} = \frac{\pi}{2}. \tag{14.16}$$

Thus a PLL using an XOR gate as a phase detector will lock to a phase difference of $\pi/2$. Of course, $\theta_f - \theta_{ref} = -\frac{\pi}{2}$ also satisfies (14.15). However, due to the even symmetry of the phase detector, the slope of its response around $-\frac{\pi}{2}$ is opposite to the slope around $\frac{\pi}{2}$, meaning if the latter is a stable equilibrium point, the former is not.

To make the connection between an XOR gate and a multiplier clearer, consider Figure 14.4, where operation of an XNOR gate is presented where a logic 0 is represented by a -1 signal. The symbol is in Figure 14.4 (a) and its truth table is in Figure 14.4 (b). The output of the gate is the product of the two inputs when we consider only full-rail signals. Parts (c), (d) and (f) show the inputs and outputs for the same phase relationships that were presented in Figure 14.3. In part (e), the two inputs are $\pi/2$ apart. The output is 50% duty cycle signal at twice the frequency, centred around 0, thus giving an average value of 0. The overall response of the XNOR phase detector is in part (g) where $\overline{V_O}$ crosses 0 at $\pm\pi/2$. The triangle wave can be viewed as a straight-line approximation of (14.14).

There are two limitations of the XOR-based (or XNOR-based) phase detector. The first is that the output has a high-frequency component even when the PLL is locked as shown in Figure 14.3 (e). Although the output is filtered, this signal will inevitably

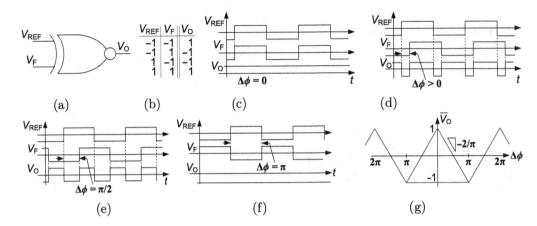

Figure 14.4 XNOR gate as a phase detector: (a) circuit symbol; (b) truth table; (c) waveforms for $\Delta\phi = 0$; (d) waveforms for $\Delta\phi > 0$; (e) waveforms for $\Delta\phi = \pi/2$; (f) waveforms for $\Delta\phi = \pi$; (g) $\overline{V_O}$ as a function of $\Delta\phi$.

modulate the VCO. The second limitation relates to initial frequency acquisition when the reference and fed-back clocks are at different frequencies. In this case the "ω difference" term in (14.12) does not have a persistent offset that can tune the oscillator in the PLL. In order to lock the PLL we need the difference in frequency to be less than the cut-off frequency of the PLL's loop filter. Since we want the loop filter's cut-off frequency to be low to filter the persistent $2\omega_{ref}$ terms, the XOR-based PLL has a very narrow lock range. To extend the range, we need a circuit that inherently detects differences in frequency as well as phase. This can be accomplished by having an additional frequency-detection loop or by a different type of phase detector that also detects frequency errors.

14.1.4   DFF-Based Phase/Frequency Detector

The schematic and operation of a circuit that can detect differences in both phase and frequency, known as a phase/frequency detector (PFD), is shown in Figure 14.5. It consists of two DFFs each with a logic "1" applied to its D terminal, as shown in Figure 14.5 (a). The clock signals are applied to the CLK terminals of the flip-flops. The FFs also have an asynchronous reset, denoted by R. Figure 14.5 (b) shows the signals when the edges of V_{REF} arrive before those of V_F. In this case, the rising edge of V_{REF} clocks the "1" at the input of the DFF to the output, Up. When the rising edge of V_F arrives, its output (Dn) rises. Once both Up and Dn are high, the AND gate generates a reset pulse ($Reset$) that resets both DFFs. The phase error is encoded in the difference between Up and Down. Even if the inputs are phase locked as shown in part (d), outputs Up and Dn still rise once per cycle for a short interval of time, denoted by T_{min}. Ideally, these minimum pulses will be exactly equal in duration, leading to zero difference between Up and Dn. The phase to voltage conversion is

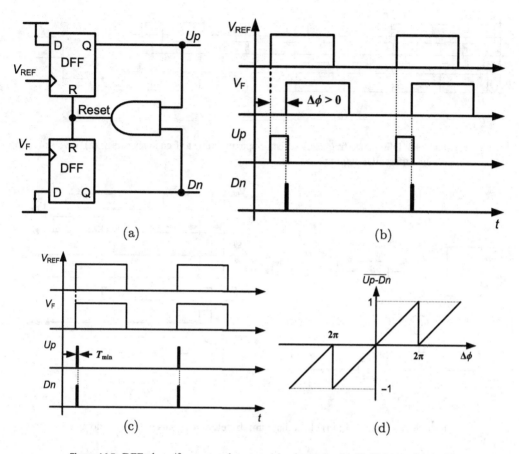

Figure 14.5 DFF phase/frequency detector: (a) schematic; (b) *Up/Dn* waveforms when V_{REF} leads V_F; (c) *Up/Dn* waveforms when V_{REF} and V_F are in phase; and (d) average differential output $(\overline{Up - Dn})$ vs phase error.

shown in Figure 14.5 (d). Over the range of phase difference $-2\pi < \Delta\phi < 2\pi$, the phase detector has a gain of $1/2\pi$.

The behaviour of the PFD for $\Delta\phi$ beyond $\pm 2\pi$ is what makes the circuit useful for frequency detection. Consider the waveforms in Figure 14.6 (a). Near $t = 0$, V_F lags V_{REF} by about one-eighth of the period. However, V_F is also at a lower frequency. We can see the *Up* pulse grows each cycle until it lasts for almost the entire period. The longest *Up* pulse, shown in the dashed ellipse, sees two rising edges of V_{REF} between rising edges of V_F. After this "cycle slip," the *Up* pulse may be quite short. According to the transfer characteristic in Figure 14.5 (d), it can decrease to zero when the phase difference is exactly 2π. Then, due to the lagging frequency of V_F, the *Up* pulse grows in duration, before returning to zero. However, throughout this the *Dn* pulse has been a minimum-length pulse. While the DFF PFD is linear for phase offsets, it is nonlinear for frequency errors. As presented in Figure 14.6 (b), on average, the output will be

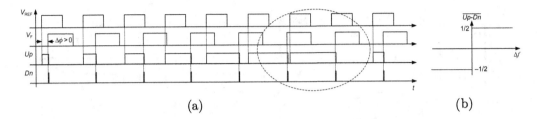

Figure 14.6 DFF phase/frequency detector operation due to frequency offset: (a) waveforms; (b) average differential output vs Δf.

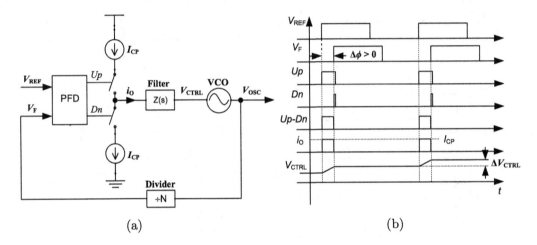

Figure 14.7 Analog PLL: (a) block diagram; (b) relevant signals in the time domain.

1/2 when the frequency of the fed-back signal is lower than the reference and $-1/2$ when the fed-back signal is at a higher frequency.

Thus far, the implementation details of the rest of the PLL have been glossed over. The next section presents analog phase-locked loops in detail, showing the circuitry that connects the phase detector to the VCO.

14.2   Analog Phase-Locked Loops

The previous section presented several phase detectors. In this section, the complete implementation and modelling of analog PLLs using the DFF-based PFD will be discussed. Various nonidealities will also be considered.

Figure 14.7 (a) shows a block diagram of a PFD-based analog PLL. The feedback path includes a divider that divides the frequency of the VCO's output by N and feeds it back to the PFD.

Between the PFD's outputs and the oscillator are a charge pump consisting of two switches and two current sources and a filter, denoted by $Z(s)$. This combination of circuits converts the difference in duration of the Up and Dn pulses to a smooth control

Table 14.1 Charge pump operation.

Up	Dn	i_O
0	0	0
0	1	$-I_{CP}$
1	0	I_{CP}
1	1	0

voltage (V_{CTRL}) to be applied to the oscillator. The operation of the charge pump is summarized in Table 14.1. When Up is high, the upper switch closes, connecting the upper current source (I_{CP}) to the charge pump's output, i_O. Likewise, when Dn is high the lower switch closes and the charge pump sinks a current of I_{CP}. When both switches are closed the matched current sources "cancel" and no current flows to the filter; instead, the current flows from V_{DD} to GND.

Figure 14.7 (b) shows example waveforms for the case in which V_{REF} leads the fed-back signal V_F by some $\Delta\phi$. The Up pulse is longer than the minimum-length Dn pulse. For the interval in which Up is high but Dn is low, a current I_{CP} flows out of the charge pump. In this example, the filter, $Z(s)$ is a grounded capacitor (C), meaning that the control voltage is the integral of the charge pump's output. Each cycle, the control voltage increases by ΔV_{CTRL} due to the longer Up pulse. This change in voltage is:

$$\Delta V_{CTRL} = \frac{I_{CP}\Delta T}{C}, \tag{14.17}$$

where ΔT is the difference between the duration of the Up and Dn pulses. I_{CP} is the charge pump current and C is the capacitance. When the reference and fed-back clocks are in phase, the Up/Dn pulses have equal duration and V_{CTRL} is constant. The average output of the combined PFD/CP, defined as $\overline{i_O}$, can be expressed in terms of time difference or phase difference as:

$$\overline{i_O} = \frac{\Delta T}{T_{ref}}I_{CP} = \frac{\Delta\theta}{2\pi}I_{CP}. \tag{14.18}$$

The gain of the overall PFD/CP is thus $\frac{I_{CP}}{2\pi}$. This is valid for the range of $-2\pi < \Delta\theta < 2\pi$. In general the loop filter will be a more complex circuit than a single capacitor. It will nevertheless convert i_O to V_{CTRL} and apply it to the VCO. Once we have a model of the PLL, we will consider various implementations of the loop filter.

14.2.1 Linear Time Invariant Modelling of a PLL

Phase-locked loops are complex systems whose transient simulations can be slow, particularly when done at the transistor level. This is because the simulator must take small time steps, determined by the time constants in high-speed blocks such as the VCO, but simulation over a long interval of time is needed to show the PLL's settling

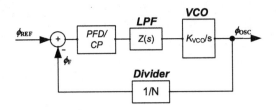

Figure 14.8 Block diagram of a linear time-invariant PLL with signals in the Laplace domain.

dynamics, which are set by comparatively long time constants in the loop filter. We can, however, gain some insight into the PLL's dynamics when it is locked by considering a linear time-invariant model in the "s" domain. Such a model is shown in Figure 14.8.

Whereas Figure 14.7 has the actual transient voltages as input and output signals, the model in Figure 14.8 has the input and output phase. The combined PFD/CP has a gain of $\frac{I_{CP}}{2\pi}$. The as yet unknown loop filter has a transfer function of $Z(s)$. The VCO has a tuning characteristic with a slope of K_{VCO}. This gain maps tuning voltage changes to frequency changes. However, we want to convert this to phase changes. Since phase is the integral of frequency, the VCO is modelled as the product of K_{VCO} and $\frac{1}{s}$ for an overall transfer function of $\frac{K_{VCO}}{s}$.

The feedback divider generates an output edge every N edges of its input. If the output of the oscillator shifts by ΔT, so too does the output of the divider. However, because the period of the divided signal is N times longer than the oscillator's, a shift of ΔT represents a proportionally smaller shift in phase for the divider's output. Thus the divider has a gain of $\frac{1}{N}$.

The LTI model extracts the essence of the blocks, mapping the relevant signal quantities. Notice that the units of the signals vary, meaning that each block has different units for its gain. For example:

- PFD-CP: Output is current. Input is phase. Gain has units $[\frac{A}{rad}]$.
- Loop filter: Output is voltage. Input is current. Gain has units $[\frac{V}{A}] = [\Omega]$.
- VCO: Output is phase. Input is voltage. Gain has units $[\frac{rad}{V}]$.
- Divider: Both input and output are phase. Gain is unitless.

As with any feedback system the loop gain (i.e., the product of all gains forming the feedback loop) is unitless. The transfer function of the PLL, $H(s)$, can be found using standard control theory as:

$$H(s) = \frac{\phi_{OSC}}{\phi_{REF}} = \frac{G(s)}{1 + G(s)\beta(s)}, \tag{14.19}$$

where

$$G(s) = \frac{I_{CP}}{2\pi} Z(s) \frac{K_{VCO}}{s}, \tag{14.20}$$

$$\beta = \frac{1}{N}. \tag{14.21}$$

Therefore,

$$H(s) = \frac{\frac{I_{CP}}{2\pi}Z(s)\frac{K_{VCO}}{s}}{1 + \frac{I_{CP}}{2\pi}Z(s)\frac{K_{VCO}}{s}\frac{1}{N}},$$ (14.22)

$$H(s) = N\frac{\frac{I_{CP}}{2\pi}Z(s)K_{VCO}}{Ns + \frac{I_{CP}}{2\pi}Z(s)K_{VCO}}.$$ (14.23)

At this point we investigate choices for $Z(s)$ and each resultant $H(s)$. For Figure 14.7 we discussed feeding the charge pump output to a capacitor, C. If we set $Z(s) = Z_C(s) = \frac{1}{sC}$ (14.23) becomes:

$$H_C(s) = N\frac{\frac{I_{CP}}{2\pi}\frac{1}{sC}K_{VCO}}{Ns + \frac{I_{CP}}{2\pi}\frac{1}{sC}K_{VCO}},$$ (14.24)

$$H_C(s) = N\frac{\frac{I_{CP}}{2\pi C}K_{VCO}}{Ns^2 + \frac{I_{CP}}{2\pi C}K_{VCO}}.$$ (14.25)

The denominator of $H_C(s)$ is of the form $as^2 + c$. Therefore, it has poles on the $j\omega$ axis, meaning that this is not a good choice for $Z(s)$ since it will lead to persistent ringing in the step response. By considering the forward transfer function of:

$$G_C(s) = \frac{I_{CP}}{2\pi}Z_C(s)\frac{K_{VCO}}{s},$$ (14.26)

$$G_C(s) = \frac{I_{CP}}{2\pi C}\frac{K_{VCO}}{s^2},$$ (14.27)

we see a forward transfer function with phase that is always $-180°$, and therefore has $0°$ of phase margin. To address this, we modify $Z(s)$. We still want it to have very high, ideally infinite dc gain, but have a phase closer to $-90°$ at the unity-gain frequency of $G(s)$. By adding a series R to the loop filter, $Z(s)$ becomes:

$$Z_{RC}(s) = R + \frac{1}{sC} = \frac{1 + sRC}{sC}.$$ (14.28)

With this loop filter, $G(s)$ becomes:

$$G_{RC}(s) = \frac{I_{CP}(1 + sRC)}{2\pi C}\frac{K_{VCO}}{s^2}.$$ (14.29)

Now the overall transfer function becomes:

$$H_{RC}(s) = \frac{\frac{K}{C}(1 + sRC)}{s^2 + \frac{KR}{N}s + \frac{K}{NC}},$$ (14.30)

where $K = \frac{I_{CP}K_{VCO}}{2\pi}$. This can be compared to the standard second-order transfer function to yield

$$\omega_n = \sqrt{\frac{K}{NC}},$$ (14.31)

$$\zeta = \frac{R}{2}\sqrt{\frac{KC}{N}}.$$ (14.32)

ζ	ω_z	ω_{p1}	ω_{p2}
1	0.5	1	1
$\sqrt{2}$	$\frac{1}{2\sqrt{2}}$ $=0.35$	$\sqrt{2}-1$ $=0.41$	$\sqrt{2}+1$ $=2.41$
2	0.25	0.27	3.73
3	0.167	0.172	5.83

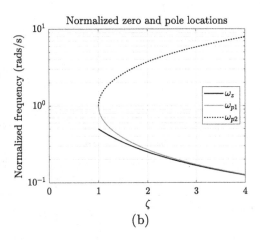

(a)   (b)

Figure 14.9 Magnitude of normalized poles and zero frequencies for various values of ζ: (a) table showing sample values; (b) plot of poles and zero for $\zeta \geq 1$.

Unlike many second-order systems that are designed to have poles that are critically damped ($\zeta = 1$) or a Butterworth response ($\zeta = \frac{1}{\sqrt{2}}$), this PLL is usually designed to have a high damping factor, which results in near cancellation of the pole and the zero. For this transfer function, the zero is given by:

$$\omega_z = -\frac{1}{RC} = -\frac{\omega_n}{2\zeta}. \tag{14.33}$$

The poles of any second-order system are given by:

$$\omega_{p1,2} = -\zeta\omega_n \pm \omega_n\sqrt{\zeta^2 - 1}. \tag{14.34}$$

As $\zeta \to \infty$

$$\omega_{p1} \to -\frac{\omega_n}{2\zeta}, \tag{14.35}$$

$$\omega_{p2} \to -2\zeta\omega_n = -2\frac{R}{2}\sqrt{\frac{KC}{N}}\sqrt{\frac{K}{NC}} = -R\frac{K}{N} = -R\frac{I_{CP}K_{VCO}}{2\pi N}. \tag{14.36}$$

Figure 14.9 shows the magnitude of the poles and zero as ζ is increased, starting at 1, for $\omega_n = 1$. As ζ increases, the lower frequency pole approaches the zero and the higher frequency pole increases. We choose a design point with large ζ to avoid peaking in the frequency response that would otherwise occur. If we assume the lower frequency pole and zero cancel, the 3 dB bandwidth of the PLL is set by ω_{p2} and becomes:

$$\omega_{bw} \approx |\omega_{p2}| = R\frac{I_{CP}K_{VCO}}{2\pi N}. \tag{14.37}$$

As a counter example, if $\zeta = 1$, the transfer function has two real poles with magnitude ω_n and a zero with magnitude $\frac{\omega_n}{2}$. Although this $H(s)$ is generally low-pass, it will increase about 3 dB between the zero and the poles, meaning that reference clock jitter is amplified in this region.

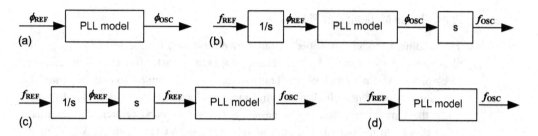

Figure 14.10 Conversion of phase-based PLL model to frequency-based model.

It might appear viable to design the PLL with $\zeta = \frac{1}{2}$, giving $\omega_z = \omega_n$. However, the real zero does not cancel the effect of a complex pole. A unity-gain all-pole transfer function with $\zeta = \frac{1}{2}$ reaches a peak gain of 1.15 at $\omega = \frac{1}{\sqrt{2}}\omega_n$. The addition of a zero at ω_n only adds to the peaking, raising the peak gain to 1.47 at $1.71\omega_n$.

With a loop filter $Z(s)$ that integrates, the forward transfer function is Type II, meaning it has two poles at $s = 0$. This is a common implementation, as Type II systems can track ramps. Although we often think about the inputs and outputs being phase offsets, they are actually the total phase of input and output signals which are ramp signals in the time domain. Therefore, to control the oscillator so that the fed-back signal has zero phase difference, we need a Type II system.

Since we often want to consider the PLL's operation under small frequency shifts, we need a model with input and output variables in units of frequency. Figure 14.10 shows how we can add integration and differentiation to the input and output (b), respectively, to use the PLL model with inputs and outputs with units of frequency. Using the commutative property of LTI systems, the order of the blocks can be rearranged (c). Clearly the differentiation undoes the integration and we are left with the original model in part (d). This allows us to use the previously derived transfer function to contemplate small variations in reference frequency.

The continuous-time LTI model is an approximation of PLL dynamics. The real PLL is a hybrid system with charge-pump pulses occurring synchronously with the clock signals. The LTI model assumes we are applying an average current to the loop filter and that subsequent changes in phase of the VCO appear instantly through the divider. A general rule of thumb is that the LTI model is reasonable when the reference clock is at least 10x higher than the bandwidth of the closed-loop system. A second limitation, as noted in Section 14.1.4, is that the gain of the PFD/CP under large frequency changes is not predicted by the same linear model of the PFD we use for phase changes.

Despite the conditions attached to the model, it nevertheless allows us to:

1. Determine if the system is stable when locked.
2. Determine output phase noise due to reference clock phase noise.
3. Determine output phase noise due to oscillator phase noise.

Point 14.2.1 is self-explanatory. Noise modelling is investigated in the next section.

14.2.2   Phase-Noise Modelling of PLLs

The s-domain model developed in the previous section can be used to explore the phase-noise performance of a PLL. Here, $H(s)$ is the transfer function from reference clock phase to output clock phase. This transfer function applies to both deterministic variations in reference clock phase and random fluctuations in phase due to noise. Since the transfer function is low-pass, we see that in-band reference clock phase noise appears at the output. The control voltage of the VCO adjusts the VCO's phase to track the in-band reference clock jitter. If the reference clock is a clean signal, we can tolerate a wide PLL bandwidth. Note as well that $H(s)$ has a dc gain of N. Therefore, in-band phase noise is amplified by N. If we have a large divide ratio and a noisy reference clock we favour a lower PLL bandwidth.

Example 14.1   Consider a reference clock at 100 MHz with an rms random jitter of 10 ps, due to phase noise within a bandwidth of 2 MHz. Assume it is the reference clock for a PLL with a bandwidth of 4 MHz. What is the output clock jitter due to reference clock jitter if $N = 25$ (i.e., $f_{OSC} = 2.5$ GHz)?

Solution
We can answer this question in two ways. In the first, we note that the edges of the reference clock move around with 10 ps jitter. The phase noise/jitter is within the bandwidth of the PLL. Therefore, the VCO tracks this jitter. The output clock jitter is also 10 ps. The jitter is not 25 times worse in an absolute sense. However, in most applications, 10 ps of jitter on a 2.5 GHz clock is worse than in a system operating with a 100 MHz clock, since at 2.5 GHz, 10 ps is a larger fraction of the clock period than at 100 MHz. The second approach to relate input and output jitter is by considering how the input phase noise is amplified by the PLL. Phase noise and jitter are related by:

$$J_{rms} = \frac{T}{2\pi}\sqrt{\int_{f_1}^{f_2} S_\phi(f)df}.$$ (14.38)

For the reference clock, we have:

$$J_{ref,rms} = \frac{T_{ref}}{2\pi}\sqrt{\int_{f_1}^{f_2} S_{\phi,ref}(f)df} = 10 \text{ ps}.$$ (14.39)

Since this phase noise is in band, $S_{\phi,out} = N^2 S_{\phi,ref}$. However, $T_{out} = \frac{T_{ref}}{N}$. Therefore,

$$J_{out,rms} = \frac{T_{out}}{2\pi}\sqrt{\int_{f_1}^{f_2} S_{\phi,out}(f)df},$$ (14.40)

$$= \frac{T_{ref}}{2\pi N}\sqrt{\int_{f_1}^{f_2} N^2 S_{\phi,ref}(f)df},$$ (14.41)

$$= \frac{T_{ref}}{2\pi} \sqrt{\int_{f_1}^{f_2} S_{\phi,ref}(f) df} = 10 \text{ ps.} \tag{14.42}$$

In-band phase noise is amplified by N, but phase noise is relative to a period of 2π. To convert phase noise to jitter, we multiply by the period T, which is N times smaller at the output. In units of time, in-band jitter is unchanged.

While in-band reference clock jitter modulates the VCO and appears as output jitter, in-band VCO phase noise is suppressed. This can be modelled by considering the block diagram in Figure 14.11 where phase noise is added at the output of the VCO. The transfer function from the noise input to the output is:

$$H_n(s) = \frac{\phi_{OSC}}{\phi_n} = \frac{1}{1 + \frac{G_{RC}(s)}{N}}, \tag{14.43}$$

$$H_n(s) = \frac{\frac{K}{NC}s^2}{s^2 + \frac{KR}{N}s + \frac{K}{NC}}. \tag{14.44}$$

$H_n(s)$ is a second-order high-pass system, with the same poles as $H_{RC}(s)$. Figure 14.12 shows $H_n(s)$ and how it shapes the typical VCO phase-noise spectrum. Due to the high-pass nature of $H_n(s)$, phase noise outside of the PLL's bandwidth (f_{HIGH}) is unfiltered. Between the two poles, VCO phase noise sees suppression of 20 dB/dec. Since VCO phase noise decreases at 20 dB/dec, between the two poles the output phase noise is flat. Below the lower pole, VCO phase noise sees suppression of 40 dB/dec, leading to phase noise that decreases at 20 dB/dec. The overall output phase noise is the sum of reference clock and VCO contributions shaped by their respective transfer functions:

$$S_{\phi,tot} = |H_{RC}(s)|^2 S_{\phi,ref} + |H_n(s)|^2 S_{\phi,vco}$$

$$= \left| \frac{\frac{K}{C}(1 + sRC)}{s^2 + \frac{KR}{N}s + \frac{K}{NC}} \right|^2 S_{\phi,ref} + \left| \frac{\frac{K}{NC}s^2}{s^2 + \frac{KR}{N}s + \frac{K}{NC}} \right|^2 S_{\phi,vco}. \tag{14.45}$$

From a phase-noise point of view, we can choose the PLL bandwidth to minimize the integral of (14.45) over some bandwidth. The overall phase-noise PSD is given by:

$$S_{\phi,tot} = \frac{\left((1 + (2\pi f\, RC)^2)\, S_{\phi,ref} + \frac{1}{N^2} (2\pi f)^4\, S_{\phi,vco} \right)}{(\frac{K}{NC} - (2\pi f)^2)^2 + (2\pi f \frac{KR}{N})^2} \frac{K^2}{C^2}. \tag{14.46}$$

In general both the reference clock and the VCO will have frequency ranges where flicker noise is significant. However, in most cases, flicker noise of the VCO will be suppressed, leaving the flicker noise of the reference as the main flicker noise component. In the subsequent analysis, all flicker noise is ignored so as to show simpler relationships. The following phase-noise spectra are assumed:

$$S_{\phi,ref} = \frac{S_{0,\phi,ref}}{f^2}, \tag{14.47}$$

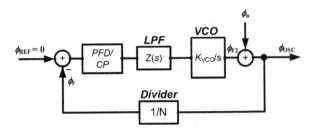

Figure 14.11 PLL block diagram to model VCO phase noise. Phase noise (ϕ_n) is added at the output of the VCO.

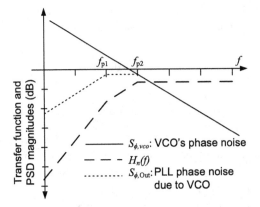

Figure 14.12 Graphical depiction of VCO phase noise contribution in a Type II PLL. The $1/f^3$ region of VCO phase noise has been ignored.

$$S_{\phi,vco} = \frac{S_{0,\phi,vco}}{f^2} = N^2 K_j \frac{S_{0,\phi,ref}}{f^2}, \qquad (14.48)$$

where K_j is a measure of how much noisier the VCO is compared to an equal-jitter reference clock. We expect the higher frequency VCO to be noisier than the reference. As presented in Example 14.1, if it has equal jitter, it would have N^2 times more phase noise, and K_j would be equal to 1. Thus, $K_j > 1$ models the situation where a crystal oscillator is used for the reference for a PLL or where an LC-based PLL is distributed to several ring-based PLLs.

To further simplify the analysis, the PLL, designed with high damping factor, is approximated as a first-order system, giving:

$$H_1 = N \frac{\omega_{bw}}{s + \omega_{bw}} \quad \text{and} \quad H_{n1} = \frac{s}{s + \omega_{bw}}, \qquad (14.49)$$

where H_1 is a first-order approximation of the input/output transfer function of the PLL and H_{n1} is a first-order approximation of the noise transfer function for the VCO's phase noise. With the relation in (14.48) and the simplifications in (14.49) overall phase noise becomes:

$$S_{\phi,tot} = |H_1(s)|^2 S_{\phi,ref} + |H_{n1}(s)|^2 S_{\phi,vco}$$

$$= \frac{S_{0,\phi,ref}}{f^2 |Den(s)|^2} \left(|N\omega_{bw}|^2 + |s|^2 N^2 K_j \right)$$

$$= N^2 \frac{S_{0,\phi,ref}}{f^2} \frac{\omega_{bw}^2 + K_j(2\pi f)^2}{\omega_{bw}^2 + (2\pi f)^2}. \tag{14.50}$$

For the particular case in which $K_j = 1$, (14.50) simplifies further allowing the jitter to be readily calculated as:

$$S_{\phi,tot} = N^2 \frac{S_{0,\phi,ref}}{f^2}, \tag{14.51}$$

$$J_{\phi,rms}^2 = \int_{f_1}^{f_2} S_{\phi,tot} df = \int_{f_1}^{f_2} N^2 \frac{S_{0,\phi,ref}}{f^2} df = -N^2 \frac{S_{0,\phi,ref}}{f} \bigg|_{f_1}^{f_2}, \tag{14.52}$$

$$J_{\phi,rms}^2 = N^2 S_{0,\phi,ref} \left(\frac{1}{f_1} - \frac{1}{f_2} \right). \tag{14.53}$$

From (14.51) to (14.53) we conclude that if $K_j = 1$, the PLL bandwidth has no impact on overall jitter, when we consider the reference and VCO phase. Bandwidth would be selected based on other considerations such as the suppression of VCO modulation from periodic disturbances to the control voltage, charge-pump and loop-filter noise, and loop settling constraints.

Often, the reference clock has lower jitter than the oscillator in the PLL in question, meaning $K_j > 1$. In this case, the simplification done in (14.51) is not possible and (14.50) must be integrated. This can be approached by performing a partial-fraction expansion to (14.50), to give:

$$S_{\phi,tot} = N^2 S_{0,\phi,ref} \left(\frac{1}{f^2} + \frac{(K_j - 1)(2\pi)^2}{\omega_{bw}^2 + (2\pi f)^2} \right). \tag{14.54}$$

This can be integrated to compute jitter:

$$J_{\phi,rms}^2 = \int_{f_1}^{f_2} S_{\phi,tot} df = \int_{f_1}^{f_2} N^2 S_{0,\phi,ref} \left(\frac{1}{f^2} + \frac{(K_j - 1)(2\pi)^2}{\omega_{bw}^2 + (2\pi f)^2} \right) df$$

$$= N^2 S_{0,\phi,ref} \left\{ \left(\frac{1}{f_1} - \frac{1}{f_2} \right) + \frac{K_j - 1}{f_{bw}} \tan^{-1} \left(\frac{f}{f_{bw}} \right) \bigg|_{f_1}^{f_2} \right\}. \tag{14.55}$$

Assuming $f_2 \gg f_1$ the $\frac{1}{f_2}$ term is negligible. Also, if $f_2 \gg f_{bw} \gg f_1$, the $\tan^{-1}$ term reduces to $\frac{\pi}{2}$, further simplifying (14.55) to:

$$J_{\phi,rms}^2 = N^2 S_{0,\phi,ref} \left(\frac{1}{f_1} + \frac{\pi}{2} \frac{K_j - 1}{f_{bw}} \right). \tag{14.56}$$

The conclusions that can be drawn from (14.56) are:

- If $K_j = 1$, (14.56) simplifies to (14.53).
- If $K_j > 1$ the second term adds to overall jitter. Its contribution is minimized by maximizing f_{bw}. This cannot be done indefinitely for reasons of loop stability.
- if $K_j < 1$ the second term subtracts from overall jitter. Therefore, we reduce the jitter by making f_{bw} as small as possible. Intuitively, this makes sense. If the VCO is cleaner then the reference clock we want to filter more reference-clock phase noise. However, shrinking the bandwidth may cause lock-time to increase too much.

14.2.3   Practical Considerations for Charge Pumps

Figure 14.7 (b) showed examples of waveforms for V_{CTRL} for the case of a phase error. When the PLL has settled with zero phase error, V_{CTRL} is constant since no current leaves the charge pump. Even when there is a phase error and V_{CTRL} is updated each reference clock cycle, it increased in smooth steps. In this section, we consider V_{CTRL} waveforms due to the resistor, R, in the loop filter and due to various nonidealities in the PFD/CP. In particular, we consider the effect of:

- A timing offset between Up and Dn pulses.
- A mismatch between the two charge-pump current sources.
- Incomplete settling of charge-pump currents, motivating a minimum duration of charge-pump pulses.

14.2.3.1   **Timing Offsets Between** Up **and** Dn **Pulses**
In Figure 14.13 (a) we see the loop filter composed of a series RC circuit. When there is a net output of the charge pump, each loop filter element develops a corresponding voltage. The voltage across the capacitor integrates i_O, as was shown in Figure 14.7 (b). In addition, as shown in Figure 14.13 (b), a voltage of $I_{CP}R$ develops across the resistor and adds to the integral voltage to form the overall control voltage. When the PLL is locked, we would like to have a constant V_{CTRL}. Various phenomena discussed here and in the following sections can lead to a periodic signal on the control voltage, leading to deterministic jitter in the output clock. To estimate this jitter we must consider both the integral (C) and proportional (R) current-to-voltage conversion of the loop filter.

In Figure 14.13 (c) we show a scenario in which the PLL is locked but the Dn pulse arrives ΔT after the Up pulse. This happens when one signal is driven from the DFF whereas the other is inverted and therefore delayed. In this case, although the charge injected from the Up pulse is equal to the charge removed by the Dn pulse, the net current i_O has a positive pulse and a negative pulse. The resulting control voltage is shown below. Although at the end of the Up/Dn pulses it returns to the voltage before the phase comparison, this brief period of time when V_{CTRL} is disturbed leaves the phase of the VCO advanced. Therefore, the steady-state V_{CTRL} will be slightly lower than it would be without the ΔT mismatch so as to return the VCO's phase to match the reference clocks. More importantly, the periodic signal on V_{CTRL} modulates the VCO, giving rise to deterministic jitter.

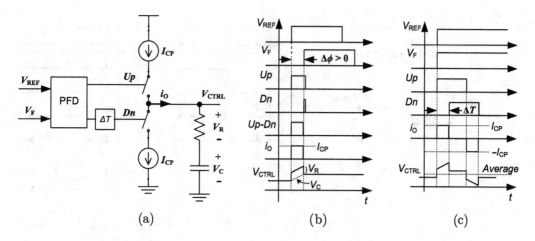

Figure 14.13 Effect of a timing mismatch, ΔT between Up and Dn pulses: (a) schematic with ΔT delay for the Dn pulse; (b) waveforms showing operation of the RC loop filter when there is a phase offset $\Delta \phi > 0$ but $\Delta T = 0$; (c) waveforms for $\Delta \phi = 0$ but with $\Delta T > 0$.

14.2.3.2   Mismatch Between Charge-Pump Current Sources

The situation depicted in Figure 14.14 is similar. Here, the Up current is larger than the Dn current by ΔI_{CP} (Figure 14.14 (a)). If the fed-back and reference clocks are phase-aligned, the Up/Dn pulses are matched in timing, but due to current mismatch, there is an equal-duration net current into the loop filter of ΔI_{CP} that is integrated to give ΔV_{CTRL}. The steady-state waveforms are shown in Figure 14.14 (c). In order to get equal charge each period, the lower magnitude Dn pulse must last a proportionally longer time, accomplished through a static phase error. Assuming a minimum pulse duration of T_{min} the static time offset ΔT is found by equating the short negative charge pump current pulse to the longer (T_{min}) positive pulse. That is:

$$\Delta T I_{CP} = T_{min} \Delta I_{CP}, \tag{14.57}$$

$$\Delta T = \frac{\Delta I_{CP}}{I_{CP}} T_{min}. \tag{14.58}$$

The static phase error is found by normalizing ΔT to the period of oscillation and scaling by 2π:

$$\Delta \phi_{ss} = \frac{\Delta I_{CP}}{I_{CP}} \frac{2\pi T_{min}}{T_{osc}}, \tag{14.59}$$

where $\Delta \phi_{ss}$ is the steady-state phase error between the fed-back signal and the reference. The signal on the control voltage also leads to deterministic jitter.

14.2.3.3   Incomplete Settling of Charge-Pump Currents

Although we have stated that PFD pulses must have a minimum duration, thus far we have not justified this need. Figure 14.15 demonstrates this requirement. In part (a) the Up/Dn currents are shown with exaggerated rise and fall times. Notice that I_{Dn} has the minimum pulse width, but nevertheless rises to reach the peak value of charge-pump

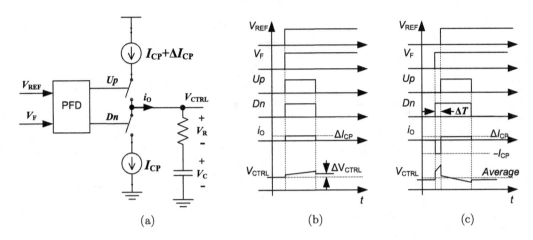

Figure 14.14 Effect of mismatched Up and Dn current: (a) schematic showing the Up current ΔI_{CP} larger than the Dn current; (b) waveforms for $\Delta\phi = 0$; (c) resulting steady-state waveforms.

current, I_{CP}. Therefore, when I_{Up} lasts for ΔT_1 longer, the net charge delivered to the loop filter, denoted by the shaded region, is given by:

$$\Delta q = I_{CP}\Delta T_1. \tag{14.60}$$

Equation (14.60) holds for arbitrarily small ΔT. This gives rise to a constant PFD/CP gain as shown in Figure 14.15 (e), where the amount of charge (q) delivered to the loop filter each cycle grows linearly with ΔT. This is contrary to the case in parts (c) and (d). The Up and Dn currents have been aligned to each other to facilitate their subtraction. In both, I_{Dn}, due to a minimum pulse, rises, but does not reach I_{CP}. The peak value of I_{Up} depends on the time offset. The area of shaded regions shown for small and large offsets should differ by only their duration. However, when the time offset is longer (d), the height of I_{Up} and i_O is larger than in part (c). This leads to a nonlinear phase detector gain. At small time offsets, we have a lower slope than at large time offsets. In the worst case, if Up/Dn pulses are very small in amplitude, for small time offsets we may get no net current into the loop filter, as shown in part (f). This leads to a dead zone in the PFD/CP characteristic, which in turn leads to wander in the VCO's output phase, since phase deviations are only detected/corrected when they grow beyond the horizontal region of the PFD/CP transfer characteristic. This situation can be avoided by adding additional delay to the AND gate in the PFD of Figure 14.5 (a).

14.2.3.4   Finite Output Resistance of the Charge Pump

In the analysis above, we have modelled the charge-pump current as an ideal current source. However, due to the finite output resistance of transistors, we will not have an ideal current source. Two phenomena occur. The first is that, as the control voltage moves toward one extreme, due to the output resistance of the charge pump, one charge-pump current increases and the other decreases. This smaller current must be

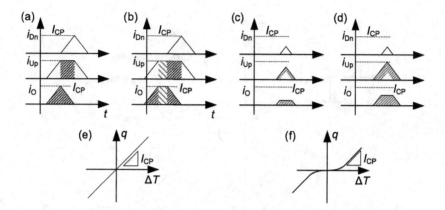

Figure 14.15 Dead zone induced by incomplete settling of charge-pump currents: (a) sufficient T_{min}, Up longer than Dn by ΔT_1. i_O reaches I_{CP}; (b) sufficient T_{min}, Up longer than Dn by ΔT_2. Additional charge to loop filter relative to (a) of $I_{CP}(\Delta T_2 - \Delta T_1)$; (c) insufficient T_{min}, Up longer than Dn by ΔT_1 but also taller. i_O does not reach I_{CP}; (d) insufficient T_{min}, Up longer than Dn by ΔT_2. Additional charge to loop filter relative to (a) less than $I_{CP}(\Delta T_2 - \Delta T_1)$ since $i_O < I_{CP}$; (e) charge-pump response when T_{min} is long enough. Linearly increasing charge delivered to loop filter each reference clock cycle; (f) charge-pump response when T_{min} is too short. Small increases in charge delivered to loop filter for small ΔT, increasing to slope in (e) for larger time offsets.

supplied by a longer pulse leading to a static phase error similar to that shown in Figure 14.14. The second phenomena is that charge can leak from the loop filter in between Up/Dn pulses, also leading to static phase error.

14.2.4   Modified Loop Filters

Any periodic signal on V_{CTRL} due to the various nonidealities above modulates the VCO leading to frequency content at a frequency of $f_{osc} \pm N f_{REF}$, known as reference spurs. To reduce spurs, an extra capacitor (C_P) is often added to the loop filter as shown in Figure 14.16 (a). The extra capacitor gives a loop filter impedance of:

$$Z_p(s) = \left(R + \frac{1}{sC}\right) \| \frac{1}{C_P},$$
(14.61)

$$= \frac{1 + sRC}{s^2 RCC_P + s(C + C_P)}.$$
(14.62)

The magnitude and phase response of this filter's impedance is plotted in Figure 14.16 (b). At low frequency, it acts like a capacitor of $C + C_P$. It has a zero at $\omega_z = -\frac{1}{RC}$ and a pole at $\omega_p = -\frac{1}{RC_{eq}}$ where C_{eq} is the equivalent capacitance of C in series with C_P. Between the zero and the pole, the impedance is set by R. The dashed line shows the impedance of the loop filter without C_P. It is critical that C_P is not too large leading to too small of a distance between ω_p and ω_z.

Qualitatively, the need for sufficiently wide spacing between ω_p and ω_z is explained as follows:

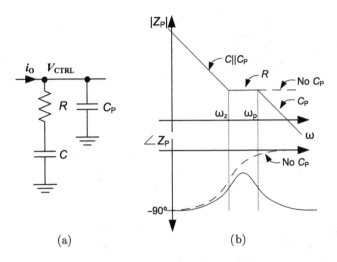

Figure 14.16 Modified loop filter: (a) schematic with additional parallel capacitor, C_P; (b) Bode plot with and without C_P.

- It was noted in Section 14.2.1 that if the loop filter has only capacitor C, the loop gain, $G_C(s)$ has two poles at $s = 0$ and a phase of $-180°$, leading to no phase margin.
- The series R was added to allow the loop gain, $G_{RC}(s)$ to reach a phase of $-90°$ at frequencies larger than ω_z. Note that at $\omega = \omega_z$, $G_{RC}(s)$ has a phase of $-135°$.
- If the open-loop unity gain frequency, ω_T is much larger than ω_z, the phase margin can be almost $90°$. If $\omega_T = \omega_z$, we have only $45°$ of phase margin.
- In Figure 14.16 (b) we see that the phase of the loop filter's impedance does not reach $0°$, unless the pole and zero move very far apart, thus the loop gain's phase margin will not reach $90°$.

For the case of $C_P < 0.1C$, $\omega_p = 11\omega_z$, the phase of the loop filter's impedance reaches a peak of $-33.6°$, giving a maximum phase margin of $56°$, assuming the unity-gain frequency is exactly at the geometric mean of the zero and pole frequencies. To enable reasonable phase margin in the presence of variation in the PLL constants such as charge-pump current and VCO gain, we need $C > 10C_P$.

14.2.5   Split Loop Filters and VCO Tuning Ports

In the RC loop filter shown in Figure 14.17 (a) we see that the control signal, V_{CTRL}, is the sum of the voltage across the capacitor, V_C, and the voltage across the resistor, V_R. This total control voltage can be converted to a tuning current, i_{Tune}, in the VCO using a single transconductor, G_m. However, we can convert these voltages separately. In Figure 14.17 (b) each voltage is applied to a separate V–I converter whose outputs are summed to form i_{Tune}. In part (c) the tuning paths are further split to include two charge pumps. In this case the combined PFD/CP/LF/VCO gain is:

$$G_{Split}(s) = \frac{K_{i,VCO}}{2\pi s} \left(I_{CP_P} R G_{mP} + \frac{I_{CP_I} G_{mI}}{sC} \right), \tag{14.63}$$

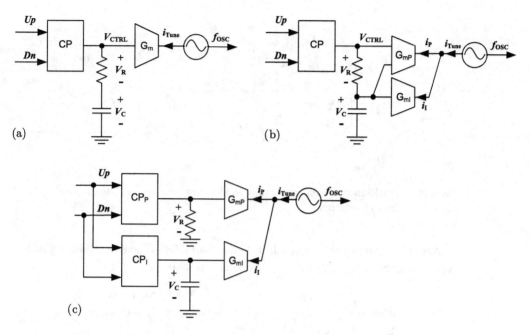

Figure 14.17 Split loop filters and VCO tuning ports: (a) conventional single charge pump with a series RC loop filter and a single V-to-I converter; (b) summing the effect of the series RC in the current domain using two V-to-I converters; and (c) separate charge pumps, loop filter components and V-to-I converters.

where $K_{i,VCO}$ is the VCO gain in units of rads/s per A and I_{CP_P} and I_{CP_I} are the charge pump currents for CP_P and CP_I, respectively. The voltage developed across the resistor will be proportional to the phase difference, motivating the subscript P while the voltage across the capacitor is the integral of phase errors, motivating the subscript I. Thus, G_{mP} and G_{mI} are the proportional and integral path transconductors that generate currents i_P and i_I, respectively. These sum to produce an overall current i_{Tune}.

Splitting the tuning path allows for more degrees of freedom in the design. It also allows easier tuning of loop dynamics. For example, the zero, that was simply $\omega_z = \frac{1}{RC}$ for the system in Figure 14.17 (a) becomes:

$$\omega_z = \frac{1}{RC} \frac{I_{CP_I} G_{m_i}}{I_{CP_p} G_{m_p}}. \tag{14.64}$$

Adjustment of the ratio of charge-pump current or transconductance can easily tune the zero location that would be subject to the at least $\pm 10\%$ variation of each passive component.

The proportional path is responsible setting for the bandwidth. Equation (14.37) is modified to give:

$$\omega_{bw} \approx R \frac{I_{CP_P} G_{mP} K_{i,VCO}}{2\pi N}. \tag{14.65}$$

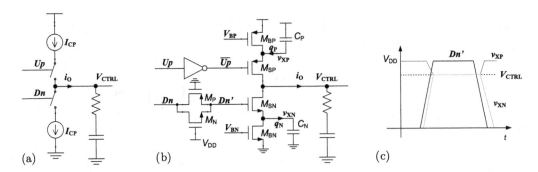

Figure 14.18 Single-ended charge pump: (a) with ideal components; (b) transistor-level schematic; and (c) waveforms.

Another feature of the approach in Figure 14.17 (c) is increased linearity of the proportional path V-to-I converter, as explained as follows:

- In part (a), V_{CTRL} moves up and down as the VCO's frequency is tuned.
- Due to nonlinearity in the transconductor, the value of G_m and hence the bandwidth of the PLL will change as the VCO is tuned.
- With two tuning paths in (b) and (c), V_C moves up and down as the VCO's frequency is tuned, but in steady-state operation, V_R remains close to 0 (assuming differential operation in (b)).
- Thus, bandwidth, now dependent only on G_{mP}, is no longer sensitive to G_m nonlinearity, since the input to G_{mP} remains close to 0.

14.2.6   Charge-Pump Circuits

A single-ended charge pump is shown in Figure 14.18. Part (a) shows the circuit with ideal switches and current sources while part (b) gives transistor-level details. Gate voltages V_{BP} and V_{BN}, generated by current mirrors, set the charge pump currents. Since the switch M_{SP} is a PMOS transistor, conducting when its gate is pulled low, an inverter generates $\overline{Up}$ from Up. To match the delay of the signal applied to M_{SN}, the signal Dn is delayed by an always-on pass-gate to produce Dn'.

A disadvantage of the single-ended design is that charge sharing between the output and the intermediate nodes v_{XN} and v_{XP} leads to static phase offset in the PLL. To illustrate this, consider the waveforms in part (c). Here, V_{CTRL} is above $V_{DD}/2$. When Dn' is low, the node v_{XN} has been discharged to ground by M_{BN}. When Dn' is high, the switch has high conductance and v_{XN} rises drawing a charge q_N from the output node. Likewise, when $\overline{Up}$ is low, v_{XP} reduces from V_{DD}, sending a charge q_P to the output node. A net charge q_{Tot} is sent to the loop filter. These charges are:

$$q_N = C_N \Delta v_{XN}, \tag{14.66}$$

$$q_P = C_P \Delta v_{XP}, \tag{14.67}$$

$$q_{Tot} = q_P - q_N, \tag{14.68}$$

where C_P and C_N are the parasitic capacitance at nodes v_{XP} and v_{XN}, respectively. These charges are independent of pulse width and are redistributed even when the PLL is locked, with minimum pulse widths. When these charges are shared with the loop filter, the resulting voltage change is much smaller owing to the much larger loop filter capacitor, but they nonetheless disturb V_{CTRL}. Therefore, the charge pump is designed to minimize q_{Tot}. If $C_P = C_N = C_X$, achieved by matching transistor dimensions in the PMOS and NMOS transistors, the total charge becomes:

$$q_{Tot} = q_P - q_N = C_X \left(\Delta v_{XP} - \Delta v_{XN} \right). \tag{14.69}$$

The waveforms in part (c) are drawn assuming $M_{SN/P}$ are in deep triode with small on-conductance such that $v_{XN/P} \rightarrow V_{CTRL}$ when the switches conduct. In this case, the total charge becomes:

$$q_{Tot} = C_X \left(V_{DD} - 2V_{CTRL} \right), \tag{14.70}$$

which is non-zero except when $V_{CTRL} = V_{DD}/2$.

The effect of the extra charge added or subtracted each reference clock cycle, as predicted by (14.70), is explained as follows. Suppose the VCO tuning curve requires a particular V_{CTRL} assumed to be less than $V_{DD}/2$. Therefore $q_{Tot} = C_X \left(V_{DD} - 2V_{CTRL} \right)$ is added to the loop filter capacitor each reference clock cycle, increasing V_{CTRL}. To prevent this, the Dn pulse must be extended by ΔT so as to remove an equal amount of charge, giving rise to:

$$I_{CP}\Delta T = q_{Tot} = C_X \left(V_{DD} - 2V_{CTRL} \right). \tag{14.71}$$

This corresponds to a static phase offset ($\Delta \phi_{ss}$) at the phase detector of:

$$\Delta \phi_{ss} = 2\pi \, \Delta T f_{ref}, \tag{14.72}$$

$$= 2\pi f_{ref} \frac{C_X \left(V_{DD} - 2V_{CTRL} \right)}{I_{CP}}. \tag{14.73}$$

This error can be mitigated using a differential charge-pump circuit. Since the error predicted in (14.73) is proportional to the voltage excursion (relative to $V_{DD}/2$) it acts similarly to finite output resistance of the charge pump.

Figure 14.19 shows a differential charge pump. Signals $\overline{Up}$ and Dn' direct current to the loop filter, while Up' and $\overline{Dn}$ direct current to a dummy node (V_{REF}) when the charge-pump pulses are inactive. If the switching transistors operate in saturation, v_{XN} is the same when Dn' or $\overline{Dn}$ is high, eliminating the charge sharing seen in the single-ended example. However, low (high) V_{CTRL} will push M_{SN} (M_{SP}) into triode. If we make the output voltages of the differential structure equal to one another, the symmetry of the circuit keeps $v_{XN/P}$ constant even if the switches are in triode. This is accomplished by driving V_{REF} to V_{CTRL} using a buffer. If V_{REF} is left floating, it will be driven in the opposite direction as V_{CTRL}. For example, if V_{CTRL} is increased by $\overline{Up}$ lasting longer than Dn', $\overline{Dn}$ is therefore shorter in duration than Up', pulling V_{REF} low. This circuit appears to have no common-mode feedback to compensate for mismatch in the currents conducted by $M_{BP/N}$. However, any mismatch results in current into

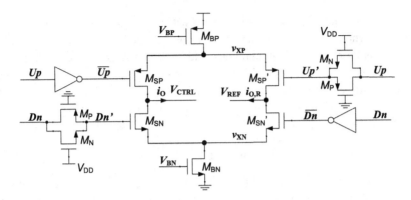

Figure 14.19 Differential charge-pump circuit.

the output of the buffer, and a static phase error in the VCO which, when detected by the phase detector, provides feedback.

The disadvantage of the differential structure is that it consumes dc power, unlike the single-ended version which only drew current on pulses. Since in lock, pulses have a very short duty cycle, the differential charge pump will dissipate significantly more power. However, even a differential charge pump dissipates less power than each of the VCO and divider, meaning that the increase in power dissipation of the charge pump is a small relative increase in overall power dissipation.

14.3   Divider Implementation

A DFF connected with $D = \overline{Q}$ acts as a divide-by-two block. Higher exponents of two can be realized by cascading circuits of this type. For example, a divide by 16 can be implemented with four of these circuits as shown in Figure 14.20. The advantage of this approach is that high-speed DFFs can operate at high frequency. The disadvantage of this approach is that power-supply-induced delay variation accumulates through the circuit, since the output of one divider acts as the clock input of the next. For example, each edge of v_F arrives after four τ_{CLK-Q} delays. If each divider has a delay variation of ΔT, the four-stage divider will have a delay variation of $4\Delta T$. This problem can solved by retiming the divider's output using a DFF clocked by v_{OSC}.

To generate divide ratios not of the form 2^n, a counter-based approach can be used in which edges of v_{OSC} increment a counter. The output of the counter is compared against the divide ratio. When it is reached, an output edge is toggled. A PLL that divides by an integer value of N is known as an integer-N PLL. Its output frequency is constrained to be integer multiples of the reference clock. This is usually sufficient for wireline applications. However, if finer steps are needed, a fractional-N scheme can be used. By varying the divide ratio between two consecutive integers, (e.g., 63 and 64), fractional dividing ratios are possible. This only works if the rate by which the divide ratio is switched is higher than the PLL's bandwidth, otherwise the PLL will pull the VCO back and forth between the two divide ratios.

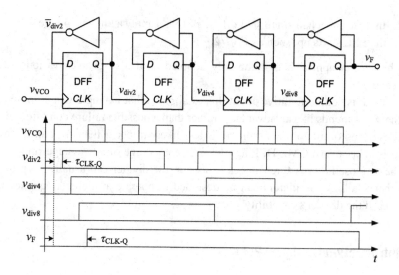

Figure 14.20 Feedback divider constructed using cascaded divide-by-two circuits. Schematic (top) and waveforms (bottom).

A pitfall in divider design occurs if the frequency range of the VCO exceeds the divider's. This is illustrated in Example 14.2.

Example 14.2 Consider a PLL with 200 MHz reference clock and a feedback divider of $N = 32$. The VCO can tune as high as 8 GHz while the divider works up to 7.5 GHz. Is this a problem? What can go wrong?

Solution

If no extra circuitry is included to force V_{CTRL} to a particular voltage we cannot make any assumptions about its value when the PLL powers on. If it were to power on at the top of the tuning range, the VCO would operate at 8 GHz. Under this condition, we want the fed-back clock, operating at $\frac{8 \text{ GHz}}{32} = 250$ MHz to be compared against the reference clock. Since the fed-back clock is at a much higher frequency than the reference clock, we expect the PFD's Dn pulse to be asserted for longer intervals than the Up pulse. However, for this to occur, edges of the fed-back clock must arrive at the PFD. In this case, if the divider fails to operate at such a high frequency, no edges will arrive at the PFD. With only reference clock edges, the Up pulse will be continuously asserted, giving the opposite signal to the charge pump to what we want. The VCO will be kept at the top of the tuning range.

There are two solutions that can help address the problem outlined in Example 14.2:

1. Ensure that the ends of the operation range of the divider exceed the lower and upper end of the VCO's tuning range.

2. Include a mechanism to shut off the PFD and apply a known voltage to V_{CTRL} when the PLL starts up, such as $V_{DD}/2$.

The second approach is commonly adopted as it also enables various calibration schemes.

What is counterintuitive about this example is that having a wide tuning range for the VCO sounds like an advantage, rather than a possible failure condition. If the PLL uses a ring oscillator, this scenario is less likely to occur than with an LC oscillator. This is because a CML DFF has the frequency range of a two-stage ring. So long as the ring oscillator is three stages or longer, the divider can usually keep up. On the other hand, LC oscillators can be designed to operate at higher frequency and may exceed the divider's capability.

14.4   Introduction to Digital PLLs

Phase-locked loop designers and the system architects who floorplan wireline and wireless transceivers have long bemoaned the die area of passive components, with the loop filter capacitor being the largest, even larger than the inductor in an LC VCO. Loop filter capacitors also present reliability challenges, since the probability of a pinhole in the oxide increases with capacitor area. Even non-catastrophic capacitor leakage causes static phase error. The analog behaviour of PLLs presents device modelling challenges as well. As each new CMOS technology is developed, circuits are being designed before the process is mature. The more "analog" the circuits, the more the designers must hit a moving target by designing high-performance circuits with varying device models. Even as a technology matures, if most of the products it supports are purely digital, analog structures, such as current mirrors or high-gain opamps, that rely on accurate modelling of the transistors will be the first to suffer a yield reduction if the process drifts.

In the early 2000s, PLL designers started developing digital PLLs for both wireless and wireline applications. With a knowledge of analog PLLs a designer needs to do the following to convert it to a digital PLL:

1. The voltage-controlled oscillator is modified to make it a *digitally* controlled oscillator (DCO).
2. The phase detector must output signals compatible with standard digital approaches.
3. The analog loop filter is replaced with a digital filter whose output is applied to the DCO.

These modifications are discussed one by one below.

14.4.1   Converting a VCO to a DCO

In the case of a ring oscillator, this conversion amounts to replacing whatever transistor or transistors that were tuned gradually by V_{CTRL} with a collection of smaller, parallel transistors that are enabled or disabled using digital signals. For example, the analog control of V_{PM} in Figure 13.17 is replaced by a parallel array of PFETs, where each gate is raised to V_{DD} to lower the frequency or pulled to GND to increase the frequency. In general, the total width of the digitally controlled PFETs will be approximately equal to that of the original PFET. Meanwhile, LC oscillators are tuned digitally by subdividing the tuning varactors into smaller varactors and similarly controlling their tuning voltages using $[V_{DD}, GND]$ rather than a continuously variable voltage.

Noise that is coupled into the loop filter of an analog PLL can modulate the VCO. We have already discussed periodic disturbances to V_{CTRL} stemming from several charge-pump nonidealities. However, we can also have power-supply noise or substrate coupling. Given the high gain of the VCO, this can induce deterministic jitter. In a DCO, each control bit fully switches an element. Once a given control voltage is switched, noise on that control voltage sees a very low gain. For example, in an analog LC PLL, at least one varactor is biased such that variations in tuning voltage lead to variations in C and hence frequency. In its digital counterpart, all varactors are fully switched, meaning that small variations in digital tuning signal voltages do not change capacitance by nearly as much. Hence, digital control is less noise sensitive. However, since control of the DCO is associated with discrete steps in tuning, the output phase will have quantization effects, discussed in Section 14.4.3.

14.4.2   Phase Detection in Digital PLLs

Similar to their analog counterparts, digital PLLs either use a nonlinear phase detector (bang-bang) or linear time-to-digital converters (TDC). The operation of the nonlinear phase detector in the digital PLL context is the same as that shown in Figure 14.2. This scheme gives a 1-bit signal to the loop filter that can be filtered over several reference clock cycles and supplied to the DCO as a multi-bit signal.

On the other hand, TDCs produce a multi-bit estimate of phase error.

14.4.3   All-Digital Loop-Filter Design

Digital loop filters convert the low-resolution phase detector output to a higher resolution DCO control, typically using proportional and integral paths. Broadly speaking, digital PLLs draw on many of the techniques used in oversampled data converters that allow quantization noise to be reduced in a signal band of interest.

14.5   Injection Locking and Injection-Locked Oscillators

An increasingly important technique in wireline systems is the use of injection-locked oscillators (ILOs). In Figure 14.21 a four-stage ring oscillator receives a signal v_{INJ} from outside the ring. If $v_{INJ} = 0$, the oscillator operates at its free-running frequency (f_0). Its four outputs, if taken with both polarities, give eight phases separated by 45°. If $v_{INJ} = A_{inj} \cos(\omega_{INJ} t)$, the oscillator is pulled from f_0 to f_{INJ}, provided the two

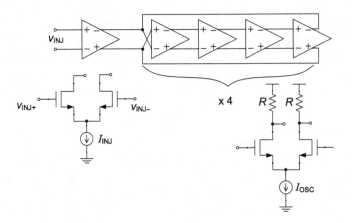

Figure 14.21 Ring-based injection-locked oscillator.

signals are close enough in frequency. The phase offset between the ILO's output and the input signal is determined by the offset between f_{INJ} and f_0, as well as the relative "strength" of the injection signal, expressed as the ratio I_{INJ}/I_{OSC}. Thus we can tune f_0 away from f_{INJ} in order to tune the static phase difference between the signals. This is useful in a clock-forwarded link.

Even when there is a static phase offset between the injection signal and the ILO's output, the phase of the ILO tracks the phase *changes* of the reference clock, within a bandwidth, known as the ILO's jitter tracking bandwidth (JTB). This means that the phase noise of the ILO is suppressed down to the level of the reference clock's within this bandwidth. This feature allows the use of a relatively noisy ring ILO for multiphase generation so long as it is driven by a clean clock, probably generated by an LC PLL. The power dissipation of this PLL is amortized over multiple transmit and receive lanes.

A thorough treatment of ILOs for clock buffering and deskew is found in [91]. Some key relationships are summarized as follows. The ILO can lock to the injection signal only within a lock range given by:

$$\omega_{LOCK} = \frac{K}{A\sqrt{1 - K^2}}, \tag{14.74}$$

where $K = \frac{I_{INJ}}{I_{OSC}}$ and $A \approx \frac{n}{2\omega_0} \sin\left(\frac{2\pi}{n}\right)$. Here, A models the frequency to phase conversion of an n-stage ring oscillator. The steady-state phase offset between a small amplitude injection signal and the oscillator's output is:

$$\theta_{ss} = \sin^{-1}\left(\frac{K}{A}\Delta\omega\right), \tag{14.75}$$

where $\Delta\omega = \omega_0 - \omega_{INJ}$. The jitter characteristics of an ILO are modelled as a first-order PLL, meaning that up to a bandwidth ω_P, oscillator phase noise is filtered and

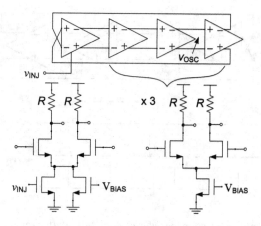

Figure 14.22 Ring-based injection-locked oscillator with $\omega_{OSC} = \omega_{INJ}/2$.

the injection signal's phase noise is transferred to the output. This bandwidth, referred to as the ILO's jitter tracking bandwidth, can be expressed in the following forms:

$$\omega_P = \sqrt{\frac{K^2}{A^2} - \Delta\omega^2} = \frac{K}{A}\cos\theta_{SS}. \qquad (14.76)$$

From (14.76) we conclude that if we want oscillator phase-noise suppression, we should ensure that θ_{SS} is kept away from $\pm90°$, otherwise the JTB drops to zero and oscillator phase noise is not suppressed. Although such deskew angles are achievable, the ILO's phase-noise performance will be degraded.

The ILO in Figure 14.21 locks to an injection signal near the oscillator's free-running frequency. That is, the ILO locks to the first harmonic of the injection signal; ILOs can also lock to harmonics and sub-harmonics of the injection signal. For example, if the injection signal is fed into the tail current source of the differential pair as in Figure 14.22, $\omega_{OSC} = \omega_{INJ}/2$. In this circuit, three of the four stages are conventional differential pairs. In the fourth, the current source is split into two branches. The larger branch is controlled by a fixed bias voltage, V_{BIAS}, while the smaller receives the injection signal. To understand why this works, we recall the signal at the tail node of a differential pair. Under small-signal excitation, the tail node is a virtual ground, although as the amplitude of a differential pair grows to the level typical in a ring oscillator, the tail node rises for large differential inputs. That is, it rises and falls twice per cycle (i.e., at the second harmonic). Thus, the tail node has a strong second harmonic component as does current flowing through the current source's output resistance. By feeding in a second harmonic we can push the zero crossings (phase) of the ring, giving us a dividing ILO.

Figure 14.23 shows the operation of a multiplying ILO. If pulses extracted from injection-signal edges occurring every j^{th} oscillator period are injected into an oscillator, the oscillator can be locked to the jth harmonic of the injection signal [92]. These short pulses are applied to a MOSFET switch across the differential output of one ILO

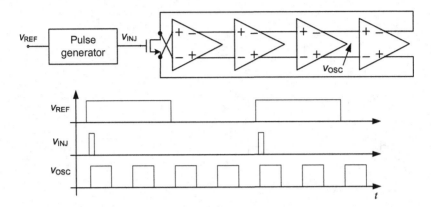

Figure 14.23 Ring-based multiplying injection locked oscillator (MILO): (top) circuit schematics; (bottom) representative waveforms.

stage. This shorts the outputs together every j^{th} oscillator period thereby resetting the phase of the ILO (every j^{th} oscillator period) to match that of the injection signal.

14.6   Summary

We began this chapter with the relationship between frequency and phase, stating that angular frequency is the rate of change (derivative) of phase. Methods for comparing the phase of two periodic signals were considered, starting with the nonlinear comparison offered by a DFF. The XOR gate or a multiplier is promising for high-speed operation, but will not support large frequency errors. The DFF-based phase-frequency detector (PFD) was linear for phase offsets and nonlinear for frequency differences, making it a common choice.

Phase detectors and charge pumps give time-varying signals. The notion that it is the time-averaged output that is relevant in modelling the circuits was introduced. However, this is only valid when the dynamics of the system are slow. We gave the rule of thumb that linear time-invariant modelling is appropriate when the bandwidth of the PLL is no higher than 10% of the PLL's reference frequency.

A typical, Type II analog PLL consisting of a PFD, charge pump, loop filter, VCO and divider was modelled. This modelling showed that if a loop filter only has a capacitor, the closed-loop response has $\zeta = 0$. Thus a resistor was added in series. Unlike all-pole responses that are often designed for $\zeta = 1/\sqrt{2}$, to avoid any peaking in the PLL's response, we usually choose a large ζ. In this case, the bandwidth is set by the higher-frequency pole. The LTI model allowed us to consider both an input to output transfer function and a VCO phase noise to output transfer function. We saw that when we have a noisy VCO, wider loop bandwidth is preferred and when the VCO has less phase noise than the reference we prefer a narrow loop bandwidth. The term "noisy" was made more precise in the context of references and VCOs at different frequencies, i.e., when the divide ratio is not equal to 1.

Various nonideal scenarios were considered: charge-pump mismatch, timing offset, finite charge-pump output resistance and slow settling of charge-pump currents.

Charge-pump and divider implementations were given. In the case of the divider, we saw that its output can show a large cumulative power-supply-induced delay variation when we cascade many divide-by-two circuits.

With digital PLLs becoming commonplace, the modifications needed to convert an analog PLL to a digital PLL were briefly introduced.

Ring-based injection-locked oscillators were presented. These circuits are becoming commonplace as low-jitter signals from an LC PLL are distributed around a chip. Though smaller than an LC PLL, a ring-oscillator-based PLL may still have large passive components in its loop filter. An ILO provides first-order PLL behaviour without an explicit loop filter. It operation was described, giving expressions for its lock range and its jitter tracking bandwidth.

Problems

14.1   Compare the design choice of a 1 MHz vs 10 MHz PLL bandwidth from the point of view of jitter considering phase-noise contributions from the reference and VCO. Assume a noise parameter $K_j = 5$ when looking at (14.56). Use $f_1 = 200$ kHz.

14.2   Consider the phase-locked loop shown in Figure 14.7 (a). The frequency of the oscillator is given by:

$$f_{osc} = 16V_{CTRL}^2 \qquad (14.77)$$

for a range of V_{CTRL} from 0.25 to 0.75 V. $I_{CP} = 100\mu A$. $f_{REF} = 250$ MHz, $N = 16$.

(a)   Choose suitable components (and their values) to implement the filter $Z(s)$. Choose a bandwidth that maximizes VCO phase-noise suppression while keeping the bandwidth low enough that the LTI assumptions of the PLL model are valid across the tuning range.

(b)   Give the poles, zeros and damping factor of the transfer function given the element values you have chosen. Consider the upper and lower ends of the tuning range.

(c)   Draw a sketch of the VCO's contribution to PLL output phase noise when operating at the lower end of the tuning range, taking into consideration a typical VCO phase-noise behaviour (thermal noise only) and the noise transfer function from the VCO to the output. Clearly label all relevant frequencies.

14.3   Assume a Type II PLL is designed with $N = 20$ and $\zeta = 3$. If a lower reference frequency is used with the same target VCO output frequency, raising N to 40, explain what happens to the damping factor, poles, zeros and PLL bandwidth. What happens to the phase margin, assuming no other changes to loop parameters? If the designer does not like this new PLL response, what other parameter can be adjusted to return the PLL to its original ζ?

14.4   Design a ring-based CML ILO. Compare its jitter tracking bandwidth when $\Delta\omega = 0$ to that predicted by theory. How much difference between the injection frequency and the free-running frequency is needed to cause a 45° steady-state phase offset?

15 Clock and Data Recovery

With PLLs and ILOs introduced in Chapter 14, this chapter presents the systems that synchronize clocks to incoming data, known as clock and data recovery (CDR) systems. The chapter starts with an introduction and discussion of the metrics of CDRs. Phase detection is done differently in CDRs compared to PLLs. This is explained before the most common approaches are described. Several options are available to the designer for how phase comparisons should be acted on. These are presented and compared next. The chapter continues with an introduction to baud-rate phase detection schemes built on Mueller–Muller phase detection.

15.1   CDR Introduction and Metrics

Clock and data recovery refers to the control loop that synchronizes a clock signal to an incoming data stream to allow sampling of the data where the eye is most open. An example of this is shown in Figure 15.1. The receiver consists of an analog front-end (AFE) and a decision circuit. The output of the AFE is applied to the clock recovery block. This block compares the phase of the data stream to a clock signal with the objective of locking the clock to the data stream. This recovered clock is then used to sample the data. The sampling and regeneration of the data is the data recovery aspect of the CDR. The phase alignment of the clock is the clock recovery aspect. In some instances the recovered clock is then used to relaunch the data at a high data rate. In other cases, once received, the recovered data stream is demultiplexed to a lower data rate.

Some CDRs use an external reference clock. In this case, the output of the phase detector is used to control the phase of the reference clock. This approach requires a PLL to generate the reference clock. When a chip supports several parallel lanes, the PLL can be shared among the lanes. The PLL shown in Figure 1.4 (b) can supply clock signals to multiple transmitters, shown in Figure 1.4 (c), and receivers (a).

Without an external reference, the CDR controls the phase/frequency of a local oscillator. These so-called reference-less CDRs are common when a chip supports only one data lane. Reference-less CDRs are similar to PLLs, having charge pumps and loop filters when implemented in the analog domain.

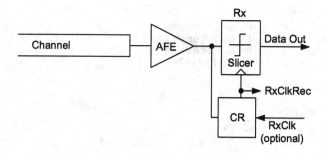

Figure 15.1 High-level block diagram of a clock and data recovery system.

The jitter performance of CDRs is described by three metrics:

1. Jitter tolerance;
2. Jitter transfer; and
3. Jitter generation.

Each of these is described in the following subsections.

15.1.1   Jitter Tolerance

Jitter tolerance is a measure of the maximum peak-to-peak sinusoidal jitter imposed on the data that the CDR can track with BER less than some specified threshold. An example of jitter tolerance is shown in Figure 15.2. To measure jitter tolerance, the clock signal used to launch data is modulated as:

$$v_{clk}(t) = A\sin(2\pi f_b t + \pi S_{J_{PP}}\sin(2\pi f_j t)), \qquad (15.1)$$

where f_b is the data rate, $S_{J_{PP}}$ is the peak-to-peak jitter added to the data in units of UI and f_j is the frequency of the jitter added to the data. Measurements are made by sweeping f_j and, for each f_j, finding the maximum $S_{J_{PP}}$ at which the BER is better than a predefined level.

Figure 15.2 shows measured results for a system's four receivers. The dynamics of a CDR can be similar to those of a PLL. Tolerating jitter means that the CDR's clock tracks input jitter. Therefore, we expect tracking up to the bandwidth of the system. This bandwidth corresponds to the frequency at which the decreasing jitter tolerance levels off. In the measurements, this occurs at approximately 20–30 MHz. Within this bandwidth, the CDR fails if the maximum *rate-of-change* of jitter (denoted by $\frac{dj}{dt}_{max}$) is too high. This is found by differentiating the jitter added in 15.1, simplified as:

$$\frac{dj}{dt}_{max} = \pi S_{J_{PP}} f_j. \qquad (15.2)$$

The slope of the jitter tolerance below the CDR's bandwidth is such that $S_{J_{PP}} \propto \frac{1}{f_j}$. Above the CDR bandwidth, jitter tolerance is not zero, simply because the eye has some horizontal opening within which the data can move relative to the sampling clock. Figure 15.2 also shows jitter tolerance requirements of the communication

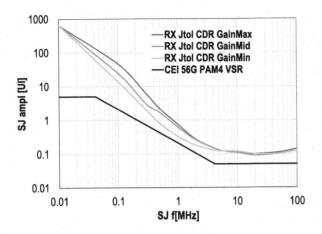

Figure 15.2 Measured jitter tolerance. Bottom trace shows the minimum requirement from the targeted standard (CEI 56G PAM4 VSR). Jitter tolerance of the receiver lane for different CDR gain settings are shown above it. Maximum gain has the highest jitter tolerance. ©2019 IEEE. Reprinted, with permission, from [3].

standard, shown in black below the measured performance, referred to here as the template. It is sometimes referred to as the mask.

15.1.2 Jitter Transfer

Jitter transfer refers to the transfer function from input data jitter to recovered clock jitter. In the optical link transceiver example in Figure 1.6 the recovered clock from the input data on the optical side relaunches the data on the electrical side. Within the CDR bandwidth, jitter on the optical data will be translated to jitter on the recovered clock ultimately appearing on the relaunched data, along with jitter generated in the transceiver. Any peaking in the jitter transfer characteristic of the CDR above unity gain will amplify jitter around a certain frequency. In applications where several CDRs are used in cascaded repeaters, peaking in the CDR's response cannot be tolerated. Therefore, we will scrutinize the jitter transfer of CDRs, both in terms of bandwidth and peaking.

15.1.3 Jitter Generation

The jitter in the Rx CLK when jitter-free data are applied corresponds to the jitter generation of the CDR. When the CDR has a local oscillator, the feedback loop can suppress LO phase noise within the loop bandwidth, as was the case in a PLL.

15.2 Phase Detection

Given the emphasis on digital PLLs over the last two decades, this section will focus on phase detection approaches that are amenable to digital implementation. The main

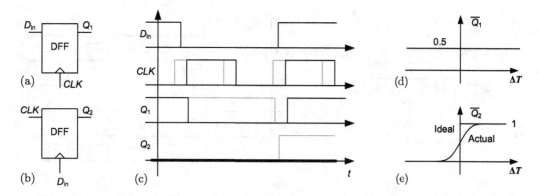

Figure 15.3 A D flip-flop as a phase detector in a CDR: (a) configuration in which the clock we are aligning drives the clock port of the DFF; (b) configuration in which the data stream drives the clock port of the DFF; (c) example waveforms in which the grey (black) clock edge arrives before (after) the data edge; (d) average output $\overline{Q}_1$ for the configuration in (a); and (e) average output $\overline{Q}_2$ for the configuration in (b).

challenge in the design of phase detectors for CDRs is that the data stream is not periodic. Whereas phase detectors in PLLs compare two periodic signals, in a CDR the phase detector compares a periodic clock signal and a non-periodic data signal. Since we represent data using non-return-to-zero modulation, at most, there is a transition once per UI. For example, when "10101010...." is transmitted, the data look like a half-rate clock. A typical data pattern will have CIDs, leading to fewer transitions in the data stream.

15.2.1   DFF as a Phase Detector

The simplest method of phase detection is to use a single DFF. Consider the two options for connection shown in Figure 15.3 (a) and (b). If the input data (D_{in}) are fed to the data port and the DFF is clocked using the full-rate clock as in (a), the output Q_1 will capture the data. Changes in clock phase only change which UI is sampled by a given edge. The average value of the DFF's output $\overline{Q}_1$ is simply the average value of the data stream. For balanced data $\overline{Q}_1 = 0.5$, regardless of the clock phase (time offset ΔT between clock and data edges).

In Figure 15.3 (b) the connections to the DFF are reversed. Now the data stream (D_{in}) is used to sample the full-rate clock. In this case, if the data edge arrives before the rising edge of the clock (black clock trace), $Q_2 = 0$, whereas if the data edge arrives after the rising edge of the clock (grey clock), $Q_2 = 1$ (also shown in grey). This gives rise to the transfer characteristic shown in Figure 15.3 (e). The ideal operation is a step at $\Delta T = 0$. However, due to metastability and clock jitter, the actual transition from 0 to 1 is smoother. This approach can be used to lock the clock edge to the data edges. One would then use the falling edge of the clock to sample the input data, since this edge will be aligned with the middle of the data eye. The disadvantage of this approach is that in the absence of data edges, the previous measure of phase

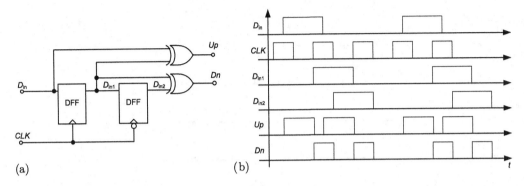

Figure 15.4 Hogge phase detector: (a) implementation; (b) waveforms when rising edge of the clock lags behind the middle of the data eye.

error is held, which can push the phase of the sampling clock too far. The other issue with this scheme is that it requires a large enough data signal to correctly operate the DFF's clock port. Whether CML- or sense-amplifier-based, DFFs can resolve smaller signals on the D input provided the clock signal is large. However, the clock signal cannot be shrunk as much. This approach would require more gain on the receive path than an approach that applied the input data to only the D input of DFFs.

15.2.2   Linear Phase Detection: Hogge Phase Detector

A Hogge phase detector, shown in Figure 15.4 (a) is a linear phase detector suitable for random data. This implementation, designed for a full-rate clock, consists of two DFFs with the second one being negative edge-triggered. Here, D_{in} and the output of the first DFF (D_{in1}) are applied to an XOR gate, generating Up. As presented in Figure 15.4 (b), whenever there is a data transition, $D_{in} \neq D_{in1}$ before the clock's edge and $D_{in} = D_{in1}$ after the clock edge. Therefore, Up is a pulse whose width is equal to the time difference between the D_{in} edge and the clock edge. Note that DFF timing is considered later on. On the rising edge of the clock, D_{in1} changes. Therefore a bit transition leads to $D_{in1} \neq D_{in2}$ until D_{in1} is clocked through to D_{in2}. Since the second DFF is negative edge-triggered, D_{in2} changes $\frac{1}{2}$UI after D_{in1}. Therefore Dn is a pulse of $\frac{1}{2}$UI whenever D_{in} changes, regardless of the phase difference between clock and D_{in}. If the rising edge of the clock occurs at the mid-point of the UI, Up also lasts for $\frac{1}{2}$UI. In the example waveforms, the rising edge of the clock is before the centre of the UI. Thus, Up lasts for a longer duration than Dn. Pulses Up and Dn, when connected to a charge pump and loop filter, will advance the phase of an oscillator, bringing the edge of the clock to the centre of the UI.

When connected to a charge pump/loop filter, the relevant output quantity of the PD becomes the difference in area under the Up/Dn pulses. Clearly when the falling edge of the clock is aligned with the data transition, both pulses last $\frac{1}{2}$UI or π in units of phase. If the data edges are delayed such that the Up pulse shrinks to zero, the difference between Up/Dn is a Dn pulse lasting $\frac{1}{2}$UI or π in units of phase. For a

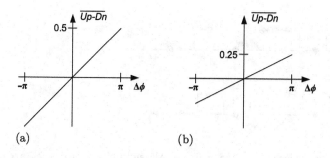

Figure 15.5 Hogge phase detector transfer characteristic for: (a) 101010 pattern and (b) random data.

repeating 101010 pattern, where Up/Dn pulses are generated every UI, this gives an average output amplitude of $\frac{\pi}{2\pi} = \frac{1}{2}$ as shown in Figure 15.5 (a). However, Up/Dn pulses are generated only when there are transitions in data. Therefore, when D_{in} is a random data stream, the actual average output is that shown in (a), scaled by the transition density, assumed to be $\frac{1}{2}$, plotted in Figure 15.5 (b). The gain of the Hogge PD is:

$$K_{pd} = \frac{1}{4\pi},\qquad(15.3)$$

when $\frac{1}{2}$ transition density is accounted for.

15.2.3 Non-Idealities of the Hogge Phase Detector

In the discussion thus far of the Hogge PD, the clock-to-Q delay (t_{clkQ}) of the flip-flops has been ignored. Pulses Up and Dn both last for $\frac{1}{2}$UI if the rising edge of the clock is aligned with the middle of the eye. Assuming the clock-to-Q delay of FF_1 and FF_2 are equal and the XOR gates have matched rising and falling delays, Dn remains a $\frac{1}{2}$UI pulse. However, when the rising edge of the clock is aligned to the centre of the eye, Up will last:

$$t_{Up} = \frac{UI}{2} + t_{clkQ}.\qquad(15.4)$$

If left uncorrected, a static phase error of t_{clkQ} will accumulate until the steady-state durations of Up and Dn match, occurring when:

$$t_{Up_{ss}} = t_{Dn_{ss}} = \frac{UI}{2}.\qquad(15.5)$$

One solution is to add a delay to the data path before the XOR gate, as shown in Figure 15.13. This delay must be matched to t_{clkQ} of the first flip-flop.

Another challenge in using the Hogge PD is that the variable width Up pulse can disappear entirely as its width shrinks. Even if it does not disappear, the circuit's operation is nevertheless sensitive to changes to the duty cycle of pulses. On the other hand,

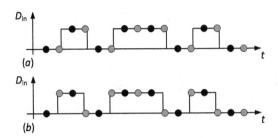

Figure 15.6 Data and edge samples captured in an Alexander phase detector for (a) early and (b) late clock. Data samples are shaded black and edge samples are shaded grey.

no analog information is encoded in the widths of the pulses generated by the Alexander phase detector described in Section 15.2.4, making it less sensitive to analog effects.

15.2.4   Alexander Phase Detector

An Alexander phase detector applies an input data stream to the D-input of DFFs, sampling the data twice per UI. When the recovered clock is aligned to the data, these samples are at the data's edges and in the middle of the eye. Samples are shown in Figure 15.6 for early and late clocks, where data samples are shown in black and edge samples are shaded grey.

Applying a data stream to the D-input of a DFF was previously considered and rejected as a viable phase detector for a CDR in Section 15.2.1, shown in Figure 15.3 (a), (c) and (d). Rather than consider only the average value of the DFF's output, an Alexander phase detector compares a given edge sample, e_0, with the data samples immediately before (d_0) and after (d_1). We can see in Figure 15.6 (a) that when the clock is early, $e_0 = d_0$. From Figure 15.6 (b), we see that for a late clock, $e_0 = d_1$. Of course, we can only draw a conclusion about clock timing when there is a data transition (i.e., $d_0 \neq d_1$). All combinations of edge and data samples are summarized in Table 15.1.

An early clock should give rise to a Dn signal while a late clock an Up signal. These conditions may be expressed using a sum of products as:

$$Dn = \overline{d_0}\,\overline{e_0}d_1 + d_0 e_0 \overline{d_1}, \tag{15.6}$$

$$Up = \overline{d_0}e_0 d_1 + d_0 \overline{e_0}\,\overline{d_1}. \tag{15.7}$$

Since Up/Dn signals are generated based on how the edge sample is resolved, this phase detector is also a nonlinear phase detector. However, in the absence of data transitions, no phase adjustment action is taken since both Up and Dn are low. Therefore, this phase detector addresses one deficiency of the stand-alone DFF presented in Figure 15.3 (b). Recall that in the presence of CIDs, that PD held the previous phase decision, possibly pulling the clock's phase more than needed.

Due to jitter and metastability, when the clock is aligned such that edge samples are taken at the transitions, the phase detector's abrupt transfer characteristic is smoothed

Table 15.1 Interpretation of edge and data samples in an Alexander phase detector. Note that e_0 is the edge sample between data samples d_0 and d_1.

d_0	e_0	d_1	Conclusion
0	0	0	None (CID = 0)
0	0	1	Clk early
0	1	0	None (not expected)
0	1	1	Clk late
1	0	0	Clk late
1	0	1	None (not expected)
1	1	0	Clk early
1	1	1	None (CID = 1)

out. Additional averaging takes place based on how the Up/Dn samples are acted on. For example, Up/Dn signals could be computed based on eight consecutive edge samples. An overall phase-adjustment signal can be computed as:

$$\Delta\phi = \sum_{i=0}^{7} Up_i - \sum_{i=0}^{7} Dn_i. \tag{15.8}$$

If edge samples are taken far from transitions all data transitions will produce the same Up or Dn signal. In this case, $\Delta\phi$ will have a magnitude equal to the number of data transitions over the eight UIs. However, as the edge samples move closer to data transitions, jitter gives rise to differing Up/Dn signals and a smaller $\Delta\phi$ control signal. Thus, although under ideal (jitter-free and uniform clock phase spacing) the Alexander PD is nonlinear, in practice, due to jitter and nonuniform spacing, the input/output behaviour is smoothed out.

Taking a closer look at Table 15.1, equations (15.6) and (15.7) can be rewritten as:

$$Dn' = e_0 \oplus d_1, \tag{15.9}$$
$$Up' = d_0 \oplus e_0, \tag{15.10}$$

where $\oplus$ denotes the exclusive OR operation. This formulation of Up'/Dn' produces high signals for both Up'/Dn' should the "unexpected" rows of Table 15.1 occur. Since phase updates are based on differences between Up' and Dn' we can ignore this effect.

15.2.5 Baud-Rate Phase Detection

Despite the widespread use of the Alexander PD, it has one important disadvantage, in that it requires two samples per UI and thus requires twice as many clock phases as a system needing only one sample per UI, known as a baud-rate CDR. In the case of a differential full-rate clock, one might argue that sampling twice per UI can

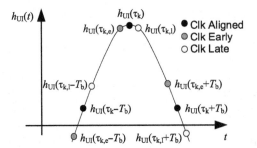

Figure 15.7 Mueller–Muller phase detector concept, based on [93].

be achieved without adding clock phases, since a differential clock signal is effectively two phases. However, CDRs usually operate with sub-rate clocks. For example, a quarter-rate Alexander PD requires eight samples per clock period, requiring the generation and distribution of eight clock phases, spaced by 45°. On the other hand, baud-rate sampling requires phases spaced by 90°. The extra phases of the Alexander PD are not simply the complements of the quadrature phases, but rather are halfway in between quadrature phases.

The question naturally arises as to whether we can infer clock alignment with only one sample per UI. The principle behind this is depicted in Figure 15.7 where a channel's pulse response, ($h_{UI}(t)$), assumed to be symmetrical about its peak, has a peak located at $t = \tau_k$. Due to the assumed symmetry, samples of the pulse response T_b on either side of the peak are equal. That is, $h_{UI}(\tau_k - T_b) = h_{UI}(\tau_k + T_b)$ or, alternatively, using notation introduced in Chapter 1, $h_{-1} = h_1$. The solid black circles show h_{-1}, h_0 and h_1 when the clock is aligned to sample at the peak of the pulse response. If the clock advances, sampling early at $\tau_{k,e} < \tau_k$ then $h_{-1} < h_1$. Similarly, if the clock is late, sampling at $\tau_{k,l} > \tau_k$, then $h_{-1} > h_1$. These observations are summarized using the notation of Figure 15.7 as:

$$h_{UI}(\tau_k - T_b) < h_{UI}(\tau_k + T_b), \text{ Conclusion: Clock early.} \qquad (15.11)$$

$$h_{UI}(\tau_k - T_b) > h_{UI}(\tau_k + T_b), \text{ Conclusion: Clock late.} \qquad (15.12)$$

Thus we can update τ_k using:

$$\Delta \tau_k = \mu \left(h_{UI}(\tau_k + T_b) - h_{UI}(\tau_k - T_b) \right), \qquad (15.13)$$

where μ a gain term mapping differences in the pulse response samples to shifts in sampling point.

In practice, we do not have direct access to an individual pulse response in the receiver, unless an isolated 1 or 0 is transmitted. Rather, the received signal $x(t)$ is a linear combination of pulse responses weighted by the transmitted data. That is,

$$x(t) = \sum_m A_m h_{UI}(t - mT_b), \qquad (15.14)$$

where A_m is the mth symbol. The kth sample of $x(t)$ is given by:

$$x_k = x(kT_b + \tau_k) = \sum_m A_m h_{UI}((k - m)T_b + \tau_k) = \sum_i A_{k-i} h_{UI}(iT_b + \tau_k). \quad (15.15)$$

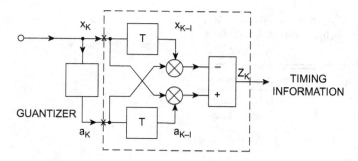

Figure 15.8 Mueller–Muller phase detector implementation. ©1976 IEEE. Reprinted, with permission, from [94].

The Mueller–Muller phase detector [94] published in 1976 implements (15.13) using the block diagram in Figure 15.8. Consider first the -ve input of the differencing block. Assuming equal probability data that is uncorrelated from sample to sample, the expectation of $A_{k-1}x_k$ can be shown to be given by:

$$E[A_{k-1}x_k] = \sum_i E[A_{k-i}A_{k-1}h_{UI}(iT_b + \tau_k)] \approx A^2 h_{UI}(\tau_k + T_b) \qquad (15.16)$$

Similarly,

$$E[A_k x_{k-1}] \approx A^2 h_{UI}(\tau_k - T_b). \qquad (15.17)$$

Therefore, the expectation of the overall output of the phase detector is:

$$E[A_{k-1}x_k - A_k x_{k-1}] \approx A^2 \left(h_{UI}(\tau_k + T_b) - h_{UI}(\tau_k - T_b)\right). \qquad (15.18)$$

The output (z_k), can be used to update the sampling point, τ_k. Note that since (15.18) is framed in terms of expectation, updates of τ_k must be done gradually, in order to average z_k.

To arrive at (15.18) requires either an analog delay to generate x_{k-1} from x_k, or a high enough resolution ADC to generate x_k, allowing straightforward digital delay. In this case, x_k/x_{k-1} are typically multi-bit representations of the continuous-valued input. With ADC-based links being more common, baud-rate CDRs are also becoming common place. The question naturally arises whether baud-rate phase detection can be done without an ADC or analog delay/multipliers.

An alternative baud-rate phase detector is a sign-sign Mueller–Muller phase detector. Consider the waveform and sample points in Figure 15.9. The input data (d_n) are compared against symmetrical reference voltages $(\pm V_{REF})$ to generate binary error signals (e_n) that are 1 when d_n is outside the range between $\pm V_{REF}$ and -1 inside this range. Black circles correspond to data and clock being properly aligned. If the clock is too early, the grey-shaded points are sampled, while a late clock samples the light-shaded samples. Early and late scenarios are summarized in Table 15.2.

A clock phase update function can be written as:

$$\Delta T_n = d_n \times d_{n-1} \times (e_n - e_{n-1}). \qquad (15.19)$$

Table 15.2 Interpretation of data and error samples in a sign-sign Mueller–Muller phase detector.

d_n	d_{n-1}	e_n	e_{n-1}	Conclusion
1	−1	1	−1	Late
−1	1	1	−1	Late
1	−1	−1	1	Early
−1	1	−1	1	Early
All	other	cases		Unknown

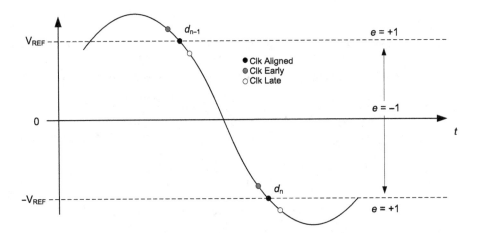

Figure 15.9 Waveform for sign-sign Mueller–Muller phase detector [95]. Input data d_n compared against target levels $\pm V_{REF}$ to generate error signals e_n.

Early/late signals (ΔT_n) can be averaged over several observations and then update the sampling clock's phase.

15.3   Referenceless Clock and Data Recovery Using a Hogge Phase Detector

The Hogge phase detector can be connected to a charge pump, loop filter and oscillator as shown in Figure 15.10.

15.3.1   Transfer Functions and Jitter Performance of a Hogge Phase Detector-Based Referenceless CDR

With a phase detector gain that takes into account the transition density, the loop can be analyzed to give equations similar to those derived in Chapter 14 for a Type II PLL:

$$H(s) = \frac{\frac{K}{C}(1 + sRC)}{s^2 + KRs + \frac{K}{C}}, \tag{15.20}$$

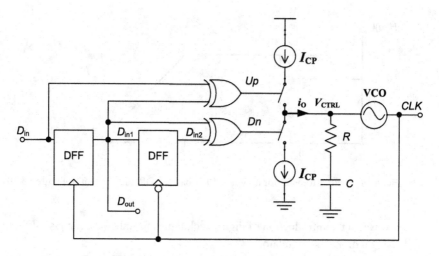

Figure 15.10 Overall block diagram of a reference-less (PLL-based) CDR using a Hogge phase detector.

where $K = \frac{I_{CP}K_{VCO}}{4\pi}$. This can be compared to the standard second-order transfer function to yield:

$$\omega_n = \sqrt{\frac{K}{C}}, \tag{15.21}$$

$$\zeta = \frac{R}{2}\sqrt{KC}. \tag{15.22}$$

The zero is given by:

$$\omega_z = -\frac{1}{RC} = -\frac{\omega_n}{2\zeta} \tag{15.23}$$

The poles are:

$$s_{1,2} = -\zeta\omega_n \pm \omega_n\sqrt{\zeta^2 - 1}, \tag{15.24}$$

As $\zeta \to \infty$

$$s_1 \to -\frac{\omega_n}{2\zeta} \to -\frac{1}{RC}, \tag{15.25}$$

$$s_2 \to -2\zeta\omega_n \to -2\frac{R}{2}\sqrt{KC}\sqrt{\frac{K}{C}} \to -KR. \tag{15.26}$$

As $\zeta \to \infty$, the zero and s_1 nearly cancel. Therefore, $H(s)$ can be approximated by a first-order transfer function given by:

$$H_{1st} = \frac{1}{\frac{s}{KR} + 1}. \tag{15.27}$$

Interestingly, the capacitor C does not determine the bandwidth, so long as ζ is large. However, to increase ζ without increasing the bandwidth we need a large C as it is the only parameter in ζ that can be adjusted without also changing s_2. Therefore, PLL-based CDRs usually have large capacitors, often implemented off-chip. This has

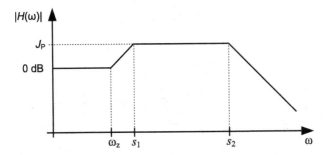

Figure 15.11 Jitter transfer for a Type II PLL-based CDR such as that in Figure 15.10.

motivated the introduction of digital techniques for integrating phase-detector outputs using a digital accumulator.

Meanwhile, $H(s)$ gives the jitter transfer of the CDR. Recall that jitter transfer describes the conversion of data jitter to recovered clock jitter. Since the recovered clock is used to retime the data, jitter in the recovered clock also appears in the retimed data. The need to significantly increase ζ stems from a more detailed consideration of $|H(\omega)|$ between ω_z and s_2. A piece-wise linear sketch is shown in Figure 15.11. Between the zero and the first pole, $|H(\omega)|$ rises to a value denoted as J_P:

$$J_P = \frac{s_1}{\omega_z} = \frac{\zeta \omega_n \left(1 - \sqrt{1 - \frac{1}{\zeta^2}}\right)}{\frac{\omega_n}{2\zeta}} = 2\zeta^2 \left(1 - \sqrt{1 - \frac{1}{\zeta^2}}\right). \tag{15.28}$$

To approximate the amount of jitter peaking the radical can be expanded using a Taylor-series expansion, as suggested in [63]:

$$J_P = 2\zeta^2 \left(1 - \left(1 - \frac{1}{2\zeta^2} - \frac{1}{8\zeta^4}\right)\right) = 1 + \frac{1}{4\zeta^2}. \tag{15.29}$$

As $\zeta \to \infty$, $J_P \to 1$, which corresponds to a flat transfer function. For $\zeta > 2$, (15.29) estimates $J_P - 1$ to within 10%.

Example 15.1   Estimate the minimum value of ζ to keep $J_P < 0.2$ dB.

Solution
Rearranging (15.29) gives:

$$\zeta = \sqrt{\frac{1}{4(J_P - 1)}}. \tag{15.30}$$

With J_P expressed in dB, we have:

$$\zeta = \sqrt{\frac{1}{4\left(10^{J_{PdB}/20} - 1\right)}} = \sqrt{\frac{1}{4\left(10^{0.2/20} - 1\right)}} = 3.28. \tag{15.31}$$

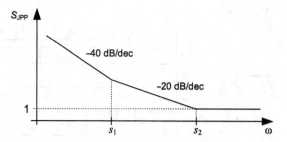

Figure 15.12 Jitter tolerance for a Type II PLL-based CDR.

Such a large ζ leads to widely spaced poles. In fact, for large ζ,

$$\frac{s_2}{s_1} \rightarrow 4\zeta^2. \tag{15.32}$$

Thus in this example the poles are spaced by a ratio greater than 36.

The jitter tolerance of the CDR can be found as follows. If we assume an eye that is very open, the phase of the recovered clock (θ_{CK}) can differ from the centre of the eye by at most 0.5 UI. In units of phase, this is π. Therefore,

$$\theta_{D_{in}} - \theta_{CK} < \pi \tag{15.33}$$

To find the largest modulation to $\theta_{D_{in}}$, using notation in (15.1), we set $\theta_{D_{in}} = \pi S_{Jpp}$. We can relate θ_{CK} to $H(s)$ as:

$$\theta_{CK} = H(s)\theta_{D_{in}} = H(s)\pi S_{Jpp}. \tag{15.34}$$

Therefore (15.33) becomes:

$$\pi S_{Jpp}(1 - H(s)) < \pi, \tag{15.35}$$

$$S_{Jpp} < \frac{1}{1 - H(s)} = \frac{s^2 + KRs + \frac{K}{C}}{s^2}. \tag{15.36}$$

Equation (15.36) is sketched in Figure 15.12. The frequency at which the jitter tolerance drops to 1 UI (maximum) is s_2 which is also the bandwidth of the jitter transfer of the CDR. Therefore, reducing the jitter transfer bandwidth also reduces the CDR's jitter tolerance.

The jitter generation of the CDR shown in Figure 15.10 is similar to that of the Type II PLL implemented with an XOR gate PD. It consists of random contributions from the VCO's phase noise and from the thermal noise of the loop filter. The VCO's phase noise sees a transfer function of:

$$H_{VCO} = \frac{s^2}{s^2 + KRs + \frac{K}{C}}. \tag{15.37}$$

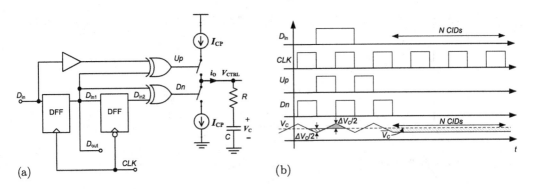

Figure 15.13 Hogge CDR: (a) schematic and (b) waveforms showing pattern-dependent capacitor voltage assuming a data transition every UI.

15.3.1.1   Deterministic Jitter Generation in a Hogge PD-Based Referenceless CDR

In addition to random jitter (phase noise), there is also a deterministic component due to the periodic modulation of the loop filter's voltage. However, the edge density in a CDR is not constant (unlike a PLL). Fluctuations on the control voltage that occur only on a data edge lead to pattern jitter, which is explained in the following discussion.

Figure 15.13 shows the Hogge phase detector and its waveforms when the phase of the oscillator has settled to its steady-state alignment with the incoming data. The capacitor voltage (V_C) is shown in the lowermost trace. Since the switches Up/Dn are active one after another, V_C is charged and discharged each cycle. The excursion on V_C is calculated as:

$$\Delta V_C = \int_0^{T_b/2} \frac{I_{CP}}{C} dt = \frac{I_{CP}T_b}{2C}. \tag{15.38}$$

The average value of V_C denoted as $\overline{V_C}$ is such that the oscillator's output frequency is f_{bit}. During a long 101010 pattern (transition density = 1) assumed to have occurred before $t = 0$ in Figure 15.13, $\overline{V_C}$ is halfway up the ramp while V_C fluctuates by $\pm\frac{\Delta V_C}{2}$. Since the VCO's phase is the integral of the control voltage, the maximum deviation in VCO output phase can be found by finding the triangular area ($\frac{bh}{2}$) above $\overline{V_C}$, scaled by the VCO gain, given by:

$$\Delta\phi_{VCO} = K_{VCO}\frac{1}{2}\frac{T_b}{2}\frac{\Delta V_C}{2} = K_{VCO}\frac{I_{CP}T_b^2}{16C}. \tag{15.39}$$

The output clock is modulated with a bounded phase deviation of $\pm K_{VCO}\frac{I_{CP}T_b^2}{16C}$.

The analysis above assumes a repeating 1010 pattern. A more realistic scenario is to assume an average transition density of 0.5 as depicted in Figure 15.14. In this case, the triangular area of $\Delta V_C\frac{T_b}{2}$ is spread over 2 UI on average, meaning that $\overline{V_C}$ sits one-quarter of the way up the triangle. If data transitions stop due to a long run of CIDs, the control voltage stays $\frac{\Delta V_C}{4}$ below its average when the transition density was

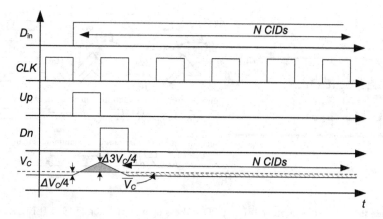

Figure 15.14 Hogge CDR waveforms showing pattern-dependent capacitor voltage assuming 0.5 transition density before a run of CIDs

0.5. Since the CDR is a low-pass system, designed to have dynamics that respond after several UI, a phase error between the VCO and the data accumulates each UI. After N CIDs, the phase offset is given by:

$$\Delta\phi_{VCO} = NK_{VCO}T_b\frac{\Delta V_C}{4} = NK_{VCO}\frac{I_{CP}T_b^2}{8C}. \tag{15.40}$$

This phase error is referred to as pattern-dependent jitter.

15.3.1.2 Improvement to the Hogge Phase Detector to Address Pattern-Dependent Jitter Generation

To address the per-data-transition modulation described in (15.39) an extra capacitor can be used in the loop filter, which smooths out fluctuations on the time-scale of a UI, similar to what was proposed in Chapter 14. To address pattern jitter an improved Hogge phase detector is shown in Figure 15.15 [96]. When phase-locked, for each data edge, the conventional Hogge phase detector gives:

$$\int i_{CP}(t)dt = 0, \tag{15.41}$$

indicating that its control voltage is returned to the value it was before each edge. However, due to non-zero i_{CP}, the control voltage is perturbed by the charge pump current. Further, the control-voltage perturbation has non-zero average (i.e., the area under a triangle in Figure 15.13). That is,

$$\int \frac{1}{C}\int i_{CP}(t)dt = \int V_C \neq 0, \tag{15.42}$$

showing that the per-edge control-voltage perturbation has non-zero average value, leading to pattern jitter.

The circuit in Figure 15.15 (a) generates four $\frac{1}{2}$UI "Phase Out" pulses referred to as x_{1-4}. Input data are applied to a cascade of level-sensitive latches, the first two of

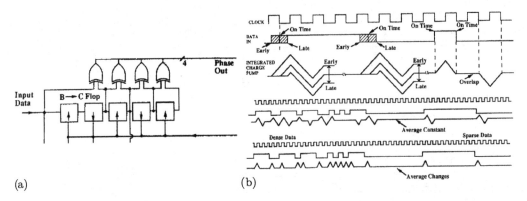

Figure 15.15 Devito's improved Hogge PD addressing pattern jitter. ©1991 IEEE. Reprinted, with permission, from [96].

which form a flip-flop. Pulses x_1 and x_2 correspond to *Up* and *Dn* in Figure 15.13, where the width of x_1 depends on the clock phase but x_2 is fixed in width. Meanwhile, x_3 is a second fixed-width *Dn* pulse while x_4 is a fixed-width *Up* pulse. These four pulses are applied to a charge pump giving currents that last the duration of the pulses. The sequence of *Up, Dn, Dn* and *Up* gives a sawtooth voltage on the capacitor as shown in Figure 15.15 (b), denoted by "integrated charge pump." Not only does this pattern return the capacitor voltage to its previous value (satisfying (15.41)) but the integral in (15.42) is also 0. This means that the average value of the control voltage is independent of edge density and pattern jitter is eliminated. Additional explanation of the approach can be found in [97].

15.4   Phase Control Approaches for CDRs with a Frequency Reference

An alternative to the referenceless CDR presented in Section 15.3 is to supply a CDR with a reference clock that is at a frequency suitable for sampling the incoming data. The CDR uses the reference clock and phase error information from a phase detector to generate a clock whose phase matches that of the incoming data stream. So long as the phase of the generated clock can be updated over a range of at least 360°, its frequency, and hence the data rate, can be different from the reference clock by some offset. The offset is usually specified as a "parts-per-million offset."

Once a phase error between the local clock and the incoming data has been computed we need a method for adjusting the phase of the reference clock. Two main methods are used, the most common of which uses phase interpolators to generate clocks of any phase from the reference clock. The second method uses a phase-locked loop where interpolating phase detectors select the phase of the VCO's output to be compared with the reference clock by the PLL's phase detector. This method is discussed first.

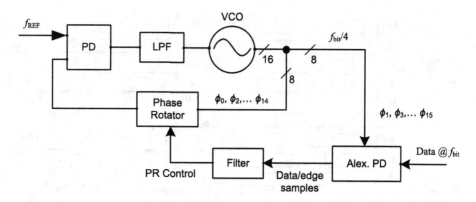

Figure 15.16 Phase tracking through using a PLL where the signal that is applied to the PLL's phase detector is adjusted via a phase rotator based on clock/data phase comparisons [98].

15.4.1   CDR Using an Interpolating Phase Detector in a PLL

As data rates move higher, most CDRs operate with a sampling clock at half or a quarter of the data rate. To sample twice per UI, in order to generate data and edge samples, a quarter-rate CDR requires eight clock phases. An N-stage differential ring oscillator generates $2N$ clock phases when we allow the output of a given stage to be taken with both signal polarities.

In the scheme shown in Figure 15.16 [98] a PLL is wrapped around such an eight-stage ring oscillator. The PLL is a conventional, Type II PLL using a static-CMOS ring oscillator, nominally tuned to 10 GHz. It is used as a quarter-rate clock for a 40 Gb/s CDR. The PLL has no divider in the feedback path, meaning both the VCO's output signal and the reference clock are at 10 GHz. At such a high frequency, an XOR-based phase detector is used. If a single, fixed phase from the VCO were fed to the phase detector, all 16 VCO phases would be locked to the reference clock's phase. Instead, eight phases of the VCO ($\phi_0, \phi_2, \ldots \phi_{14}$) are fed to a phase interpolator block that can generate an arbitrary output phase from its inputs, the output of which is applied to the phase detector. By adjusting which phase is applied to the phase detector, the VCO's phase relative to the reference clock is controlled. The other eight phases of the VCO ($\phi_1, \phi_3, \ldots \phi_{15}$) drive an Alexander phase detector.

The PLL is a relatively fast loop. The phase selection input to the phase interpolator is controlled by a secondary slower loop consisting of an Alexander phase detector and a digital loop filter. Based on averaged early/late signals computed from four data and four edge samples, the phase of the signal fed to the XOR phase detector is selected. Figure 15.17 shows examples of the phase relationships based on different phase selection by the phase interpolator. Remember that an XOR phase-detector-based PLL locks so that the fed-back clock is 90° from the reference clock. In Figure 15.17 (a) the phase interpolator selects ϕ_2, which settles to 90° behind ϕ_{REF}. Part (d) shows the relationship among the even VCO phases and the reference. The grey arrow

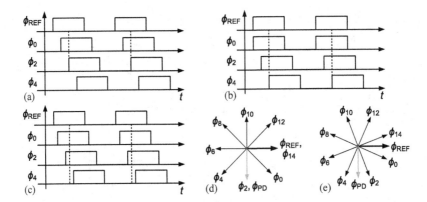

Figure 15.17 Phase relationships for different phase interpolator output: (a) phase interpolator selects ϕ_2; (b) phase interpolator outputs ϕ_4; (c) phase interpolator outputs a phase half-way between ϕ_2 and ϕ_4; (d) relationship among the even VCO phases and the reference corresponding to (a). The grey arrow indicates the phase that is applied to the PLL's phase detector; (e) VCO's phases in part (c).

indicates the phase that is locked by the PLL. In part (b) the phase interpolator is adjusted to output ϕ_4. All VCO phases advance by 45° bringing ϕ_4 in quadrature with ϕ_{REF}. The phase interpolator not only can output signals matched to the phase of one of its inputs; it can interpolate between two phases. For example, part (c) shows the scenario in which the phase interpolator outputs a phase halfway between ϕ_2 and ϕ_4. Therefore, ϕ_2 is less than 90° from ϕ_{REF} and ϕ_4 is more. Part (e) shows the VCO's phases in this situation. The grey line represents the interpolated phase that is applied to the phase detector.

Odd VCO phases ($\phi_1, \phi_3, \ldots \phi_{15}$) drive the Alexander phase detector. The edge and data samples are used to determine if the sampling clocks are early or late relative to the data. The digital filter generates control signals for the phase interpolator.

The implementation of the combination of the phase interpolator and detector is shown in Figure 15.18 [99]. Conceptually, the circuitry above the horizontal dashed grey line implements phase detection while the circuitry below performs phase selection and interpolation. At most, two consecutive coarse control bits are active. This selects the phases between which interpolation takes place. In the example in Figure 15.17 (c), *coarse*<1> and *coarse*<2> would be high, selecting ϕ_2 and ϕ_4.

When *coarse*<0> is active, transistors M_1 to M_6 form a multiplier (XOR gate) with a bias current controlled by bits *fine*<0:8>. This multiplies the differential phases of $\phi_{REF/REFB}$ and $\phi_{0/0B}$. By adding only transistors M_7, M_8 and M_{10}, *coarse*<4> can select the multiplication of $\phi_{REF/REFB}$ and $\phi_{0B/0}$, i.e., ϕ_8.

The circuit consisting of M_1 through to M_{10} is duplicated, for a total of four instances. Interpolation takes place when two consecutive circuits are activated by their respective *coarse* bit. Interpolation is controlled by *fine*<0:8> bits where *fine* is applied to the first and third circuits, while *fineb* is applied to the second and fourth circuits. Returning to the example in Figure 15.17 (c) with *coarse*<1>and

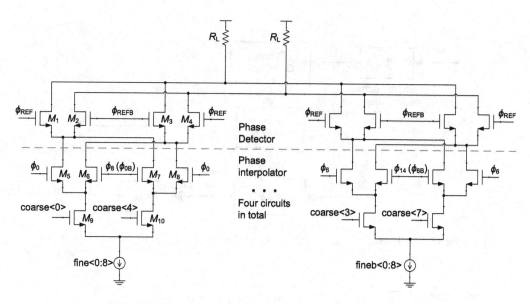

Figure 15.18 Phase interpolator/detector implementation [99].

coarse<2> active, to move the phase applied to the phase detector further toward ϕ_4, *fineb* bits are lowered and *fine* bits are raised. This happens until all *fine* bits are high. To move the phase beyond ϕ_4 toward ϕ_6, *coarse*<1> is lowered and *coarse*<3> is raised.

The advantage of this architecture is that all data and edge samples are moved earlier or later while only having one phase interpolator active at a time. This is opposite to the approach in the next section where each sampling phase is generated by a separate phase interpolator.

If the reference clock and the data have a ppm frequency difference, the phase control of the phase detectors cycles through the VCO phases continuously so that the effective frequency of the signal applied to the samplers is at a higher or lower frequency than the reference clock.

15.4.2   Phase Interpolator-Based Approaches

Figures 15.19 and 15.20 [99] show the other main approaches to phase tracking. In Figures 15.19 a shared PLL generates a reference clock that is supplied to a phase generator circuit, which generates sufficient phases for the CDR. In this example of a quarter-rate Alexander phase detector, it generates eight phases. These phases are applied to eight phase rotators controlled by a phase control code. Outputs of the phase rotators drive an Alexander phase detector. Data/edge samples are used to determine if the applied phases are early or late with respect to data. A suitably filtered phase control signal is generated and applied to the eight phase rotators. This approach would allow for per-phase calibration by applying different signals to each phase rotator. However, having eight phase rotators significantly increases its power dissipation.

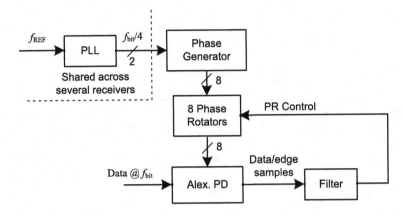

Figure 15.19 Phase tracking through using one phase rotator per phase detector input phase.

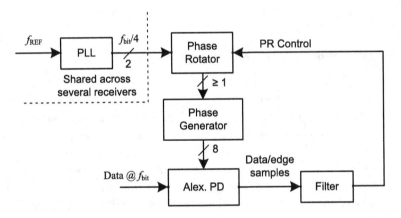

Figure 15.20 Phase tracking through using a single phase rotator to control the input to a phase generator [99].

Figure 15.20 [99] shows a simpler approach to phase tracking. In this approach, the PLL's output is applied to a single phase rotator. Its output is applied to a phase generator block such as an injection-locked oscillator that generates eight phases for the phase detector. The advantage of this system is that it has only one phase rotator. Similar to the design in Figure 15.19, data and edge samples are used to control the phase rotator via a digital filter.

15.5   CDR Design Decisions and Considerations

The following design decisions are discussed briefly:

1. **CDR rate:** for a given data rate, moving from $\frac{1}{2}$ to $\frac{1}{8}$ rate relaxes settling time requirements of latches, but may require additional circuitry to independently adjust each clock phase. With a longer clock period a given phase misalignment (in radians) translates to a larger time offset (in ps) in the sampling position.

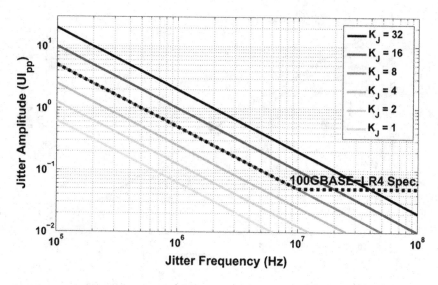

Figure 15.21 Jitter tolerance for different values of phase correction gain. ©2015 IEEE. Reprinted, with permission, from [4].

2. **Loop filter clock frequency:** higher clock frequency favours jitter tolerance since more frequent updates to the phase rotator allow the recovered clock to keep up with faster changes in the phase of the data stream.
3. **Phase interpolator step size:** larger step size can favour jitter tolerance, at the cost of larger steady-state jitter. Trade-offs due to gain along the phase detection/correction path are illustrated by Figure 15.21. Here, a larger K_j improves jitter tolerance, by allowing larger steps in phase interpolators. However, this is at the cost of larger phase noise, as shown in Figure 15.22.

15.6 Clock-Forwarded Links

Figures 12.2 and 12.3 show block diagrams of clock-forwarded links. Here, the circuitry responsible for aligning the receiver-side clock to the data streams is considered. Relevant block diagrams are presented in Figure 15.23 [84] and operate as follows:

- The differential forwarded clock, $fwdclock_B$, is received by each of three receiver lane bundles.
- An expanded view of the circuitry inside $rxbundle_A[0]$ is shown.
- The main phase tracking loop is inside the block labelled, "phase rotator." This block aligns the forwarded clock for all 10 of the $rxlane[9:0]$.
- The contents of the phase rotator block are presented in Figure 15.23 (b). It consists of a delay-locked loop to generate appropriated spaced clock phases. The CML MUX, Polarity sel and Mixer perform interpolation to generate data and edge clock phases.
- Phase detector outputs from all lanes are fed to the Error Counter block to perform global phase alignment.

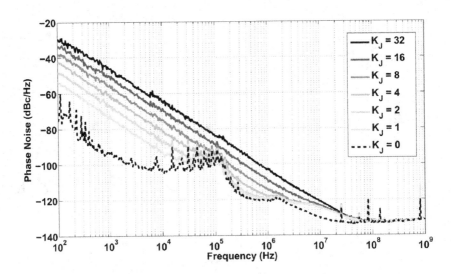

Figure 15.22 Phase noise for different values of phase correction gain. ©2015 IEEE. Reprinted, with permission, from [4].

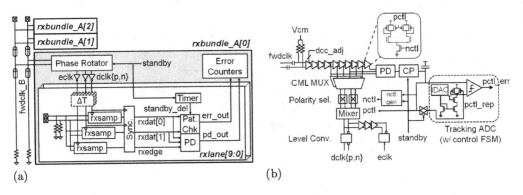

(a)   (b)

Figure 15.23 Clock-forwarded receiver synchronization showing (a) global synchronization for a bundle of Rx lanes and (b) details of the per-bundle phase rotator. ©2010 IEEE. Reprinted, with permission, from [84].

• Referring back to part (a), per-lane timing control is offered by the ΔT block, but for well-controlled channels is not needed.

Recall that in this approach, the clock has no frequency offset relative to the data. The receiver amplifies/buffers this clock and applies a phase offset to it before sampling the incoming data. Unlike a CDR for a system with a frequency offset between Tx and Rx clocks where phase rotators must adjust phases continuously around a phase circle, phase shifters in clock-forward systems must only introduce phase offsets between $-\pi$ and π; further, even with significant jitter on the clock and the data, the *differential* jitter between the signals is usually small, meaning that the phase difference need not be adjusted quickly.

In other designs, the DLL/phase rotator blocks are replaced by phase rotator/ILO.

15.7   Summary

This chapter presented techniques for locking clock signals in a receiver to an incoming data stream, collectively known as clock and data recovery. Three jitter metrics are important for these systems, namely: jitter tolerance, jitter transfer and jitter generation.

Before a clock signal's phase can be adjusted, a phase error between the data stream and the receiver's clock must be measured using a phase detector. The simplest approach using a single DFF was presented, followed by the Hogge, Alexander and Mueller–Muller phase detectors. An example of an analog, reference-less CDR based on a Hogge phase detector was analyzed. The relationship between its damping factor (ζ) and peaking in its jitter transfer characteristic was presented, motivating a design point with a large ζ, usually leading to a large capacitor value. The modulation to the control voltage, present even under locked conditions, gives rise to data-dependent jitter. An improved Hogge phase detector eliminated this.

Various techniques for building CDRs where a reference clock is available were described. These either make use of phase rotators/interpolators to synthesize sampling clock phases from the outputs of a PLL or select a phase for phase comparison in a PLL.

Problems

15.1 Define each of jitter tolerance, jitter generation and jitter transfer.

15.2 What are the advantages and disadvantages of using a single DFF as a phase detector in a CDR as presented in Figure 15.3 (b)?

15.3 If the DFFs in a Hogge phase detector are implemented using CML latches, propose a schematic for the delay element in Figure 15.13 (a) that will delay the input signal appropriately to keep both pulses $\frac{1}{2}$UI long when the clock is aligned. The objective is to design a logic buffer that will have a delay inherently similar to that of the DFF.

15.4 Determine the value of ζ needed to keep jitter peaking to less than 0.3 dB for the CDR in Figure 15.10.

15.5 Show the schematic for an Alexander phase detector-based analog CDR. That is, adapt Figure 15.10 to use an Alexander phase detector.

15.6 Explain why a variable delay element with finite-delay range cannot be used to align clock and data in the presence of a ppm frequency offset.

15.7 Design a phase interpolator. Explore the linearity of its digital control to phase delay transfer characteristic.

15.8 What are the advantages and disadvantages of a Mueller–Muller phase detector as presented in Figure 15.8 compared to an Alexander phase detector? Which of the disadvantages are mitigated by the sign-sign version presented in Figure 15.9?

Appendix A: Frequency Domain Analysis

This appendix reviews frequency domain circuit analysis techniques. For circuits with only two or three nodes, hand analysis can provide useful insight into the circuit's frequency response. From an analysis perspective, a circuit simulator or hand analysis can determine a transfer function or the numerical values of poles and zeros. However, hand analysis can often give expressions for bandwidth or poles/zeros that can provide some design insight. For example, which capacitor is most responsible for limiting a circuit's bandwidth.

We start with a review of circuit analysis in the s-domain by way of nodal analysis and then the solution of the linear equations using Cramer's rule. Methods for estimating the bandwidth from a transfer function are given.

A.1   Cramer's Rule for Solving s-Domain Nodal Equations

In this section, a technique for computing transfer functions from nodal equations in the s-domain is presented, by way of an example.

Figure A.1 (a) shows a common-source amplifier. The voltage source, v_{in}, is assumed to provide both the ac signal and the necessary dc biasing voltage. Figure A.1 (b) shows the ac small-signal model. The resistor R'_L is the parallel combination of R_D, the transistor's r_o and any other load resistance added. Likewise, C_O is the sum of the transistor's C_{db} and any wiring capacitance or loading from a follow-on circuit.

The nodal equations presented below have been written with the convention that currents leaving a node are positive. To eliminate fractions, voltage differences have been multiplied by conductances, rather than divided by resistances. That is, a given conductance G_x is equal to R_x^{-1}. Kirchhoff's current law equations at nodes V_g and V_o are written as:

$$(V_{gs} - V_{in})G_S + V_{gs}sC_{gs} + (V_{gs} - V_o)sC_{gd} = 0, \qquad (A.1)$$
$$(V_o - V_{gs})sC_{gd} + g_m V_{gs} + V_o G'_L + V_o sC_O = 0. \qquad (A.2)$$

The first equation is the nodal equation at V_g, while the second is at V_o. These equations can be put in matrix form as:

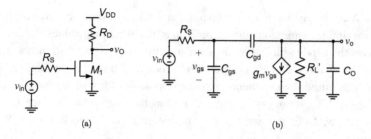

Figure A.1 Common-source amplifier: (a) schematic (bias details omitted); (b) small-signal model with relevant capacitances included.

$$\begin{bmatrix} G_S + s(C_{gs} + C_{gd}) & -sC_{gd} \\ g_m - sC_{gd} & G'_L + s(C_{gd} + C_O) \end{bmatrix} \begin{bmatrix} V_{gs} \\ V_o \end{bmatrix} = \begin{bmatrix} G_S V_{in} \\ 0 \end{bmatrix}$$

$$[G + sC]\,V = J, \tag{A.3}$$

where $G+sC$ is the system matrix, V is the vector of unknowns and J is the source vector. Solving these equations using Gaussian elimination can be rather tedious. Instead, Cramer's rule is a convenient method for systems up to second or third order.

Given: $[G + sC]\,V = J$, Cramer's rule tells us:

$$V_i = \frac{|G + sC|_i}{|G + sC|}, \tag{A.4}$$

where V_i is the ith entry in V, $|\ |$ denotes the determinant operator and $|\ |_i$ means the determinant of the matrix $G + sC$ where the source vector J has been put in the ith column. The determinant of the system matrix is computed as:

$$|G + sC| = (G_S + s(C_{gs} + C_{gd}))(G'_L + s(C_{gd} + C_O)) - (-sC_{gd})(g_m - sC_{gd})$$
$$= s^2(C_{gs}C_O + C_{gs}C_{gd} + C_{gd}C_O) + s(C_{gs}G'_L + C_{gd}(g_m + G_S + G'_L) + C_O G_S) + G_S G'_L.$$

Since we want to compute V_o, which is the second entry in V, $i = 2$. We put the source vector into the second column of the system matrix and compute this matrix's determinant:

$$|G + sC|_2 = \left| \begin{bmatrix} G_{sig} + s(C_{gs} + C_{gd}) & G_{sig}V_{in} \\ g_m - sC_{gd} & 0 \end{bmatrix} \right|$$
$$= (G_{sig} + s(C_{gs} + C_{gd}))(0) - (g_m - sC_{gd})(G_{sig}V_{in})$$
$$= -(g_m - sC_{gd})G_{sig}V_{in}. \tag{A.5}$$

Therefore, $\frac{V_o}{V_{in}}$ is:

$$\frac{V_o}{V_{in}} = \frac{-(g_m - sC_{gd})G_S}{s^2(C_{gs}C_O + C_{gs}C_{gd} + C_{gd}C_O) + s(C_{gs}G'_L + C_{gd}(g_m + G_S + G'_L) + C_O G_S) + G_S G'_L}. \tag{A.6}$$

From (A.6) we see that there are two poles and a right-half plane zero. However, we do not have simple expressions for the poles in terms of element values and it is not obvious what the bandwidth is. Finally, since a common objective is to increase a circuit's bandwidth, we would like an expression for bandwidth in terms of element values so we can determine what elements are most responsible for limiting the bandwidth.

When looking at pole/zeros and transfer functions, we often look for RC time constants. Below, (A.6) is rewritten after multiplying above and below by $R'_L R_S$. This leads to:

$$\frac{V_o}{V_{in}} = \frac{-(g_m - sC_{gd})R'_L}{s^2 R_S R'_L (C_{gs}C_O + C_{gs}C_{gd} + C_{gd}C_O) + s(R_S C_{gs} + C_{gd}[(g_m R'_L + 1)R_S + R'_L] + R'_L C_O) + 1}.$$

In general, we cannot factor it symbolically, though we could manipulate the quadratic formula symbolically. We can also substitute element values and compute the roots. As a side note, if $C_{gd} = 0$, it is factorable.

$$\frac{V_o}{V_{in}} = \frac{-g_m R'_L}{s^2 R_S R'_L (C_{gs}C_O) + s(R_S C_{gs} + R'_L C_O) + 1}$$

$$= \frac{-g_m R'_L}{(sR_S C_{gs} + 1)(sR'_L C_O + 1)}. \tag{A.7}$$

In this case, we would have two real poles, one set by the input side ($R_S C_{gs}$), the other set by the output side ($R'_L C_O$). In fact, if $C_{gd} = 0$ we would have written this almost by inspection by writing:

$$\frac{V_o}{V_{in}} = \frac{V_o}{V_{gs}} \frac{V_{gs}}{V_{in}}. \tag{A.8}$$

Although interesting to consider what would happen if $C_{gd} = 0$, the effect of non-zero C_{gd} is often significant. Thus, an estimate for the bandwidth from the detailed analysis is useful. This is the topic of the next section.

A.2   Bandwidth Estimation

This section presents techniques for estimating the bandwidth of transfer functions. The first is known as the dominant pole approximation. The second requires expressions or values of the poles.

A.2.1   Dominant Pole Approximation

Assume a second-order system with real poles with magnitudes $\alpha \ll \beta$. The denominator, $D(s)$, can be written in factored and expanded form as:

$$D(s) = (s + \alpha)(s + \beta) = s^2 + (\alpha + \beta)s + \alpha\beta = s^2 + bs + c. \tag{A.9}$$

When we derive a transfer function we usually have expressions for b and c but not α and β. However, we can approximate α, the pole of smaller magnitude, as follows:

$$\frac{c}{b} = \frac{\alpha\beta}{\alpha + \beta} \approx \alpha \tag{A.10}$$

The lower frequency pole (α) can be approximated using the coefficients of the denominator. When $\alpha \ll \beta$ the $\omega_{-3db} \approx \alpha$. As β approaches α, $\frac{c}{b}$ may give a poor approximation of α but it is still a reasonable approximation of ω_{-3dB} if we use $\omega_{-3dB} = \frac{c}{b}$.

The larger pole (β) can be approximated from the coefficient of "s" as:

$$b = \alpha + \beta \approx \beta \tag{A.11}$$

If we have a non-zero coefficient for s^2, say a, we approximate β as $\frac{b}{a}$.

A.2.2   Dominant Pole Approximation Applied to Common-Source Amplifier

From earlier analysis:

$$D(s) = s^2 R_S R_L'(C_{gs}C_O + C_{gs}C_{gd} + C_{gd}C_O) + \dots$$
$$s(R_S C_{gs} + C_{gd}[(g_m R_L' + 1)R_S + R_L'] + R_L'C_O) + 1$$

$$\alpha \approx \frac{\text{constant term}}{\text{coefficient of } s}$$

$$\omega_{-3dB} \approx \frac{1}{R_S C_{gs} + C_{gd}[(g_m R_L' + 1)R_S + R_L'] + R_L'C_O},$$

where ω_{-3dB} can be evaluated by substituting in values for the Rs and Cs. Each RC term in the denominator lowers the upper 3 dB frequency. We have a term from the input RC, the output RC and then extra terms due to C_{gd}. The extra term containing C_{gd} is:

$$R_S C_{gd}(1 + g_m R_L'). \tag{A.12}$$

Note that $g_m R_L'$ is the magnitude of the gain from one side of the gate-drain capacitor to the other side. As gain is increased by increasing R_L', $R_L'C_O$ increases and so does $R_S C_{gd}(1 + g_m R_L')$. (A.12) can also be explored considering the Miller effect.

A.2.3   Extension of the Dominant Pole Approximation to Higher-Order Systems

In Section A.2.1, the estimation of a dominant pole and the system's bandwidth was presented for a second-order system. This expression can be applied to higher-order systems as well. That is, $\omega_{-3dB} \approx \frac{c}{b}$ holds for a higher-order transfer function with real poles. Circuit analysis of higher-order circuits can become time-consuming. Rather than derive a circuit's entire transfer function, $\frac{c}{b}$ can be calculated from the circuit using the method of open circuit time constants. Please see [100] for more details. A summary is as follows:

$$\omega_{-3dB} \approx \frac{c}{b} = \frac{1}{\Sigma_i^M R_i C_i}, \tag{A.13}$$

where C_i is each of the circuit's M capacitors and R_i is the dc resistance "seen" by C_i when the other capacitors are treated as open circuits. Note that the computation of $\frac{c}{b}$ is exact, although this is only approximately equal to the circuit's bandwidth. In this section, the short hand $\frac{c}{b}$ is interpretted as $\frac{\text{constant term}}{\text{coefficient of } s}$.

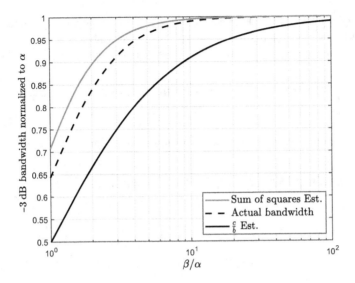

Figure A.2 Actual bandwidth and bandwidth estimates for second-order systems as a function of the ratio between the non-dominant and dominant poles.

A.2.4   Estimate Requiring Poles and Zeros

If we have values for the two poles, the bandwidth can be approximated using α and β as:

$$\frac{1}{\omega_{-3dB}} \approx \sqrt{\frac{1}{\alpha^2} + \frac{1}{\beta^2}}. \tag{A.14}$$

The quality of these two approximations is shown in Figure A.2. The horizontal axis is the ratio between the non-dominant pole (β) and the dominant pole (α). The vertical axis shows the normalized -3 dB bandwidth relative to the dominant pole. The actual -3 dB bandwidth drops to 64% of α when the two poles are identical. The approximation in (A.10) labelled as "$\frac{c}{b}$ Est." is pessimistic, putting the bandwidth at 50% of the pole magnitude. The approximation in (A.14), labelled as "Sum of Squares Est." is 71% of the repeated pole. Although this is closer to the actual bandwidth, it is an over-estimate of the bandwidth. Given that any unaccounted for parasitic capacitor will tend to further reduce the bandwidth, the pessimistic estimate provided by (A.10) is usually better.

Appendix B: Noise Analysis

This appendix reviews noise models of circuit elements and techniques for noise analysis of linear time invariant circuits. Special thanks are due to Peter Kinget of Columbia University whose lecture notes shaped the author's approach to teaching this topic.

The term "noise" can be used to describe the inherent noise of circuits as considered in this appendix as well as interference phenomena such as fluctuations in power supply voltage, induced by switching activity in neighbouring circuits. In this text, noise will be used only for signal fluctuations in a circuit caused either by thermal noise, shot noise or flicker noise.

This appendix reviews mathematical descriptions of random signals, noise models for circuit elements and the overall steps for computing the noise at the output of a linear time-invariant circuit. The appendix ends with a discussion of impedance scaling and how it affects noise performance of a voltage-to-voltage circuit as well as a transimpedance circuit.

B.1   Time and Frequency Domain Descriptions of Noise

The randomness of a noise signal $x(t)$ prevents us from specifying it with a specific function of time. We describe a noise signal indirectly, by calculating its mean and variance or by its probability density function (PDF). For a signal $x(t)$, its mean, mean-squared and root-mean-squared values are defined as:

$$\overline{x(t)} = \frac{1}{T} \int_0^T x(t)dt, \tag{B.1}$$

$$X_{MS} = \frac{1}{T} \int_0^T x^2(t)dt, \tag{B.2}$$

$$X_{RMS} = \sqrt{\frac{1}{T} \int_0^T x^2(t)dt}, \tag{B.3}$$

as $T \to \infty$. If $x(t)$ has zero mean, X_{MS} and X_{RMS} correspond to the signal's variance and standard deviation, respectively. The mean-squared value of a signal is often referred to as its normalized power since it represents the power delivered to a $1\ \Omega$ resistor if $x(t)$ is a voltage applied across the resistor or the current through it.

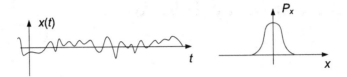

Figure B.1 Example of a noise signal, $x(t)$ and its PDF, P_x.

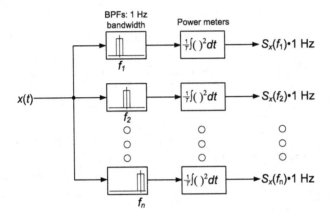

Figure B.2 Conceptual block diagram of power-spectral density measurement.

The probability density function (PDF) of $x(t)$, $P_x(x)$ is defined as

$$P_x(x)dx = P(x < x(t) < x + dx). \tag{B.4}$$

An example of a signal along with its PDF is shown in Figure B.1. Notice that none of the PDF, mean or mean-squared value gives any description of how a signal evolves in time: that is, does it vary slowly having only low-frequency fluctuations or can it also vary quickly? The most common mathematical tool for characterizing the transient behaviour of noise signals is the power-spectral density (PSD) of the signal. A conceptual block diagram of how the PSD is measured from a signal is shown in Figure B.2. The PSD at a frequency f is the normalized power that would be measured at the output of an ideal band-pass filter centred at f, having a bandwidth of 1 Hz. The units of PSD in the case of a voltage signal are V^2/Hz. In the case of a current signal the units are A^2/Hz. PSDs of real-valued signals are real-valued functions of frequency, unlike Fourier and Laplace transforms which are complex functions of frequency. By describing noise in the frequency domain, we can determine the effect of frequency-dependent systems on noise signals. For example, as shown in Figure B.3, if a noise signal $x(t)$ with PSD $S_x(f)$ is applied to the input of a linear time-invariant system with a transfer function of $H(f)$, the PSD of the output signal y can be found as:

$$S_y(f) = |H(f)|^2 S_x(f). \tag{B.5}$$

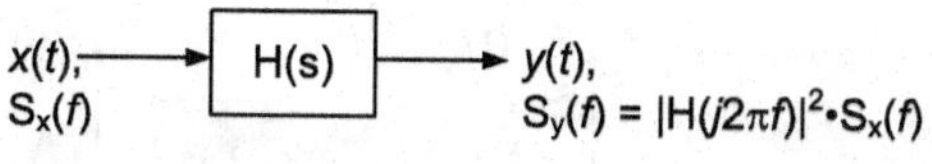

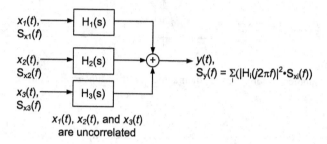

Figure B.3 Block diagram showing an input PSD shaped by a transfer function.

Figure B.4 Summation of noise signals processed by multiple transfer functions.

This is similar to what happens when a deterministic signal with a frequency response $X(f)$ is processed by $H(f)$. However, since noise signals are described by PSDs, with units V^2/Hz, we scale the input PSD by $|H(f)|^2$ rather than simply $H(f)$.

The MS value of a signal can be found from the PSD as:

$$Y_{MS} = \int_0^\infty S_y(f)df. \qquad (B.6)$$

The limits of this integration can be changed in order to compute the mean-squared power over a particular bandwidth.

The circuits we analyze in this book usually have many noise sources that each contribute noise to the output via their frequency-dependent transfer function, referred to as a noise transfer function. Assuming the sources of noise are uncorrelated, the overall output PSD can be found by adding the individual contributions from each source.

In Figure B.4 uncorrelated noise sources $x_1(t)$, $x_2(t)$ and $x_3(t)$ described by individual PSDs $S_{x_1}(f)$, $S_{x_2}(f)$ and $S_{x_3}(f)$ each pass through an LTI transfer function to the output, where the signals add. Therefore, the overall PSD at the output y is given as

$$S_y(f) = |H_1(f)|^2 S_{x_1}(f) + |H_2(f)|^2 S_{x_2}(f) + |H_3(f)|^2 S_{x_3}(f). \qquad (B.7)$$

This is generalizable to N noise signals and transfer functions. For LTI systems, we expect the principle of superposition to hold. In the case of noise signals, we add the power of the signals.

The PSD is a frequency-domain description of a noise signal. Thus far, we have not addressed how we compute it from a time-domain signal, aside from the block diagram in Figure B.2, involving ideal BPFs. The PSD is the Fourier transform of the noise signal's autocorrelation function. While circuit designers readily analyze circuits and think in terms of PSD, comparatively few compute or think in terms of

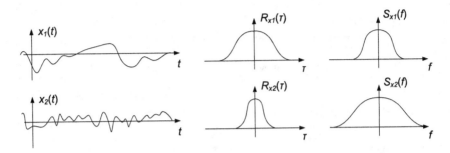

Figure B.5 Two examples of noise signals, their autocorrelation function and power-spectral density.

autocorrelation functions. Nevertheless these relationships are given below. The autocorrelation function $R_X(\tau)$ is defined as

$$R_X(\tau) = \overline{x(t)x(t+\tau)} = \frac{1}{T} \int_0^T x(t)x(t+\tau)dt. \tag{B.8}$$

$R_X(\tau)$ is a measure of the correlation of the signal $x(t)$ at one time location to the same signal, offset in time by τ. Two examples of noise signals, autocorrelation functions and PSDs are shown in Figure B.5. Qualitatively, the top noise signal varies more slowly than the lower noise signal. The integral in (B.8) increases whenever $x(t)$ and $x(t+\tau)$ have the same sign and decreases when they have opposite sign. Since the top signal varies more slowly the autocorrelation function is greater than zero for larger values of τ. On the other hand, the lower noise signal varies more quickly and changes sign along a shorter timescale. Assuming that these sections of the signals are representative of the entire signals, we expect that the autocorrelation of the lower signal will decay to zero for a smaller value of τ compared to that for the upper signal. The autocorrelation functions are shown in the middle. As the autocorrelation shrinks in τ (a measure of time), it expands in frequency when we consider its Fourier transform. Hence in the frequency domain, the PSD of x_2 is wider than the PSD of x_1.

Finally, the autocorrelation function shows the mean-squared signal value. By considering $\tau = 0$, we see that

$$R_X(0) = \overline{x(t)x(t)} = \overline{x^2(t)} = \frac{1}{T} \int_0^T x^2(t)dt = X_{MS}. \tag{B.9}$$

B.2 Noise Models of Circuit Elements

This section presents the noise models of various circuit elements.

B.2.1 Resistors

Resistors generate thermal noise, also known as Johnson noise, which can be modelled with either a series voltage source or parallel current source as shown in Figure B.6.

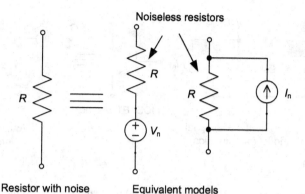

Noiseless resistors

Resistor with noise   Equivalent models

Figure B.6 Noise models of a resistor.

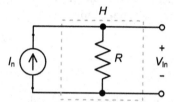

Figure B.7 Schematic for finding a resistor's voltage noise from its current noise.

The noisy resistor R is modelled as a noiseless resistor of the same value in series or parallel with the noise source. The PSDs of the noise sources are

$$S_{V_n}(f) = 4kTR, \tag{B.10}$$

$$S_{I_n}(f) = \frac{4kT}{R}, \tag{B.11}$$

where k is Boltzmann's constant and T is the absolute temperature in Kelvin. For frequencies of interest in wireline we can take the PSDs as being flat (i.e., white noise). Notice that (B.6) evaluated to infinite frequency for a white noise PSD gives infinite power. In reality, at THz frequencies (B.10) decreases, leading to finite power when integrated.

To show that these two sources are equivalent and to illustrate the use of (B.5), consider the PSD of the noise voltage that is measured across the terminals of the current source model, V_{I_n}. Although our description of the current source is in terms of its current PSD (in units of A^2/Hz), we determine the transfer function $H(f)$ by analyzing the circuit in Figure B.7, treating the noise source as we would any source. In this case, H is the function that maps I_n to V_{I_n}. Therefore $H(f) = R$. Applying (B.5) we have

$$S_{V_{I_n}}(f) = |H(f)|^2 S_{I_n}(f), \tag{B.12}$$

$$S_{V_{I_n}}(f) = |R|^2 \frac{4kT}{R} = 4kTR = S_{V_n}(f). \tag{B.13}$$

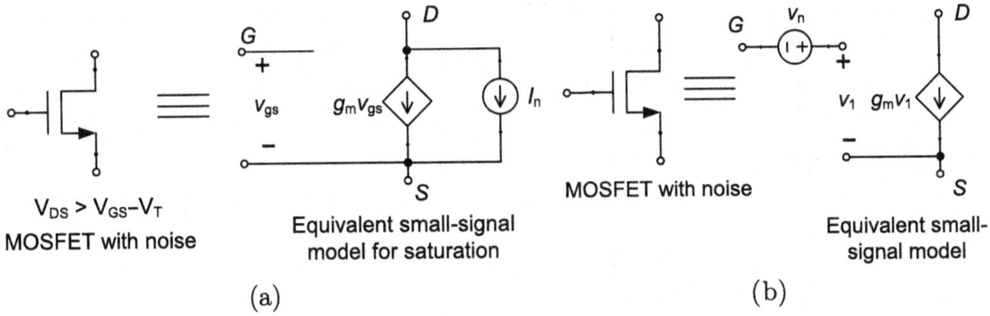

Figure B.8 Noise models of a MOSFET in saturation (a) thermal noise and (b) flicker noise.

Therefore, the parallel current source model gives the same PSD of voltage that the voltage source model does. This also serves as a template for how the various transfer functions ($H(f)$) are calculated.

B.2.2   Reactive Elements

Ideal inductors and capacitors generate no noise. However, real inductors and capacitors may generate thermal noise due to resistive losses. What complicates matters is that not all phenomena modelled with resistances in an equivalent circuit model generate noise.

B.2.3   MOSFETs

A simple noise model of a MOSFET in saturation has two noise sources, as shown in Figure B.8. The parallel noise current source in Figure B.8 (a) models channel thermal noise, whereas the voltage source in series with the gate in Figure B.8 (b) models flicker noise. The PSDs are:

$$S_{V_n}(f) = \frac{K}{C_{OX}WLf},\tag{B.14}$$

$$S_{I_n}(f) = 4kT\gamma g_m,\tag{B.15}$$

where K is a constant for a given technology's flicker noise, not to be confused with Boltzmann's constant k, and γ is a parameter for thermal noise, not to be confused with the body-effect parameter often given the same symbol. In long channel MOSFETs (i.e., those that closely follow a square-law model for I_D vs V_{GS} in saturation) γ is 2/3. However, in short-channel MOSFETs γ is higher, but usually less than 2.

Flicker noise is believed to originate from charge traps at the channel–oxide interface. Unlike thermal noise which has a flat PSD, flicker noise follows a $1/f$ relationship and is often referred to as "one over f" noise.

The overall drain-current noise PSD from both flicker and thermal noise can be found considering Figure B.9 (a) and (B.7). The PSDs of the contribution of each type of noise add to give the total drain current noise ($I_{n,tot}$). The noise current from thermal

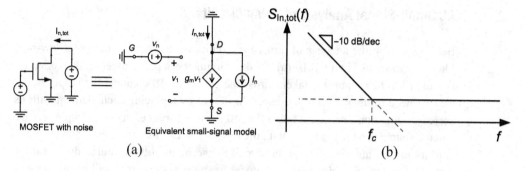

Figure B.9 Overall drain current noise (a) model and (b) plot.

noise flows directly to the output, implying $H_{Thermal} = 1$. Flicker noise contributes a term at the output through the transconductance of the MOSFET ($H_{Flicker} = g_m$). The overall output noise current PSD is given by:

$$S_{I_{n,tot}}(f) = S_{I_n}|H_{Thermal}|^2 + S_{V_n}|H_{Flicker}|^2$$
$$= 4kT\gamma g_m + \frac{K}{C_{OX}WLf}g_m^2. \tag{B.16}$$

An asymptotic plot of the overall PSD is given in Figure B.9 (b) shown on a log-log scale. Notice that at low frequencies (below f_c) the PSD is dominated by flicker noise. When expressing a PSD in dB, we take the 10log(PSD) since PSD is already in units of power. Flicker noise decreases at 10 dB/dec, unlike a single-pole low-pass transfer function that decreases at 20 dB/dec. Above f_c the PSD is dominated by thermal noise. The frequency at which thermal and flicker noise contributions are equal to one another, known as the flicker-noise corner frequency (f_c), can be found by equating the thermal and flicker noise PSDs and solving for f_c:

$$4kT\gamma g_m = \frac{K}{C_{OX}WLf_c}g_m^2$$
$$f_c = \frac{g_m K}{4kT\gamma C_{OX}WL}, \tag{B.17}$$

giving values of f_c in the kHz or MHz range, depending on sizing and biasing choices. Increasing W and L while keeping the same current will decrease the flicker-noise corner frequency. When considering this expression one must bear in mind that, depending on how a transistor is biased, changes to dimensions may also change g_m. For example, a transistor biased from a constant voltage via a current mirror will have the same f_c regardless of its width, since both g_m and the denominator will increase with increasing W.

Most circuits discussed in this book target operation in the GHz range. Flicker noise is not important in high-speed amplifiers but it is important in oscillators where a transistor's flicker noise at a given frequency, f, can be converted to phase noise at the same frequency offset ($\Delta f = f$) from the frequency of oscillation.

B.3   AC Small-Signal Analysis of Noisy Circuits

Before showing the analysis of transistor circuits we consider a first-order low-pass filter as shown in Figure B.10 (a). In this circuit the input source v_{in} is applied on the left with the output v_{out} taken across the capacitor. To compute the output noise, we first need to draw a noise model of the circuit by replacing each element with its corresponding noise model. We also shut off the input source. In this example, a noise voltage source is added in series with the resistor. Notice that the noise voltage source appears in the same location in the circuit as the input voltage source did, meaning that, in this case, the noise source's transfer function to the output will be the same as the circuit's input-output transfer function. In general, this is not the case. Each noise source will see a different noise transfer function, some being quite different from the input signal's.

Following the mathematics in (B.5) we have:

$$S_{vn,out}(f) = |H_{LPF}(f)|^2 S_{vn,R}(f)$$

$$S_{vn,out}(f) = \left| \frac{1}{j2\pi fRC + 1} \right|^2 4kTR$$

$$S_{vn,out}(f) = \frac{4kTR}{(2\pi fRC)^2 + 1}. \tag{B.18}$$

This output PSD is plotted in Figure B.10. It has a low-pass shape with a cut-off frequency of $1/(2\pi RC)$. The MS value of the noise signal can be found by integrating (B.18) up to ∞ as:

$$V_{MS} = \int_0^\infty \frac{4kTR}{(2\pi fRC)^2 + 1} df. \tag{B.19}$$

Noting that the anti-derivative of $1/(1 + x^2)$ is $\tan^{-1} x$ and making the appropriate substitution, the integration becomes the well-known result:

$$V_{MS} = \frac{kT}{C}. \tag{B.20}$$

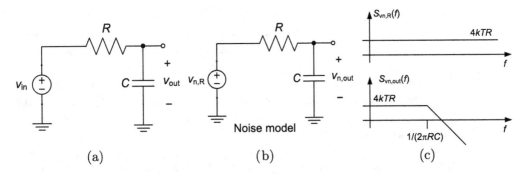

(a)   (b)   (c)

Figure B.10 Noise analysis of a first-order low-pass filter: (a) schematic; (b) noise model; and (c) PSDs.

Table B.1 Noise analysis of common-source amplifier.

Noise source	PSD	NTF	Contribution
Resistor	$\frac{4kT}{R}$ [A^2/Hz]	R [Ω]	$4kTR$ [V^2/Hz]
MOSFET	$4kT\gamma g_m$ [A^2/Hz]	R [Ω]	$4kT\gamma g_m R^2$ [V^2/Hz]
		Total	$4kTR(1+\gamma g_m R)$ [V^2/Hz]

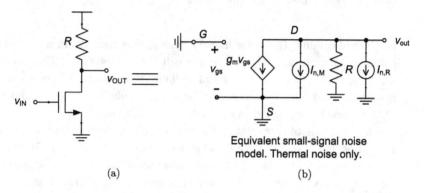

Equivalent small-signal noise model. Thermal noise only.

(a)　　　　　　　　(b)

Figure B.11 Noise analysis of a common-source amplifier considering only thermal noise: (a) schematic and (b) ac small-signal equivalent circuit with noise sources.

The disappearance of R from this expression is explained as follows: increasing R increases the intensity of the input PSD. However, the bandwidth of the filter is reduced. The result is that the area under the output PSD remains unchanged. To maintain a given cut-off frequency but with reduced mean-squared noise we must increase C and decrease R thereby reducing the impedance level of the circuit. Doing so will increase the power dissipation in the circuit. While this is a passive circuit, the circuit driving it would need to deliver and hence would dissipate more power. This illustrates a fundamental trade-off: reducing the noise in a circuit requires increasing the power dissipation. Although this is illustrated with a passive circuit, the conclusion is equally valid for an active circuit.

The schematic of a common-source amplifier is shown in Figure B.11 along with its small-signal noise model. Using this model, we can compute the overall PSD of noise at the output and also compute an equivalent input noise PSD. Note that PSDs can also be integrated to compute mean-squared noise levels, although this requires care when applied to the input PSD. Although this circuit is relatively simple, a template is used here to account for all noise sources and their respective transfer function to the output quantity of interest.

Table B.1 shows a systematic way to compute the overall output noise PSD. The first column lists the noise sources. For each noise source, the second column gives the appropriate PSD. Noise sources may be current noise or voltage noise. It is good practice to explicitly note the PSD's units. Each noise transfer function (NTF) is shown

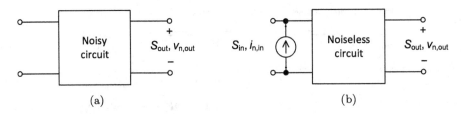

Figure B.12 Circuits illustrating the concept of input-referred noise: (a) noisy circuit and (b) noiseless circuit with equivalent input source.

in the third column. This is the transfer function that converts the noise source to the output quantity. In general, each source will have its own NTF. The contribution to the output PSD is shown in the fourth column. Note that the magnitude of the NTF is squared when computing a given noise source's contribution. The total output noise PSD is found by adding the entries in the fourth column.

Noise analysis often motivates the question as to how a design can be made less noisy. Looking at output noise PSD alone can lead to poor insight. For this example, it would appear that to reduce noise the resistor and the transconductance should be reduced. However, doing so will also reduce the signal of interest. Instead, to consider noise trade-offs, we should consider a signal-to-noise ratio (SNR), assuming we have an input signal of amplitude V_{in}. Since the gain of this circuit is $g_m R$, the output signal $V_{out} = g_m R V_{in}$. The output signal power is $(g_m R V_{in})^2$. If we consider a bandwidth B the output SNR is:

$$SNR_{OUT} = \frac{(g_m R V_{in})^2}{4kTR(1 + \gamma g_m R)B}. \tag{B.21}$$

From this we see that increasing g_m and R is good for SNR, although increasing R has an improvement that saturates since both numerator and denominator grow with R^2.

We can also compute an equivalent input-referred noise PSD and a mean-squared input-referred noise voltage.

DEFINITION B.1   Input-referred noise is the noise that, when applied to the input of the circuit, produces the same output noise as the actual, noisy circuit.

This concept is illustrated in Figure B.12 (a), which shows our original circuit that generates noise at the output with a PSD of S_{out}. In this example, it is assumed that the input to the circuit is a current, typical of the transimpedance amplifiers presented in Chapter 9. The input referred PSD (S_{in}) is the PSD that, when applied to the input of a noiseless version of the circuit from (a), produces the same output PSD. Output noise can be measured using test equipment, but input noise can only be calculated. Working with input-referred noise is attractive because it can allow us to compare the noise of amplifiers with different gains without following the signal amplitude through the system. A pitfall of this approach is that we still need to know the appropriate value of the gain, an issue discussed in [76].

Referring to (B.5) the equivalent input-referred noise PSD can be found by dividing the computed (or measured) output PSD by the frequency-dependent gain. In the common-source amplifier example, we get:

$$S_i(f) = \frac{S_o(f)}{|H(f)|^2} = \frac{4kTR(1 + \gamma g_m R)}{(g_m R)^2} = \frac{4kT}{g_m}\left(\frac{1}{A} + \gamma\right), \tag{B.22}$$

where $A = g_m R$. From this we can make the same conclusion as we did from (B.21), namely, to improve the noise performance, increase the MOSFET's transconductance. Smaller improvements can be made by increasing the gain of the amplifier.

We often compute an rms output noise using (B.6). Given the advantages in using input-referred noise for the purpose of comparing different circuits, we may want to refer the rms output noise to the input. As argued in [76], we should refer the rms output noise to input using the same gain as what the input signal sees. This issue is explored further in Chapter 10.

B.4 Equivalent Noise Bandwidth

The common-source amplifier example is greatly simplified since only the circuit's low-frequency gain is considered. Although thermal noise sources are assumed to be flat, the computed output noise is invariably shaped by the circuit's frequency response. We saw this in the RC filter example from the previous section. The integral of flat PSD shaped by a first-order LPF is a well-known result. However, the result of a flat PSD filtered by a second-order response depends on the natural frequency of the poles and their Q factor. A useful concept that captures the filtering of noise is the circuit's equivalent noise bandwidth.

DEFINITION B.2   The equivalent noise bandwidth of a system is the bandwidth (denoted as f_{ENBW}) of an ideal (brickwall) filter of equal pass-band gain that gives equal integrated noise when the input is noise with a flat PSD.

Using the first-order filter as an example, we compute the equivalent noise bandwidth. The input PSD is $4kTR$. Solving for f_{ENBW} in the following equation we compute the noise bandwidth:

$$4kTRf_{ENBW} = \frac{kT}{C}$$

$$f_{ENBW} = \frac{1}{4RC} = \frac{1}{2\pi RC}\frac{\pi}{2} = \frac{\pi}{2}f_{3dB}. \tag{B.23}$$

Figure B.13 (a) shows the output PSD for a first-order filter, normalized to its dc value. The dashed line shows the brickwall filter that gives the same integrated output noise. It has a bandwidth $\pi/2$ times larger than the first-order filter. The second-order system with $Q = \frac{1}{\sqrt{2}}$, rolls off more quickly meaning less noise beyond the bandwidth appears at the output. Its noise bandwidth is $f_{ENBW} = 1.11f_{3dB}$. This filter is shown in Figure B.13 (b). Figure B.14 summarizes the behaviour of a second-order circuit

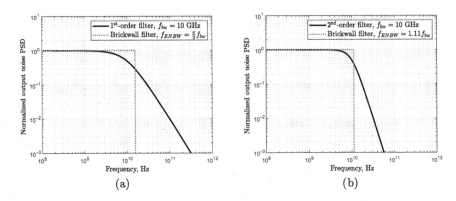

Figure B.13 Equivalent noise bandwidth for (a) a first-order system and (b) a second-order system with $\zeta = 1/\sqrt{2}$.

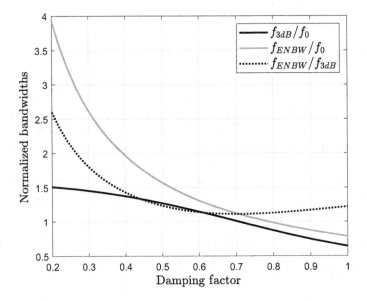

Figure B.14 Interpretation of equivalent noise bandwidth.

system as a function of ζ. The system has poles of magnitude f_0. As ζ increases the -3 dB bandwidth and equivalent noise bandwidth decrease.

B.5   Noise and Power Trade-Offs

Designers must constantly balance noise, offsets, power dissipation, chip area and several other specifications. The design which is noise optimal might be significantly larger in chip area or dissipate much more power than one that is slightly worse than optimal. To explore these trade-offs, we will focus on one, idealized design modifi-cation, namely width scaling. The starting point is a reference design (referred to as

design one) of a common-source amplifier with an input-referred noise PSD as specified by (B.22). Consider the design modification arising from widening the transistor
by 100 while maintaining the same V_{GS}. This leads to a 100x increase in drain current. In order to maintain the same voltage drop across the load resistor and the same
small-signal gain, the load resistor is reduced by 100x. These modifications are summarized below, where the subscripts 1 and 2 refer to the original and modified designs,
respectively:

$$L_2 = L_1,$$
$$W_2 = 100W_1,$$
$$R_2 = \frac{1}{100}R_1,$$
$$I_{D2} = 100I_{D1},$$
$$V_{GS2} = V_{GS1},$$
$$V_{DS2} = V_{DS1}. \tag{B.24}$$

With equal terminal voltages but 100x the width, transconductance also increases
by 100. However, the reduced load resistor preserves the gain. Design two has the
same gain, but higher power dissipation.

$$g_{m2} = 100g_{m1},$$
$$A_{v2} = A_{v1},$$
$$S_{i2} = \frac{1}{100}S_{i1},$$
$$P_2 = 100P_1,$$
$$SNR_2 = 100SNR_1,$$
$$\frac{SNR_2}{P_2} = \frac{SNR_1}{P_1}. \tag{B.25}$$

With 100x transconductance, design two has an input-referred noise PSD (S_{i2}) that
is reduced by 100x relative to the original design. This leads to a 100x improvement
in SNR at the expense of 100x the power dissipation. The ratio of SNR to power is
unchanged as a result of width scaling.

This example has thus far ignored the impact of larger transistors on the circuit
supplying the input voltage and on the circuit's ability to drive other circuits. In the
above example, all conductances grew by 100x (i.e., transconductances and the conductance of the load resistor). Notice that the capacitances of the MOSFET also scale
by 100, leading to an overall scaling of admittances by 100. If we further assume that
any load capacitance added to the circuit also scales by 100, we can extend the claim
that both designs have the same low-frequency gain to say that they have the same
frequency-dependent gain. Therefore, integrated noise will also scale by $\frac{1}{100}$.

The preceding argument was applied to a specific expression for input-referred
noise density. It would be helpful to generalize the effect of width scaling. Rather than
initiate a proof from scratch, width scaling can be viewed as an extension of impedance
scaling routinely performed in filter design. Starting with a generic ladder filter such

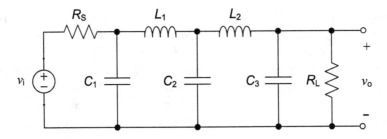

Figure B.15 Example of a fifth-order ladder filter.

as the one shown in Figure B.15, the voltage transfer function from the input to the output remains unchanged if the impedance of every element is scaled by the same factor, K. Specifically:

$$\begin{aligned} R_{new} &= R_{old}/K, \\ C_{new} &= K \cdot C_{old}, \\ L_{new} &= L_{old}/K. \end{aligned} \tag{B.26}$$

In addition to the input-to-output transfer function not changing, impedance scaling leaves all voltage-to-voltage transfer functions unchanged. For example the transfer function from the input to the voltage across L_1 is also unchanged by impedance scaling. However the driving point impedance of every node decreases by K resulting in current-to-voltage transfer functions decreasing by K. Of course wireline circuits also involve transistors. To apply impedance scaling, dc voltage sources and transistor lengths are left unchanged but dc current sources and transistor widths are also increased by K. This leaves bias voltages unchanged. With equally biased transistors now K times wider we expect the following:

$$\begin{aligned} g_{m,new} &= K \cdot g_{m,old}, \\ r_{o,new} &= r_{o,old}/K, \\ C_{i,new} &= K \cdot C_{i,old}, \end{aligned} \tag{B.27}$$

where C_i is a transistor capacitance; C_{gs}, C_{gd}, C_{sb} and C_{db} all have a linear dependence on W. Now the ac small-signal equivalent circuit of the new (scaled) circuit differs from the original (old) circuit only by impedance scaling. Voltage-to-voltage transfer functions are unchanged, while current-to-voltage transfer functions are reduced by K, but otherwise have the same frequency response. Intuitively, this is because the time constants in the circuit, which are of the form $1/RC$ or g_m/C, are unchanged.

The PSD of a resistor's thermal noise current ($4kT/R$) increases by a factor K as R is decreased by K. Likewise, the PSD of a MOSFET's thermal noise current ($4kT\gamma g_m$) also increases by K. Considering (B.7) and the approach laid out in Table B.1, the contribution of each noise source is the product of a current-noise PSD and the squared-magnitude of an impedance transfer function. This results in:

$$S_{o,i,new}(f) = |H_{new}(f)|^2 S_{i,new}(f) = \left|\frac{1}{K}H_{old}(f)\right|^2 KS_{i,old}(f)$$

$$S_{o,i,new}(f) = \frac{S_{o,i,old}}{K}, \tag{B.28}$$

where $S_{i,new}$ and $S_{i,old}$ are the PSDs of a given noise source in the scaled and original circuit, respectively. Likewise, $S_{o,i,new}$ and $S_{o,i,old}$ are the resulting contribution to output voltage noise of a given noise source in the scaled and original circuit, respectively.

Flicker noise, on the other hand, maps to the output through voltage-to-voltage transfer functions. However, due to width scaling, the PSD of the noise source is scaled by $\frac{1}{K}$, leading to the same reduction in output noise PSD.

In conclusion, the output noise PSD of a circuit is scaled by $\frac{1}{K}$ when the circuit is width scaled. With the shape of the PSD unchanged due to width scaling, the mean-squared output voltage noise is also reduced by $\frac{1}{K}$ leading to a reduction in rms output voltage noise of $\sqrt{\frac{1}{K}}$. Input-referred voltage noise is similarly scaled. As a side note, width scaling also reduces the random component of input-referred and output-referred offsets by $\sqrt{\frac{1}{K}}$.

The discussion of width scaling is concluded by a reminder that the results assume every element is impedance scaled, including source and load impedances. When a circuit's load is on-chip, it is reasonable to assume that a load capacitor could be scaled appropriately. However, when interfacing to a 50 Ω environment the real part of the load impedance is fixed, preventing exact application of width scaling.

B.5.1 Width Scaling on Transimpedance Amplifiers

Width scaling of voltage amplifiers can be applied to reduce noise at the cost of increased power dissipation. However, it is not readily applicable to transimpedance amplifiers. In the previous section it was argued that current-to-voltage transfer functions were scaled by $\frac{1}{K}$. Although a width-scaled TIA would have an rms output voltage noise scaled by $\sqrt{\frac{1}{K}}$, the input (current) to output gain would be scaled by $\frac{1}{K}$, leading to a degradation in SNR. This motivates the idea of *decreasing* the width of transistors and conductances in a TIA. To preserve the same frequency response, all capacitances must be reduced. Unfortunately, the IC designer does not have direct control of the photodiode capacitance. Decreasing the size of transistors and increasing resistor values may reduce noise, but this will occur at the expense of reduced bandwidth since some time constants will only see an increase in R without a proportional decrease in C.

Width scaling does show that, if a TIA designer were given a photodiode having a smaller capacitance, reducing the overall input capacitance consisting of photodiode and bondpad, a TIA with unchanged frequency response could be rapidly designed by width scaling with $K = \frac{C_{in,new}}{C_{in,old}}$. This new TIA would use less power, have improved noise and larger gain! This shows why reducing photodiode capacitance is so important.

B.6   Summary

This appendix reviewed noise analysis relevant for the analog circuits in this book. We began with time and frequency descriptions of noise signals, noting that in the frequency domain we use the signal's power-spectral density, a quantity with the units A^2/Hz or V^2/Hz, depending on whether the signal is a voltage or current. When noise sources are uncorrelated we find the total power-spectral density due to several sources in an LTI system, using superposition, by adding their PSDs.

To determine the output noise of a circuit, we need noise models for the circuit elements. Noise models for resistors and MOSFETs were reviewed. To analyze circuits, a systematic method was introduced, consisting of computing the noise transfer function that maps each noise source to the output quantity of interest. Total mean-squared noise can be computed by integrating the output PSD up to ∞. An RC filter was considered, giving rise to the well-known expression for mean-squared noise: kT/C. A second example considered a common-source amplifier.

Important definitions were given: input referred noise and equivalent noise bandwidth.

The trade-off between noise and the circuit's power dissipation was explored. The impedance scaling of ladder filters as well as an analogous "width" scaling in transistor circuits were considered. The results seen for voltage-to-voltage transfer functions were modified in the case of circuits whose current-to-voltage transfer function is important (i.e., TIAs).

References

[1] S. Mirabbasi, L. C. Fujino, and K. C. Smith, "Through the looking glass – the 2023 edition: Trends in solid-state circuits from ISSCC," *IEEE Solid-State Circuits Magazine*, vol. 15, no. 1, pp. 45–62, 2023.

[2] D. C. Daly, L. C. Fujino, and K. C. Smith, "Through the looking glass – 2020 edition: Trends in solid-state circuits from ISSCC," *IEEE Solid-State Circuits Magazine*, vol. 12, no. 1, pp. 8–24, 2020.

[3] E. Depaoli, H. Zhang, M. Mazzini, W. Audoglio, A. A. Rossi, G. Albasini, M. Pozzoni, S. Erba, E. Temporiti, and A. Mazzanti, "A 64 Gb/s low-power transceiver for short-reach PAM-4 electrical links in 28-nm FDSOI CMOS," *IEEE Journal of Solid-State Circuits*, vol. 54, no. 1, pp. 6–17, 2019.

[4] H. Won, T. Yoon, J. Han, J. Lee, J. Yoon, T. Kim, J. Lee, S. Lee, K. Han, J. Lee, J. Park, and H. Bae, "A 0.87 W transceiver IC for 100 Gigabit Ethernet in 40 nm CMOS," *IEEE Journal of Solid-State Circuits*, vol. 50, no. 2, pp. 399–413, 2015.

[5] M. Moayedi Pour Fard, O. Liboiron-Ladouceur, and G. E. R. Cowan, "1.23-pJ/bit 25-Gb/s inductor-less optical receiver with low-voltage silicon photodetector," *IEEE Journal of Solid-State Circuits*, vol. 53, no. 6, pp. 1793–1805, 2018.

[6] M. Raj, M. Monge, and A. Emami, "A modelling and nonlinear equalization technique for a 20 Gb/s 0.77 pJ/b VCSEL transmitter in 32 nm SOI CMOS," *IEEE Journal of Solid-State Circuits*, vol. 51, no. 8, pp. 1734–1743, 2016.

[7] F. M. Wanlass, "Low standby power complementary field effect circuitry," US Patent 3 356 858, 1967.

[8] J. R. Burns, "Self-biased field effect transistor amplifier," US Patent 3 392 341, 1968.

[9] S. T. Hsu, "Amplifier employing complementary field-effect transistors," US Patent 3 946 327, 1976.

[10] A. G. F. Dingwall, "Complementary field effect transistor differential amplifier," US Patent 3 870 966, 1975.

[11] R. Pryor, "Complementary field effect transistor differential amplifier," US Patent 3 991 380, 1976.

[12] Y. P. Tsividis and P. R. Gray, "An integrated NMOS operational amplifier with internal compensation," *IEEE Journal of Solid-State Circuits*, vol. 11, no. 6, pp. 748–753, 1976.

[13] B. Nauta and E. Seevinck, "Linear CMOS transconductance element for VHF filters," *Electronics Letters*, vol. 25, no. 7, pp. 448–450, 1989.

[14] B. Nauta, "A CMOS transconductance-C filter technique for very high frequencies," *IEEE Journal of Solid-State Circuits*, vol. 27, no. 2, pp. 142–153, 1992.

[15] U. Inan, A. Inan, and R. Said, *Engineering Electromagnetics and Waves*. Pearson, 2014.

[16] C. Trueman, *Lecture Notes for Course ELEC353*. Department of Electrical and Computer Engineering, Concordia University, 2018.

[17] T. C. Carusone, "Matlab code for ABCD-parameter analysis of channels," 2007 [Online]. https://github.com/tchancarusone/Wireline-ChModel-Matlab

[18] D. A. Frickey, "Conversions between S, Z, Y, H, ABCD, and T parameters which are valid for complex source and load impedances," *IEEE Transactions on Microwave Theory and Techniques*, vol. 42, no. 2, pp. 205–211, 1994.

[19] T. Kobayashi, K. Nogami, T. Shirotori, Y. Fujimoto, and O. Watanabe, "A current-mode latch sense amplifier and a static power saving input buffer for low-power architecture," in *1992 Symposium on VLSI Circuits Digest of Technical Papers*, pp. 28–29, June 1992.

[20] M. J. M. Pelgrom, A. C. J. Duinmaijer, and A. P. G. Welbers, "Matching properties of MOS transistors," *IEEE Journal of Solid-State Circuits*, vol. 24, no. 5, pp. 1433–1439, 1989.

[21] P. R. Kinget, "Device mismatch and tradeoffs in the design of analog circuits," *IEEE Journal of Solid-State Circuits*, vol. 40, no. 6, pp. 1212–1224, 2005.

[22] M.-J. Lee, W. Dally, and P. Chiang, "Low-power area-efficient high-speed I/O circuit techniques," *IEEE Journal of Solid-State Circuits*, vol. 35, no. 11, pp. 1591–1599, 2000.

[23] J. Kim, B. S. Leibowitz, and M. Jeeradit, "Impulse sensitivity function analysis of periodic circuits," in *2008 IEEE/ACM International Conference on Computer-Aided Design*, pp. 386–391, 2008.

[24] E. Mammei, F. Loi, F. Radice, A. Dati, M. Bruccoleri, M. Bassi, and A. Mazzanti, "Analysis and design of a power-scalable continuous-time FIR equalizer for 10 Gb/s to 25 Gb/s multi-mode fiber EDC in 28 nm LP CMOS," *IEEE Journal of Solid-State Circuits*, vol. 49, no. 12, pp. 3130–3140, 2014.

[25] F. Loi, E. Mammei, S. Erba, M. Bassi, and A. Mazzanti, "A 25mW highly linear continuous-time FIR equalizer for 25Gb/s serial links in 28-nm CMOS," *IEEE Transactions on Circuits and Systems I: Regular Papers*, vol. 64, no. 7, pp. 1903–1913, 2017.

[26] A. Agrawal, J. F. Bulzacchelli, T. O. Dickson, Y. Liu, J. A. Tierno, and D. J. Friedman, "A 19-Gb/s serial link receiver with both 4-tap FFE and 5-tap DFE functions in 45-nm SOI CMOS," *IEEE Journal of Solid-State Circuits*, vol. 47, no. 12, pp. 3220–3231, 2012.

[27] J. Sewter and A. C. Carusone, "A CMOS finite impulse response filter with a crossover traveling wave topology for equalization up to 30 Gb/s," *IEEE Journal of Solid-State Circuits*, vol. 41, no. 4, pp. 909–917, 2006.

[28] J. Sewter and A. Chan Carusone, "A 3-tap FIR filter with cascaded distributed tap amplifiers for equalization up to 40 Gb/s in 0.18-μm CMOS," *IEEE Journal of Solid-State Circuits*, vol. 41, no. 8, pp. 1919–1929, 2006.

[29] A. Momtaz and M. M. Green, "An 80 mW 40 Gb/s 7-tap T/2-spaced feed-forward equalizer in 65 nm CMOS," *IEEE Journal of Solid-State Circuits*, vol. 45, no. 3, pp. 629–639, 2010.

[30] M. H. Nazari and A. Emami-Neyestanak, "A 24-Gb/s double-sampling receiver for ultra-low-power optical communication," *IEEE Journal of Solid-State Circuits*, vol. 48, no. 2, pp. 344–357, 2013.

[31] S. Saeedi, S. Menezo, G. Pares, and A. Emami, "A 25 Gb/s 3D-integrated CMOS/silicon-photonic receiver for low-power high-sensitivity optical communication," *Journal of Lightwave Technology*, vol. 34, no. 12, pp. 2924–2933, 2016.

[32] S. Kasturia and J. H. Winters, "Techniques for high-speed implementation of nonlinear cancellation," *IEEE Journal on Selected Areas in Communications*, vol. 9, no. 5, pp. 711–717, 1991.

[33] B. Kim, Y. Liu, T. O. Dickson, J. F. Bulzacchelli, and D. J. Friedman, "A 10-Gb/s compact low-power serial I/O with DFE-IIR equalization in 65-nm CMOS," *IEEE Journal of Solid-State Circuits*, vol. 44, no. 12, pp. 3526–3538, 2009.

[34] B. Razavi, "The bridged T-coil [a circuit for all seasons]," *IEEE Solid-State Circuits Magazine*, vol. 7, no. 4, pp. 9–13, 2015.

[35] J. Paramesh and D. J. Allstot, "Analysis of the bridged T-coil circuit using the extra-element theorem," *IEEE Transactions on Circuits and Systems II: Express Briefs*, vol. 53, no. 12, pp. 1408–1412, 2006.

[36] S. C. D. Roy, "Comments on 'Analysis of the bridged T-coil circuit using the extra-element theorem,'" *IEEE Transactions on Circuits and Systems II: Express Briefs*, vol. 54, no. 8, pp. 673–674, 2007.

[37] E. L. Ginzton, W. R. Hewlett, J. H. Jasberg, and J. D. Noe, "Distributed amplification," *Proceedings of the IRE*, vol. 36, no. 8, pp. 956–969, 1948.

[38] E. Groen, C. Boecker, M. Hossain, R. Vu, S. Vamvakos, H. Lin, S. Li, M. Van Ierssel, P. Choudhary, N. Wang, M. Shibata, M. H. Taghavi, N. Nguyen, and S. Desai, "6.3 A 10-to-112Gb/s DSP-DAC-based transmitter with 1.2Vppd output swing in 7nm FinFET," in *2020 IEEE International Solid-State Circuits Conference*, pp. 120–122, 2020.

[39] Z. Wang, M. Choi, K. Lee, K. Park, Z. Liu, A. Biswas, J. Han, S. Du, and E. Alon, "An output bandwidth optimized 200-Gb/s PAM-4 100-Gb/s NRZ transmitter with 5-tap FFE in 28-nm CMOS," *IEEE Journal of Solid-State Circuits*, vol. 57, no. 1, pp. 21–31, 2022.

[40] Y. Song, R. Bai, K. Hu, H. Yang, P. Y. Chiang, and S. Palermo, "A 0.47–0.66 pJ/bit, 4.8–8 Gb/s I/O transceiver in 65 nm CMOS," *IEEE Journal of Solid-State Circuits*, vol. 48, no. 5, pp. 1276–1289, 2013.

[41] S. Shekhar, J. Walling, and D. Allstot, "Bandwidth extension techniques for CMOS amplifiers," *IEEE Journal of Solid-State Circuits*, vol. 41, no. 11, pp. 2424–2439, 2006.

[42] K. J. Zheng, "System-driven circuit design for ADC-based wireline communication links," PhD dissertation, Stanford University, CA, 2018. http://purl.stanford.edu/hw458fp0168.

[43] K. Zheng, Y. Frans, S. L. Ambatipudi, S. Asuncion, H. T. Reddy, K. Chang, and B. Murmann, "An inverter-based analog front-end for a 56-Gb/s PAM-4 wireline transceiver in 16-nm CMOS," *IEEE Solid-State Circuits Letters*, vol. 1, no. 12, pp. 249–252, 2018.

[44] G. P. Agrawal, *Lightwave Technology*. Wiley, 2004.

[45] L. Chrostowski and M. Hochberg, *Silicon Photonics Design: From Devices to Systems*. Cambridge University Press, 2015.

[46] B. Nakhkoob, S. Ray, and M. M. Hella, "High speed photodiodes in standard nanometer scale CMOS technology: A comparative study," *Optics Express*, vol. 20, no. 10, pp. 11 256–11 270, 2012.

[47] M. M. P. Fard, C. Williams, G. Cowan, and O. Liboiron-Ladouceur, "High-speed grating-assisted all-silicon photodetectors for 850 nm applications," *Optics Express*, vol. 25, no. 5, pp. 5107–5118, 2017.

[48] M. Kuijk, D. Coppee, and R. Vounckx, "Spatially modulated light detector in CMOS with sense-amplifier receiver operating at 180 Mb/s for optical data link applications and parallel optical interconnects between chips," *IEEE Journal of Selected Topics in Quantum Electronics*, vol. 4, no. 6, pp. 1040–1045, 1998.

[49] S. Huang and W. Chen, "A 10-Gbps CMOS single chip optical receiver with 2-D meshed spatially-modulated light detector," in *2009 IEEE Custom Integrated Circuits Conference*, pp. 129–132, 2009.

[50] Y. Dong and K. W. Martin, "A 4-Gbps POF receiver using linear equalizer with multi-shunt-shunt feedbacks in 65-nm CMOS," *IEEE Transactions on Circuits and Systems II: Express Briefs*, vol. 60, no. 10, pp. 617–621, 2013.

[51] S. M. Csutak, S. Dakshina-Murthy, and J. C. Campbell, "CMOS-compatible planar silicon waveguide-grating-coupler photodetectors fabricated on silicon-on-insulator (SOI) substrates," *IEEE Journal of Quantum Electronics*, vol. 38, no. 5, pp. 477–480, 2002.

[52] A. S. Ramani, S. Nayak, and S. Shekhar, "A differential push-pull voltage mode VCSEL driver in 65-nm CMOS," *IEEE Transactions on Circuits and Systems I: Regular Papers*, vol. 66, no. 11, pp. 4147–4157, 2019.

[53] V. Kozlov and A. Chan Carusone, "Capacitively-coupled CMOS VCSEL driver circuits," *IEEE Journal of Solid-State Circuits*, vol. 51, no. 9, pp. 2077–2090, 2016.

[54] S. Hu, R. Bai, X. Wang, Y. Peng, J. Wang, T. Xia, J. Ma, L. Wang, Y. Zhang, X. Chen, N. Qi, and P. Y. Chiang, "A 4 × 25 Gb/s optical transmitter using low-cost 10 Gb/s VCSELs in 40-nm CMOS," *IEEE Photonics Technology Letters*, vol. 31, no. 12, pp. 967–970, 2019.

[55] L. A. Coldren and S. W. Corzine, *Diode Lasers and Photonic Integrated Circuits*. Wiley, 1995.

[56] P. Westbergh, J. S. Gustavsson, A. Haglund, M. Skold, A. Joel, and A. Larsson, "High-speed, low-current-density 850 nm VCSELs," *IEEE Journal of Selected Topics in Quantum Electronics*, vol. 15, no. 3, pp. 694–703, 2009.

[57] H. Morita, K. Uchino, E. Otani, H. Ohtorii, T. Ogura, K. Oniki, S. Oka, S. Yanagawa, and H. Suzuki, "8.2 A 12 x 5 two-dimensional optical I/O array for 600Gb/s chip-to-chip interconnect in 65nm CMOS," in *2014 IEEE International Solid-State Circuits Conference Digest of Technical Papers*, pp. 140–141, February 2014.

[58] M. Cignoli, G. Minoia, M. Repossi, D. Baldi, A. Ghilioni, E. Temporiti, and F. Svelto, "22.9 A 1310nm 3D-integrated silicon photonics Mach–Zehnder-based transmitter with 275mW multistage CMOS driver achieving 6dB extinction ratio at 25Gb/s," in *2015 IEEE International Solid-State Circuits Conference Digest of Technical Papers*, pp. 1–3, 2015.

[59] E. Temporiti, A. Ghilioni, G. Minoia, P. Orlandi, M. Repossi, D. Baldi, and F. Svelto, "Insights into silicon photonics Mach–Zehnder-based optical transmitter architectures," *IEEE Journal of Solid-State Circuits*, vol. 51, no. 12, pp. 3178–3191, 2016.

[60] M. Kim, M. Shin, M.-H. Kim, B.-M. Yu, C. Mai, S. Lischke, L. Zimmermann, and W.-Y. Choi, "A large-signal equivalent circuit for depletion-type silicon ring modulators," 2018 Optical Fiber Communications Conference and Exposition, San Diego, CA, pp. 1–3, 2018.

[61] M. Kim, M. Shin, M.-H. Kim, B.-M. Yu, Y. Kim, Y. Ban, S. Lischke, C. Mai, L. Zimmermann, and W.-Y. Choi, "Large-signal SPICE model for depletion-type silicon ring modulators," *Photonics Research*, vol. 7, no. 9, pp. 948–954, 2019.

[62] H. Li, Z. Xuan, A. Titriku, C. Li, K. Yu, B. Wang, A. Shafik, N. Qi, Y. Liu, R. Ding, T. Baehr-Jones, M. Fiorentino, M. Hochberg, S. Palermo, and P. Y. Chiang, "A 25 Gb/s, 4.4 V-swing, AC-coupled ring modulator-based WDM transmitter with wavelength stabilization in 65 nm CMOS," *IEEE Journal of Solid-State Circuits*, vol. 50, no. 12, pp. 3145–3159, 2015.

[63] B. Razavi, *Design of Integrated Circuits for Optical Communications*. Wiley, 2012.

[64] E. Sackinger, "The transimpedance limit," *IEEE Transactions on Circuits and Systems I: Regular Papers*, vol. 57, no. 8, pp. 1848–1856, 2010.

[65] J. Rabaey, *Digital Integrated Circuits: A Design Perspective*. Prentice Hall, 1996.

[66] D. Abdelrahman, O. Liboiron-Ladouceur, and G. E. R. Cowan, "CMOS-driven VCSEL-based photonic links: An exploration of the power-sensitivity trade-off," *IEEE Access*, vol. 10, pp. 89 331–89 345, 2022.

[67] M. G. Ahmed, M. Talegaonkar, A. Elkholy, G. Shu, A. Elmallah, A. Rylyakov, and P. K. Hanumolu, "A 12-Gb/s – 16.8-dBm OMA sensitivity 23-mW optical receiver in 65-nm CMOS," *IEEE Journal of Solid-State Circuits*, vol. 53, no. 2, pp. 445–457, 2018.

[68] W. Ni, M.-A. Chan, and G. Cowan, "Inductorless bandwidth extension using local positive feedback in inverter-based TIAs," in *2015 IEEE 58th International Midwest Symposium on Circuits and Systems (MWSCAS)*, pp. 1–4, 2015.

[69] E. Sackinger, "On the noise optimum of FET broadband transimpedance amplifiers," *IEEE Transactions on Circuits and Systems I: Regular Papers*, vol. 59, no. 12, pp. 2881–2889, 2012.

[70] E. Sackinger, *Broadband Circuits for Optical Fiber Communication*. Wiley, 2005.

[71] P. P. Dash, G. Cowan, and O. Liboiron-Ladouceur, "A variable-bandwidth, power-scalable optical receiver front-end in 65 nm," in *2013 IEEE 56th International Midwest Symposium on Circuits and Systems (MWSCAS)*, pp. 717–720, 2013.

[72] J. Proesel, C. Schow, and A. Rylyakov, "25Gb/s 3.6pJ/b and 15Gb/s 1.37pJ/b VCSEL-based optical links in 90nm CMOS," in *2012 IEEE International Solid-State Circuits Conference*, pp. 418–420, February 2012.

[73] S. Galal and B. Razavi, "10-Gb/s limiting amplifier and laser/modulator driver in 0.18-μm CMOS technology," *IEEE Journal of Solid-State Circuits*, vol. 38, no. 12, pp. 2138–2146, 2003.

[74] H. Huang, J. Chien, and L. Lu, "A 10-Gb/s inductorless CMOS limiting amplifier with third-order interleaving active feedback," *IEEE Journal of Solid-State Circuits*, vol. 42, no. 5, pp. 1111–1120, 2007.

[75] I. Ozkaya, A. Cevrero, P. A. Francese, C. Menolfi, M. Braendli, T. Morf, D. Kuchta, L. Kull, M. Kossel, D. Luu, M. Meghelli, Y. Leblebici, and T. Toifl, "A 56Gb/s burst-mode NRZ optical receiver with 6.8ns power-on and CDR-lock time for adaptive optical links in 14nm FinFET CMOS," in *2018 IEEE International Solid-State Circuits Conference*, pp. 266–268, 2018.

[76] D. Abdelrahman and G. E. R. Cowan, "Noise analysis and design considerations for equalizer-based optical receivers," *IEEE Transactions on Circuits and Systems I: Regular Papers*, vol. 66, no. 8, pp. 3201–3212, 2019.

[77] J. Proesel, A. Rylyakov, and C. Schow, "Optical receivers using DFE-IIR equalization," in *2013 IEEE International Solid-State Circuits Conference Digest of Technical Papers*, pp. 130–131, February 2013.

[78] A. Sharif-Bakhtiar and A. Chan Carusone, "A 20 Gb/s CMOS optical receiver with limited-bandwidth front end and local feedback IIR-DFE," *IEEE Journal of Solid-State Circuits*, vol. 51, no. 11, pp. 2679–2689, 2016.

[79] S. Palermo, A. Emami-Neyestanak, and M. Horowitz, "A 90 nm CMOS 16 Gb/s transceiver for optical interconnects," *IEEE Journal of Solid-State Circuits*, vol. 43, no. 5, pp. 1235–1246, 2008.

[80] S. Echeverri-Chacón, J. J. Mohr, J. J. V. Olmos, P. Dawe, B. V. Pedersen, T. Franck, and S. B. Christensen, "Transmitter and dispersion eye closure quaternary (TDECQ) and its sensitivity to impairments in PAM4 waveforms," *Journal of Lightwave Technology*, vol. 37, no. 3, pp. 852–860, 2019.

[81] J. Im, K. Zheng, A. Chou, L. Zhou, J. W. Kim, S. Chen, Y. Wang, H. Hung, K. Tan, W. Lin, A. Roldan, D. Carey, I. Chlis, R. Casey, A. Bekele, Y. Cao, D. Mahashin, H. Ahn, H. Zhang, Y. Frans, and K. Chang, "6.1 A 112Gb/s PAM-4 long-reach wireline transceiver using a 36-way time-interleaved SAR-ADC and inverter-based RX analog front-end in 7nm FinFET," in *2020 IEEE International Solid-State Circuits Conference*, pp. 116–118, 2020.

[82] B. Yoo, D. Lim, H. Pang, J. Lee, S. Baek, N. Kim, D. Choi, Y. Choi, H. Yang, T. Yoon, S. Chu, K. Kim, W. Jung, B. Kim, J. Lee, G. Kang, S. Park, M. Choi, and J. Shin, "6.4 A 56Gb/s 7.7mW/Gb/s PAM-4 Wireline Transceiver in 10nm FinFET Using MM-CDR-Based ADC Timing Skew Control and Low-Power DSP with Approximate Multiplier," in *2020 IEEE International Solid-State Circuits Conference*, pp. 122–124, 2020.

[83] G. Forney, "The Viterbi algorithm," *Proceedings of the IEEE*, vol. 61, no. 3, pp. 268–278, 1973.

[84] F. O'Mahony, J. E. Jaussi, J. Kennedy, G. Balamurugan, M. Mansuri, C. Roberts, S. Shekhar, R. Mooney, and B. Casper, "A 47 × 10 Gb/s 1.4 mW/Gb/s parallel interface in 45 nm CMOS," *IEEE Journal of Solid-State Circuits*, vol. 45, no. 12, pp. 2828–2837, 2010.

[85] A. Ragab, Y. Liu, K. Hu, P. Chiang, and S. Palermo, "Receiver jitter tracking characteristics in high-speed source synchronous links," *Journal of Electrical and Computer Engineering*, 2011. doi.org/10.1155/2011/982314

[86] C. Menolfi, M. Braendli, P. A. Francese, T. Morf, A. Cevrero, M. Kossel, L. Kull, D. Luu, I. Ozkaya, and T. Toifl, "A 112Gb/s 2.6pJ/b 8-Tap FFE PAM-4 SST TX in 14nm CMOS," in *2018 IEEE International Solid-State Circuits Conference*, pp. 104–106, 2018.

[87] J. Maneatis and M. Horowitz, "Precise delay generation using coupled oscillators," *IEEE Journal of Solid-State Circuits*, vol. 28, no. 12, pp. 1273–1282, 1993.

[88] J. Maneatis, "Low-jitter process-independent DLL and PLL based on self-biased techniques," *IEEE Journal of Solid-State Circuits*, vol. 31, no. 11, pp. 1723–1732, 1996.

[89] G. E. R. Cowan, M. Meghelli, and D. Friedman, "A linearized voltage-controlled oscillator for dual-path phase-locked loops," in *2013 IEEE International Symposium on Circuits and Systems*, pp. 2678–2681, May 2013.

[90] A. Hajimiri, S. Limotyrakis, and T. H. Lee, "Jitter and phase noise in ring oscillators," *IEEE Journal of Solid-State Circuits*, vol. 34, no. 6, pp. 790–804, 1999.

[91] M. Hossain and A. C. Carusone, "CMOS oscillators for clock distribution and injection-locked deskew," *IEEE Journal of Solid-State Circuits*, vol. 44, no. 8, pp. 2138–2153, 2009.

[92] M. Hossain, K. Kaviani, B. Daly, M. Shirasgaonkar, W. Dettloff, T. Stone, K. Prabhu, B. Tsang, J. Eble, and J. Zerbe, "A 6.4/3.2/1.6 Gb/s low power interface with all digital clock multiplier for on-the-fly rate switching," in *Proceedings of the IEEE 2012 Custom Integrated Circuits Conference*, pp. 1–4, 2012.

[93] F. Musa, "High-speed baud-rate clock recovery," PhD dissertation, University of Toronto, 2008. https://tspace.library.utoronto.ca/handle/1807/11120.

[94] K. Mueller and M. Muller, "Timing recovery in digital synchronous data receivers," *IEEE Transactions on Communications*, vol. 24, no. 5, pp. 516–531, 1976.

[95] F. Spagna, L. Chen, M. Deshpande, Y. Fan, D. Gambetta, S. Gowder, S. Iyer, R. Kumar, P. Kwok, R. Krishnamurthy, C.-C. Lin, R. Mohanavelu, R. Nicholson, J. Ou, M. Pasquarella, K. Prasad, H. Rustam, L. Tong, A. Tran, J. Wu, and X. Zhang, "A 78mW 11.8Gb/s serial link transceiver with adaptive RX equalization and baud-rate CDR in 32nm CMOS," in *2010 IEEE International Solid-State Circuits Conference*, pp. 366–367, 2010.

[96] L. DeVito, J. Newton, R. Croughwell, J. Bulzacchelli, and F. Benkley, "A 52MHz and 155MHz clock-recovery PLL," in *1991 IEEE International Solid-State Circuits Conference Digest of Technical Papers*, pp. 142–306, 1991.

[97] B. Razavi, "A versatile clock recovery architecture and monolithic implementation," pp. 405–420 in *Monolithic Phase-Locked Loops and Clock Recovery Circuits: Theory and Design*. Wiley, 1996.

[98] T. Toifl, C. Menolfi, P. Buchmann, C. Hagleitner, M. Kossel, T. Morf, J. Weiss, and M. Schmatz, "A 72 mW 0.03mm^2 inductorless 40Gb/s CDR in 65 nm SOI CMOS," in *2007 IEEE International Solid-State Circuits Conference Digest of Technical Papers*, pp. 226–598, February 2007.

[99] T. Toifl, C. Menolfi, P. Buchmann, M. Kossel, T. Morf, R. Reutemann, M. Ruegg, M. L. Schmatz, and J. Weiss, "A 0.94-ps-RMS-jitter 0.016-mm^2 2.5-GHz multiphase generator PLL with 360° digitally programmable phase shift for 10-Gb/s serial links," *IEEE Journal of Solid-State Circuits*, vol. 40, no. 12, pp. 2700–2712, 2005.

[100] A. Sedra and K. C. Smith, *Microelectronic Circuits*, 3rd ed. Oxford, 2014.

Index

Printed in the United States
by Baker & Taylor Publisher Services